Legume Crop Wild Relatives

Grain legume crops are an important component of global food and nutritional security and help in maintaining agroecological systems. They fix atmospheric nitrogen via the root-inhabiting rhizobacteria, thereby minimising the harmful effects caused by the excessive application of synthetic nitrogenous fertilisers in the soil environment. There has been less focus on legume crop wild relatives for harnessing their potential traits and novel gene(s) to incorporate them into cultivated legumes for developing climate-resilient grain legumes. In this edited book, we will highlight the importance of various potential traits of crop wild relatives, which are yet to be properly harnessed for designing future climate-resilient grain legumes. We also update how advances in molecular genetics and genomics have enabled the underpinning of several candidate genes/genomic regions in various crop wild relatives harbouring adaptive traits that confer climate resilience in grain legumes.

The readers will benefit from new information on various crop wild relatives in grain legumes and how these wild relatives could be explored for novel climate resilience genes for developing future climate-resilient legume crops. They will gain an understanding of how genomic advances (genome sequences, pan genomes) have uncovered the novel genomic regions attributed to climate resilience in various grain legumes. Finally, the critical role of these wild relatives in maintaining the lost gene(s) due to the domestication process will be discussed.

Comprehensive information on conventional breeding, advanced breeding, and recent advances in genomics covering all the major crop wild relatives of legumes is not available in a single book. Thus, this book will provide readers with the latest updates on various information covering all aspects of wild species of legumes.

Legume Crop Wild Relatives

Their Role in Improving Climate Resilient Legumes

Edited by
Uday Chand Jha, Harsh Nayyar,
Kamal Dev Sharma, Eric J Bishop von Wettberg,
and Kadambot H. M. Siddique

CRC Press
Taylor & Francis Group
Boca Raton London New York

CRC Press is an imprint of the
Taylor & Francis Group, an **informa** business

First edition published 2025
by CRC Press
2385 NW Executive Center Drive, Suite 320, Boca Raton FL 33431

and by CRC Press
4 Park Square, Milton Park, Abingdon, Oxon, OX14 4RN

CRC Press is an imprint of Taylor & Francis Group, LLC

Library of Congress Cataloging-in-Publication Data
Names: Jha, Uday Chand, editor. | Nayyar, Harsh (Legume researcher), editor. |
Sharma, Kamal Dev, editor. | Von Wettberg, Eric J. Bishop, editor. | Siddique, K. H. M., editor.
Title: Legume crop wild relatives : their role in improving climate resilient legumes /
edited by Uday Chand Jha, Harsh Nayyar, Kamal Dev Sharma, Eric J Bishop-von Wettberg, and Kadambot H. M. Siddique
Description: First edition | Boca Raton, FL : CRC Press, 2025 |
Includes bibliographical references and index
Identifiers: LCCN 2024024505 (print) | LCCN 2024024506 (ebook) |
ISBN 9781032562230 (hardback) | ISBN 9781032562261 (paperback) |
ISBN 9781003434535 (ebook)
Subjects: LCSH: Legumes–Germplasm resources. | Legumes–Breeding | Wild plants,
Edible–Climatic factors
Classification: LCC SB317.L43 L4354 2025 (print) | LCC SB317.L43 (ebook) |
DDC 633.3–dc23/eng/20240715
LC record available at https://lccn.loc.gov/2024024505
LC ebook record available at https://lccn.loc.gov/2024024506

ISBN: 9781032562230 (hbk)
ISBN: 9781032562261 (pbk)
ISBN: 9781003434535 (ebk)

DOI: 10.1201/9781003434535

Typeset in Times
by codeMantra

Contents

Editor Biographies

Dr. Uday C Jha has been working in the area of grain legume breeding, genetics, and genomics for both biotic and abiotic stress tolerance since 2010 at the Indian Institute of Pulses Research, Kanpur, ICAR, India. He has more than 60 peer-reviewed international publications including two edited books published by Springer Nature. He is associated in developing 8 chickpea varieties. He also serves as a subject editor for various journals of international repute.

Dr. Harsh Nayyar is currently a Professor at Panjab University, India. Dr. Nayyar has been working on the responses of various food legumes (chickpea, lentil, beans) to drought, cold, heat, salt, and metals, for the past 15 years. He has published more than 150 research articles in peer-reviewed, high-impact scientific journals. Recently, he was rated among the top 2% of Indian scientists in a global ranking by Stanford University, USA, published in PLOS Biology.

Dr. Kamal Dev Sharma is a Professor and Head of the Department of Agricultural Biotechnology at CSKHPKV Palampur, India. His areas of expertise include plant genomics and abiotic and biotic stresses in plants, with a primary focus on Fusarium wilt and cold stress in chickpea. He has published more than 50 research and review articles in internationally reputed journals, along with several book chapters.

Dr. Eric J. Bishop von Wettberg is Professor at the University of Vermont. He has vast experience in conducting research in the areas of population genomics, domestication of legumes, symbiosis, conservation genetics, landscape genetics, and symbiont and microbial mediation of plant traits. He serves as editorial board member for various international journals. He has more than 100 peer-reviewed publications in reputed journals.

Professor Kadambot H.M. Siddique has more than 35 years of experience in agricultural research, teaching, and management in both Australia and overseas. He has developed a national and international reputation in agricultural science especially in the fields of crop physiology, production agronomy, farming systems, genetic resources, and breeding research in cereal, grain and pasture legumes, and oilseed crops.

He is the Hackett Professor of Agriculture Chair and Director of The UWA Institute of Agriculture at The University of Western Australia. Professor Siddique is a double Highly Cited Researcher 2021 in agricultural science and plant science (Web of Science).

Contributors

L. Ahmad
Dryland Agriculture Research Station
Sher-e-Kashmir University of Agriculture Science & Technology of Kashmir
Kashmir, India

Amandeep
Department of Plant Breeding and Genetics
Punjab Agricultural University
Ludhiana, Punjab, India

Asmat Ara
Dryland Agriculture Research Station
Sher-e-Kashmir University of Agriculture Science & Technology of Kashmir
Kashmir, India

Anamika Barman
Division of Genetics
ICAR-Indian Agricultural Research Institute
Pusa Campus, New Delhi, India

R. Beena
Department of Plant Physiology, College of Agriculture, Vellayani
Kerala Agricultural University
Thiruvananthapuram, Kerala, India

Sandip K. Bera
ICAR-Directorate of Groundnut Research
Junagadh, Gujarat, India

Soraya Leal-Bertioli
Department of Plant Pathology
University of Georgia
Athens, Georgia

Mohd. Yaqub Bhat
Department of Botany
University of Kashmir
Srinagar, Jammu and Kashmir, India

Ramesh Bhat
Department of Biotechnology
University of Agricultural Sciences
Dharwad, Karnataka, India

Sougata Bhattacharjee
School of Crop Science
ICAR-Indian Agricultural Research Institute
Gauria Karma, Jharkhand, 825405, India

Deekshitha Bomireddy
International Crops Research Institute for the Semi-Arid Tropics (ICRISAT)
Hyderabad, Telangana, India
and
Department of Genetics & Plant Breeding
S. V. Agricultural College
Tirupati, Andhra Pradesh, India

Subhash Chandra
ICAR-Central Institute of Cotton Research, Regional Station, Sirsa
Sirsa, Nagpur, India

S. A. Dar
Dryland Agriculture Research Station
Sher-e-Kashmir University of Agriculture Science & Technology of Kashmir
Srinagar, Jammu and Kashmir, India

Zahoor A. Dar
Dryland Agriculture Research Station
Sher-e-Kashmir University of Agricultural Sciences and Technology of Kashmir
Srinagar, Jammu and Kashmir, India

Uttarayan Dasgupta
Division of Genetics
ICAR-Indian Agricultural Research Institute
Pusa Campus, New Delhi, India

Saima Fayaz
Department of Botany
University of Kashmir
Srinagar, Jammu and Kashmir, India

Daniel Fonceka
CIRAD
UMR AGAP
Thies, Senegal, West Africa

R. K. Gill
Department of Plant Breeding and Genetics
Punjab Agricultural University
Ludhiana, Punjab, India

Sanjay Gupta
Indian Institute of Soybean Research
Indore, Madhya Pradesh, India

Sanjeev Gupta
Indian Council of Agricultural Research
New Delhi, India

M. Habib
Dryland Agriculture Research Station
Sher-e-Kashmir University of Agriculture Science & Technology of Kashmir
Srinagar, Jammu and Kashmir, India

Li Huang
Key Laboratory of Biology and Genetic Improvement of Oil Crops
Ministry of Agriculture, Oil Crops Research Institute of the Chinese Academy of Agricultural Sciences
Wuhan, Hubei, China

Shamshir ul Hussan
Dryland Agriculture Research Station
Sher-e-Kashmir University of Agriculture Science & Technology of Kashmir
Srinagar, Jammu and Kashmir, India

Uday Chand Jha
Crop Improvement Division
Indian Institute of Pulses Research (IIPR)
Kanpur, Uttar Pradesh, India

Huifang Jiang
Key Laboratory of Biology and Genetic Improvement of Oil Crops
Ministry of Agriculture, Oil Crops Research Institute of the Chinese Academy of Agricultural Sciences
Wuhan, Hubei, China

M. H. Khan
Dryland Agriculture Research Station
Sher-e-Kashmir University of Agriculture Science & Technology of Kashmir
Srinagar, Jammu and Kashmir, India

Madhavilatha Kommana
Department of Genetics & Plant Breeding
S. V. Agricultural College
Tirupati, Andhra Pradesh, India

Mackenzie K. Laverick
Plant Biology and Anthropology and Royal Botanic Garden Kew
Jhansi, Uttar Pradesh, India

Boshou Liao
Key Laboratory of Biology and Genetic Improvement of Oil Crops
Ministry of Agriculture, Oil Crops Research Institute of the Chinese Academy of Agricultural Sciences
Wuhan, Hubei, China

Ajaz A. Lone
Dryland Agriculture Research Station
Sher-e-Kashmir University of Agricultural Sciences and Technology of Kashmir
Srinagar, Jammu and Kashmir, India

Harsh Nayyar
Department of Botany
Panjab University
Chandigarh, Panjab, India

P.R. Nithya
Department of Agricultural Entomology
College of Agriculture, Vellayani, Kerala Agricultural University
Thiruvananthapuram, Kerala, India

Manish K. Pandey
International Crops Research Institute for the Semi-Arid Tropics (ICRISAT)
Hyderabad, Telangana, India

Krishnayan Paul
Division of Molecular Biology & Biotechnology
ICAR-Indian Agricultural Research Institute
New Delhi, India

Aditya Pratap
Indian Institute of Pulses Research
Kanpur, Uttar Pradesh, India

Latif Ahmad Peer
Department of Botany
University of Kashmir
Srinagar, Jammu and Kashmir, India

V. Rajesh
Indian Institute of Soybean Research
Indore, Madhya Pradesh, India

Munezeh Rashid
Dryland Agriculture Research Station
Sher-e-Kashmir University of Agriculture Science & Technology of Kashmir
Srinagar, Jammu and Kashmir, India

Muneeb Ahmad Rather
Department of Botany
University of Kashmir
Srinagar, Jammu and Kashmir, India

N. Krishan Kumar Rathod
Division of Genetics
ICAR-Indian Agricultural Research Institute
New Delhi, India

Manisha Saini
Division of Genetics
ICAR-Indian Agricultural Research Institute
New Delhi, India

Arjun Sharma
Rabindranath Tagore University
Bhopal, Madhya Pradesh, India

Parul Sharma
Department of Plant Breeding and Genetics
Punjab Agricultural University
Ludhiana, Punjab, India

Vinay Sharma
International Crops Research Institute for the Semi-Arid Tropics (ICRISAT)
Hyderabad, Telangana, India
and
Department of Genetics & Plant Breeding
Chaudhary Charan Singh University
Meerut, Uttar Pradesh, India

Kadambot H.M. Siddique
The UWA Institute of Agriculture
The University of Western Australia
Perth, WA, Australia

Anshuman Singh
Rani Lakhshmibahi Central Agricultural University
Jhansi, Uttar Pradesh, India

Gurjeet Singh
Department of Plant Breeding and Genetics
Punjab Agricultural University
Ludhiana, Punjab, India
and
India and Texas A&M University
Kanpur, Uttar Pradesh, India

Inderjit Singh
Department of Plant Breeding and Genetics
Punjab Agricultural University
Ludhiana, Punjab, India

Mohar Singh
ICAR-National Bureau of Plant Genetic Resources
Regional Station
Shimla, Himachal Pradesh, India

Dipro Sinha
Division of Bioinformatics
ICAR-Indian Agricultural Research Institute
New Delhi, India

Raghav Sood
ICAR-National Bureau of Plant Genetic Resources
Regional Station
Shimla, Himachal Pradesh, India

Akshay Talukdar
Division of Genetics
ICAR-Indian Agricultural Research Institute
New Delhi, India

Roshni Vijayan
Department of Plant Breeding and Genetics
Regional Agricultural Research Station
Mele Pattambi, Palakkad, Kerala, India

Eric J. Bishop von Wettberg
Plant and Soil Science and Gund Institute for the Environment
University of Vermont
Burlington, Vermont

Jiaping Wang
Agronomy Department
University of Florida
Gainesville, Florida

Xingjun Wang
Shandong Academy of Agricultural Sciences
Biotechnology Research Center
Jinan, China

Manu Yadav
Division of Genetics
ICAR-Indian Agricultural Research Institute
New Delhi, India

1 Wild Chickpea
Treasure of Novel Diversity for Crop Improvement

Mohar Singh and Raghav Sood

INTRODUCTION

Cultivated chickpea (*Cicer arietinum* L.) is the third most important pulse crop under cultivation in more than 59 nations contributing to global nutritional security owing to 23% protein content (FAO 2016; Singh and Pratap 2016). It is a self-pollinating true diploid ($2n = 2x = 16$) species belonging to the family *Fabaceae* and grouped into desi and kabuli types based on important distinct seed-related traits. The crop produces significant yields in drier conditions due to its deep tap root system (Singh 1997; Saraf et al. 1998). The genome size of chickpea has been estimated at 738 Mbp (Varshney et al. 2013) evolved from *C. reticulatum* progenitor species.

India, being the biggest chickpea producer in the world, accounts for ~70% of total annual pulse production of ~12 million tonnes (FAO 2018), which is below its actual strength. The main reasons underlying poor improvements are a series of biotic and abiotic stresses, which reduce yield levels and their stability (Siddique et al. 2000; Varshney et al. 2012). The non-availability of high-yielding genetically improved varieties remains a major bottleneck in achieving better productivity levels even in highly productive environments (Chaturvedi and Nadarajan 2010). Further, constraints in the breeding of cultivated chickpeas include low genetic diversity because of its single domestication event and a high rate of self-pollination (Abbo et al. 2003). Adequate sources of resilience to major biotic and abiotic stresses, including productivity-related traits, are often not present within cultivated species. This has diverted researchers' interest to using crop wild relatives (CWRs) for the genetic improvement of chickpea (Mallikarjuna et al. 2007). Therefore, to attain further breakthroughs in improving yield and stability in future crop varieties, new traits of interest need to be identified and incorporated into cultivated germplasm to develop novel breeding populations. To cater the needs of the growing human population, consolidated efforts are required to increase the yield potential and nutrition of chickpea. An immediate task is to broaden the genetic base of chickpea cultivars by introgression of wild species for diversification of the cultivated gene pool.

GENE POOL AND SPECIES DISTRIBUTION

The genus *Cicer* consists of 9 annual and 35 perennial species. Among these 44 species, *C. arietinum* is the only domesticated one (Harlan and de Wet 1971; van der Maesen 1987). The primary gene pool comprises the cultivated chickpea, *Cicer*

DOI: 10.1201/9781003434535-1

arietinum and its immediate progenitor species, *C. reticulatum*, which readily cross with cultivated chickpea with normal gene transfer. The secondary gene pool consists of *C. echinospermum*, which can cross with cultivated species but results in reduced fertility of the hybrids and subsequent progenies. The tertiary gene pool consists of the remaining 6 annual and 35 perennial *Cicer* species, which are not crossable with normal meiosis using conventional hybridization and require contemporary utilization approaches for gene transfer. As far as the distribution of *Cicer* species is concerned, most of these, including the three wild annual species *C. bijugum*, *C. echinospermum* and *C. reticulatum*, are found in West Asia and North Africa, covering Turkey in the north to Ethiopia in the south and Pakistan in the east to Morocco in the west. Moreover, chickpea genetic diversity has been identified in four centres: the Mediterranean, Central Asia, the Near East and India, along with a secondary centre of origin in Ethiopia (Vavilov 1951). Furthermore, *Cicer* species are distributed from sea level to over 5,000 m near glaciers in the inhospitable areas of the Himalayas (Chandel 1984). Although *C. arietinum* is found from the Mediterranean region to Burma, Ethiopia, Mexico and Chile, *C. bijugum*, *C. echinospermum* and *C. reticulatum* are distributed in Turkey, Syria and Iraq. Other species, such as *C. judaicum* and *C. pinnatifidum* exist in Cyprus, Lebanon, Israel and Syria, while *C. yamashitae* is distributed in Afghanistan. However, *C. microphyllum* is found in the North-west Indian trans-Himalayas and Turkey. These regions have been identified as warm spots for *Cicer* species.

CHARACTERIZATION AND EVALUATION

Characterization is the basic process of grouping the crop gene pool into distinct phenotypic categories using crop descriptor states, while evaluation is a prerequisite for understanding the value of genetic resources carrying important traits of interest for significant utilization. A large number of chickpea germplasm accessions have been characterized and evaluated over the years in various crop-based institutions across the globe. The wild *Cicer* species not only consist of useful variation in distinct agro-morphological characteristics and protein content, but they are also rich sources of resistance to various biotic and abiotic stresses (Croser et al. 2003; Sandhu et al. 2006; Mallikarjuna et al. 2007; Singh et al. 2008, 2014; Kaur et al. 2013), biochemical traits (Kaur et al. 2010) and yield components (Singh and Ocampo 1997; Singh et al. 2005, 2014). Important basic studies pertaining to the evaluation of wild *Cicer* species with respect to agro-morphological traits, major biotic and abiotic stresses and nutritional parameters have been initiated by several international and national institutions across the world.

PRODUCTION-RELATED MAJOR ISSUES

In India, the major portion of chickpea cultivation is under rainfed conditions, characterized by poor soil fertility and low moisture retention capacity. As a result, the crop often faces moisture stress at various growth and development stages. Numerous major abiotic and biotic stresses limit chickpea productivity in India. Among various diseases, Fusarium wilt, Ascochyta blight, wet and dry root rot and Botrytis gray mold are the most prominent. Likewise, chickpea pod borer, cutworms, and bruchids cause significant losses in chickpea production and productivity at various stages. Furthermoer, root-knot nematodes also pose very serious threats in parts of Rajasthan and Gujarat. Among abiotic stresses, cold, drought, heat and salinity are major issues in chickpea cultivation in the country.

ABIOTIC STRESSES

Winter Hardiness

Chickpea is known to be susceptible to cold stress. Therefore, evaluating germplasm against this trait would permit efficient selection of genotypes with cold hardiness (Singh et al. 1995, 2008). This involves rigorous screening of germplasm from wild annual species and cultivated species to identify sources of cold hardiness for use in crop improvement. In this context, reliable selection of accessions from eight annual wild *Cicer* species revealed significant levels of cold tolerance in *C. bijugum*, *C. reticulatum*, *C. echinospermum* and *C. pinnatifidum* (Singh et al. 1995, 2008). Some accessions of *C. bijugum* and *C. reticulatum* showed a high degree of cold tolerance with ratings of 1 and 2. Furthermore, Bhardwaj and Sandhu (2009) reported that cold hardiness is under the control of a single recessive gene. Screening of wild *Cicer* species has revealed promising traits for cold hardiness (Berger et al. 2012).

Drought Tolerance

Drought is the second most important abiotic factor limiting chickpea productivity. Therefore, the development of early maturing varieties coupled with early growth vigour may help to utilize the available soil moisture efficiently. The variability found in *C. judaicum* for early flowering (Robertson et al. 1997) needs to be exploited to breed early maturing genotypes. The involvement of a single recessive locus in controlling early flowering time and dominant delayed flowering has been reported (Kumar and van Rheenen 2000). This recessive allele is significant for breeding and early maturity, and molecular mapping of this locus will help chickpea breeders transfer the locus to locally adapted varieties through the backcross method more efficiently. Apart from this, a few accessions of *C. pinnatifidum* and *C. reticulatum* were found to be tolerant to drought (Toker et al. 2007). In another study, Canci and Toker (2009) reported a few accessions of *C. pinnatifidum* and *C. reticulatum* performed better under drought conditions, and these lines could be considered the best available drought-tolerant sources for breeding purposes. Similarly, drought-responsive yield traits were found to be present in *C. microphyllum* (Srivastava et al. 2016).

Heat Resistance

Yield losses due to high-temperature stress have been estimated to be around 3.3 million tonnes per annum (Ryan 1997; Canci and Toker 2009; Jha et al. 2014). Wild *Cicer* species are known to possess sources of resistance to multiple stresses including heat resistance. A study by Canci and Toker (2009) has shown that *C. reticulatum* and *C. pinnatifidum* resist heat stress up to 41.8°C, although the former should be taken into account in short-term breeding programmes since it can be crossed with cultivated chickpea.

Salinity Resilience

Chickpea production is adversely affected by salinity in arid and semi-arid regions of the world (Ali et al. 2002; Jha et al. 2014). Breeding for salt resilience relies on the assessment of allelic variation for salt resilience in the germplasm to create modern high-yielding cultivars. A perennial wild relative, *C. microphyllum*, is known to possess salinity resilience under salt stress conditions (Srivastava et al., 2016).

BIOTIC STRESSES

Disease Resistance

Chickpea is highly susceptible to *Ascochyta* blight, Botrytis gray mold (BGM), *Fusarium* wilt, nematodes and root rot. Several wild species are identified to be resistance sources for *Ascochyta* blight in chickpea, such as *C. reticulatum* and *C. echinospermum* (Collard et al. 2001). Santra et al. (2000) reported two quantitative trait loci (QTLs), QTL1 and QTL2 that confer resistance to *Ascochyta* blight in the United States of America. These QTLs accounted for an estimated 34.4% and 14.6% of the total phenotypic variance (Santra et al. 2000; Tekeoglu et al. 2002). The environmental effect of *Ascochyta* blight QTL1 and QTL2 was analyzed at Eskisehir, Turkey (Tekeoglu et al. 2004), using CRIL-7 (Santra et al. 2000), developed in the United States from the interspecific cross *C. arietinum* (FLIP84–92C) × *C. reticulatum* (PI 599072). A linkage map was constructed based on an interspecific F_2 population derived from the cross between cultivated chickpea and *C. echinospermum* (Collard et al. 2003). Thereafter, Pande et al. (2006) identified moderate levels of resistance in accessions belonging to *C. bijugum*, *C. cuneatum*, *C. judaicum* and *C. pinnatifidum*. Some accessions of *C. echinospermum* have also shown resistance to *Ascochyta* blight (Pande et al. 2006). In another study by Houari et al. (2015), *Ascochyta* blight resistance was found in accessions of *C. reticulatum*, *C. echinospermum* and *C. judaicum*.

Likewise, resistance to BGM has been identified in *C. bijugum* (Haware et al. 1992), *C. judaicum* and *C. pinnatifidum* (van der Maesen and Pundir 1984). Pande et al. (2006) also identified BGM resistance in *C. echinospermum* accessions. Stevenson and Haware (1999) attributed resistance in *C. bijugum* to BGM to high concentrations of maackiain (200–300 μg/g) compared to low concentrations of maackiain (70 μg/g) in susceptible wild and cultivated species. In a study by Kaur et al. (2013), *C. pinnatifidum* was shown to possess resistant derivative lines for BGM, which can serve as good pre-breeding material. Furthermore, *C. judaicum* has shown resistance to *Fusarium* wilt due to the presence of three isoflav-3-enes, together with two pterocarpan glycosides in its roots (Mallikarjuna et al. 2011). Initial experiments have shown that these chemical compounds may confer resistance to *Fusarium* wilt fungi (Stevenson and Veitch 1996). Later, these compounds were isolated in many annual wild *Cicer* species (Stevenson and Veitch 1998). The resistance of wild *Cicer* species to this disease has been recently documented by Jiménez-Díaz et al. (2015) and Jendoubi et al. (2017).

Pest Resistance

The legume pod borer, *H. armigera*, is an important pest of chickpeas globally, affecting crop yield considerably. Resistance to pod borers has been reported in wild *Cicer* species such as *C. bijugum*, *C. echinospermum*, *C. judaicum*, *C. pinnatifidum* and *C. reticulatum* (Sharma et al. 2005). Mallikarjuna et al. (2007) utilized *C. echinospermum* and *C. reticulatum* to obtain progeny that consistently showed low field damage (10% or less) due to pod borers. Laboratory bioassay using third instar larvae fed on the pods of resistant plants showed reduced larval weight, delayed pupation, failure to pupate or death before pupation, and in some cases, abnormal adults. This indicates that an antibiosis mechanism of resistance exists in these wild *Cicer* species, which can be transferred and exploited in breeding programmes to develop cultivars with resistance to this insect. Further, recombinant inbred lines (RILs) derived from ICC 4958 (*C. arietinum*) × PI489777 (*C. reticulatum*) were evaluated for resistance to pod borer using a detached leaf assay in the laboratory and under natural infestation in the field (ICRISAT 2008). The results indicated considerable variation in resistance to pod borer damage, growth and survival of the larvae. Several RILs (nos. 2, 13, 16, 17, 31, 40, 60, 65, 72, 81, 92, 95 and 123) showed low leaf feeding and low larval weight gain at the vegetative and/or flowering stages. These lines can be exploited in chickpea breeding programmes.

Moreover, in a recent study by Golla et al. (2018), wild relatives of chickpea were evaluated for resistance to *H. armigera* using a detached pod assay. It was observed that *C. bijugum*, *C. cuneatum* and *C. reticulatum* showed high levels of resistance to *H. armigera* due to the presence of protease inhibitors, thereby resulting in lesser growth of the pod borer.

Agro-Morphological Traits

Chickpea breeding programmes aim to develop early maturing cultivars especially to increase crop adaptation by avoiding abiotic stresses that affect crop production globally. Wild relatives of chickpea, such as *C. reticulatum*, *C. echinospermum*, *C. bijugum* and *C. pinnatifidum*, have been shown to possess early flowering and maturity rates (Singh et al. 2014).

In chickpeas, seed number and size are important traits for trade purposes. Following simple or complex crosses, large-sized advanced backcross populations are being generated using cultivated chickpea as the recipient with the objective of recovering the maximum genetic background of the cultivated types with small desirable segments introgressed from wild species. A high number of seeds per plant (up to 460 seeds) and high seed weight per plant (up to 142 g) have been obtained from advanced backcrosses between cultivated chickpea and wild species (*C. reticulatum* and *C. echinospermum*), thereby showing considerable variability for these agronomic traits (Langridge and Chalmers 2005; Velu and Singh 2013).

Interspecific hybridization has played an important role in the genetic enhancement of yield in many crops by facilitating the transfer of useful traits from wild forms. Interspecific hybridization in chickpea, involving wild species of the primary gene pool, has demonstrated the possibility of increasing seed yield. High variability for yield per plant in the F_2 generation from the *C. arietinum* × *C. reticulatum* cross was observed, and some higher-yielding F_3 progenies compared to the cultigen were isolated (Jaiswal et al. 1986; Singh et al. 2008). Transgressive segregants with early maturity were also isolated from interspecific crosses, and a high degree of heterosis for seed yield in F_1 was observed in interspecific crosses *C. arietinum* × *C. echinospermum* (153%) and *C. arietinum* × *C. reticulatum* (138.8%) (Singh and Ocampo 1993). The occurrence of varying levels of sterility from the F_2 progenies in *C. arietinum* × *C. echinospermum* crosses can be effectively eliminated by selecting completely fertile plants. Singh and Ocampo (1997) identified several transgressive segregants for agronomic traits in the F_2 population of *C. arietinum* × *C. reticulatum* interspecific cross. The F_2 recombinants with a very large number of secondary branches, high pod number and higher yield were isolated from a cross between *C. arietinum* and *C. judaicum* (Verma et al. 1995; Singh et al. 2008).

PROGRESS IN HYBRIDIZATION

About 300 wild annual *Cicer* accessions and landraces have been characterized and evaluated for various agro-morphological traits and biotic and abiotic stresses. Wild *Cicer* species have been extensively characterized and screened for *Ascochyta* blight (Collard et al. 2001; Croser et al. 2003; Pande et al. 2006), BGM (Rao et al. 2003; Pande et al. 2006), *Fusarium* wilt (Croser et al. 2003; Rao et al. 2003), *Helicoverpa* pod borer (Sharma et al. 2005), drought (Croser et al. 2003; Toker et al. 2007; Kashiwagi et al. 2008), cold (Croser et al. 2003; Toker 2005; Berger et al. 2012) and heat (Canci and Toker 2009).

In chickpea, besides tolerant sources, wild *Cicer* species harbour beneficial traits for high seed protein (Rao et al. 2003) and genetic improvement of agronomic traits in cultivated species. The productivity traits have been introgressed from *C. echinospermum*, *C. reticulatum* (Singh and Ocampo 1997; Singh et al. 2005) and *C. pinnatifidum* (Sandhu et al. 2006, 2012). Of the eight annual wild *Cicer* species, only *C. reticulatum* is readily crossable with cultivated chickpea resulting in a fertile hybrid, whereas for exploitation of the remaining seven annual wild *Cicer* species for chickpea improvement, specialized techniques such as application of growth hormones, embryo rescue, ovule culture and other tissue culture techniques have been suggested by various researchers (Lulsdorf et al. 2005; Mallikarjuna and Jadhav 2008). Promising high-yielding lines with good agronomic and seed traits such as early flowering and high 100-seed weight have also been obtained from hybrids involving *C. reticulatum* and *C. echinospermum* with cultivated chickpea (Malhotra et al. 2003; Singh et al. 2005; Upadhyaya 2008).

The first report on interspecific crosses involving *C. arietinum* with *C. reticulatum* and *C. cuneatum* was published by Ladizinsky and Adler (1976). The cross between *C. arietinum* and *C. reticulatum* was achieved successfully. Subsequently, wide hybridization was attempted between *C. arietinum* and *C. echinospermum* by various workers (Pundir and Mengesha 1995; Singh and Ocampo 1993). van Dorrestein et al. (1998) attempted crosses involving *C. arietinum, C. bijugum* and *C. judaicum*. Due to the use of *in vitro* techniques, success has been made in achieving hybrids between *C. arietinum* and *C. bijugum* and *C. arietinum* and *C. judaicum*. Badami et al. (1997) also reported successful hybridization between *C. arietinum* and *C. pinnatifidum* using embryo rescue technique.

Further, using various techniques, interspecific hybrids have been produced between *C. arietinum* × *C. cuneatum* (Singh and Singh 1989), *C. arietinum* × *C. judaicum* (Singh et al. 1999), *C. arietinum* × *C. bijugum* (Singh et al. 1999) and *C. arietinum* × *C. pinnatifidum* (Mallikarjuna and Jadhav 2008) to exploit the possibility of introgression of desirable alien genes from these wild *Cicer* species into the cultivated chickpea. Cross-combinations of cultivated chickpea cultivars with *C. reticulatum* and *C. echinospermum* exhibited higher variability for important yield-related traits (Singh et al. 2015).

Molecular Markers and Diversity Analysis

The genetic relationship among wild species of *Cicer* was analyzed using isozyme variation study (Tayyar and Waines 1996), which was later assessed by Rajesh et al. (2003) using inter-simple sequence repeat (ISSR) markers. It was found that linkage drag of undesirable traits hampered the efforts of transferring elite traits from wild species to cultivated chickpea. Therefore, Aryamanesh et al. (2010) developed a genetic linkage map of chickpea using an interspecific F_2 population between domesticated species and *C. reticulatum*. They identified a closely related marker for growth habit and QTLs for *Ascochyta* blight resistance and flowering time. Thereafter, Singh et al. (2014) made detailed characterization and evaluation of global wild annual *Cicer* accessions for target traits. Plant pigmentation showed variation in *C. reticulatum*, *C. judaicum* and *C. pinnatifidum* along with lightly pubescent leaves, except *C. yamashitae*, where it was densely pubescent. In most of the *Cicer* species, seed shape was angular, with the exception of *C. bijugum*, where it was irregular, rounded and pea-shaped. Testa texture was rough in *C. reticulatum*, *C. judaicum*, *C. pinnatifidum* and *C. yamashitae*, but appeared tuberculated in *C. bijugum* and *C. echinospermum*. Similarly, substantial variation in seed colour was observed in *C. reticulatum*, *C. judaicum* and *C. pinnatifidum*.

Although QTLs controlling agro-morphological traits in chickpeas were identified, a genome-wide scanning of wild *Cicer* accessions was revealed by the studies of Saxena et al. (2014) and Das et al. (2015). Moreover, comprehensive comparative transcriptome profiling and high-resolution QTL mapping of *C. microphyllum* revealed molecular machinery regulating agronomic traits in chickpeas (Srivastava et al. 2016), which was further followed by a detailed analysis of differentially expressed transcripts related to traits ranging from seed growth and metabolic processes to elite traits of interest in chickpea breeding programmes (Srivastava et al. 2016; Sagi et al. 2017). These findings have opened new avenues for systematic characterization and evaluation of wild *Cicer* species, which would aid in tapping the unexplored variability by trait introgression for broadening the genetic base of cultivated varieties.

Ahmad (1999) and Sudupak et al. (2002) used RAPD markers to study genetic relationships among the wild *Cicer* species. The RAPD analysis placed *C. reticulatum* and *C. echinospermum* in a single cluster; *C. chorassanicum* and *C. yamashitae* in the next cluster; *C. bijugum*, *C. judaicum* and *C. pinnatifidum* in the third cluster; and *C. cuneatum* in the fourth cluster. This investigation was further followed by AFLP analysis for the same *Cicer* species (Sudupak et al. 2004). The highest degree of conservation was observed in *C. reticulatum* (92%) and *C. echinospermum* (83%), whereas *C. cuneatum* showed the lowest (50%), which further supports the cross-ability, isozyme and RAPD studies-based grouping of species. *C. anatolicum*, a wild perennial relative of chickpea, showed 72% conservation, which strongly supports the karyotypic studies (Ahmad 1989).

CONCLUSIONS AND FUTURE PROSPECTS

The cultivated chickpea has a narrow genetic base and is susceptible to biotic and abiotic stresses. Therefore, characterization, evaluation and identification of wild *Cicer* species will be a significant aid in the discovery of target traits of interest. With the use of modern breeding approaches and pre-breeding activities, some useful traits can be introgressed into the background of the cultivated gene pool, which can be further used to enhance the yield level of chickpeas. There is an urgent need to augment germplasm collections including landraces for utilization purposes. Collaborative efforts among national and international research institutions would help in the successful evaluation of chickpea germplasm systematically at global hotspots. This will ensure the continuous supply of new genetic variability into the main breeding programmes to accelerate genetic gains.

REFERENCES

Abbo S, Jens Berger J, Turner NC (2003) Evolution of cultivated chickpea: four bottlenecks limit diversity and constrain adaptation. *Funct Plant Biol* 30:1081–1087.

Ahmad F (1989) The chromosomal architecture of *Cicer anatolicum* Alef., a wild perennial relative of chickpea. *Cytologia* 54:753–757.

Ahmad F (1999) Random amplified polymorphic DNA (RAPD) analysis reveals genetic relationships among the annual *Cicer* species. *Theor Appl Genet* 98:657–663.

Ali MY, Krishnamurthy L, Saxena NP, Rupela OP, Kumar J, Johansen C (2002) Scope for genetic manipulation of mineral acquisition in chickpea. *Plant Soil* 245:123–134.

Aryamanesh N, Nelson MN, Yan G, Clarke HJ, Siddique KHM (2010) Mapping a major gene for growth habit and QTLs for Ascochyta blight resistance and flowering time in a population between *C. arietinum* and *C. reticulatum*. *Euphytica* 173:307–331.

Badami PS, Mallikarjuna N, Moss JP (1997) Interspecific hybridization between *Cicer arietinum* and *C. pinnatifidum*. *Plant Breed* 116:393–395.

Berger JD, Kumar S, Nayyar H, Street KA, Sandhu J, Henzell JM, Kaur J, Clarke HC (2012) Temperature stratified screening of chickpea genetic resource collections reveals very limited reproductive chilling tolerance compared to its annual wild relatives. *Field Crop Res* 126:119–129.

Bhardwaj R, Sandhu JS (2009) Pollen viability and pod formation in chickpea (*Cicer arietinum* L.) as a criterion for screening and genetic studies of cold tolerance. *Indian J Agric Sci* 79:63–65.

Canci H, Toker C (2009) Evaluation of annual wild *Cicer* species for drought and heat resistance under field conditions. *Genet Resour Crop Evol* 56:1–6.

Chandel KPS (1984) A note on the occurrence of wild *C. microphyllum* Benth. and its nutrient status. *Int Chickpea Newsl* 10:4–5.

Chaturvedi SK, Nadarajan N (2010) Genetic enhancement for grain yield in chickpea-accomplishments and resetting research agenda. *Electron J Plant Breed* 14:611–615.

Collard BCY, Ades PK, Pang ECK, Brouwer JB, Taylor PWJ (2001) Prospecting for sources of resistance to *Ascochyta* blight in wild *Cicer* species. *Aust Plant Pathol* 30:271–276.

Collard BCY, Pang ECK, Ades PK, Taylor PWJ (2003) Preliminary investigation of QTLs associated with seedling resistance to *Ascochyta* blight from *Cicer echinospermum*, a wild relative of chickpea. *Theor Appl Genet* 107:719–729.

Croser JS, Ahmad CF, Clarke HJ, Siddique KHM (2003) Utilisation of wild *Cicer* in chickpea improvement-progress, constraints and prospects. *Aust J Agric Res* 54:429–444.

Das S, Upadhyaya HD, Bajaj D, Kujur A, Badoni S, Kumar V, Tripathi S, Gowda CL, Sharma S, Singh S, Tyagi AK, Parida SK (2015) Deploying QTL-seq for rapid delineation of a potential candidate gene underlying major trait-associated QTL in chickpea. *DNA Res* 22:193–203.

FAO (2016) FAOSTAT. Accessed on 20 July 2023. https://www.fao.org/faostat/en/#data/QCL

FAO (2018) FAOSTAT. Accessed on 20 July 2023. https://www.fao.org/faostat/en/#data/QCL

Golla SK, Rajasekhar P, Akbar SMD, Sharma HC (2018) Proteolytic activity in the midgut of *Helicoverpa armigera* (Noctuidae: Lepidoptera) larvae fed on wild relatives of Chickpea, *Cicer arietinum*. *J Econ Entomol* 111(5):2409–2415.

Harlan JR, de Wet MJ (1971) Towards a rational classification of crop plant. *Taxonomy* 20:509–517.

Haware MP, Nene YL, Pundir RPS, Narayana Rao J (1992) Screening of world chickpea germplasm for resistance to *Fusarium* wilt. *Field Crops Res* 30:147–154.

Houari SO, Bouteflika DA, Lamamra AA (2015) An assessment of wild *Cicer* species accessions for resistance to three pathotypes of *Ascochyta rabiei* (Pass) Labr. in Algeria. *Afr J Agron* 3:228–234.

ICRISAT (2008) International Crops Research Institute for the Semi-Arid Tropics. Accessed on 15 July 2023. https://www.icrisat.org/

Jaiswal HK, Singh BD, Singh AK, Singh RM (1986) Introgression of genes for yield and yield traits from *C. reticulatum* into *C. arietinum*. *Inter Chickpea Newsl* 14:5–8.

Jendoubi W, Bouhadida M, Boukteb A, Be´ji M, Kharrat M (2017) *Fusarium* wilt affecting chickpea crop. *Agriculture* 7:23.

Jha UC, Bohra A, Singh NP (2014) Heat stress in crop plants: its nature, impacts and integrated breeding strategies to improve heat tolerance. *Plant Breed* 133:679–701.

Jiménez-Díaz RM, Castillo P, del Mar Jime´nez-Gasco M, Landa BB, Navas-Corte´s JA (2015) *Fusarium* wilt of chickpeas: biology, ecology and management. *Crop Prot* 73:16–27.

Kashiwagi J, Krishnamurthy L, Upadhyaya HD, Gaur PM (2008) Rapid screening technique for canopy temperature status and its relevance to drought tolerance improvement in chickpea. *J SAT Agric Res* 6:4–105.

Kaur A, Sandhu JS, Gupta SK, Bhardwaj R, Bansal UK, Saini RG (2010) Genetic relationships among annual wild *Cicer* species using RAPD analysis. *Indian J Agric Sci* 80(4):309–311.

Kaur L, Sirari A, Kumar D, Sandhu JS, Singh S, Singh I, Kapoor K, Gowda CLL, Pande S, Gaur P, Sharma M, Imtiaz M, Siddique KHM (2013) Harnessing *ascochyta* blight and botrytis grey mould resistance in chickpea through interspecific hybridization. *Phytopathol Mediterr* 52(1):157–165.

Kumar J, Van Rheenen HA (2000) A major gene for time of flowering in chickpea. *J Hered* 91:67–68.

Ladizinsky G, Adler A (1976) The Origin of chickpea *Cicer arietinum* L. *Euphytica* 25:211–217.

Langridge P, Chalmers K (2005) The principle: identification and application of molecular markers. In: Horst L, Wenzel G (eds) *Biotechnology in Agriculture and Forestry, Molecular Marker Systems in Plant breeDing and Crop Improvement*. Springer, Berlin, pp. 3–21.

Lulsdorf M, Mallikarjuna N, Clarke H, Tar'an B (2005) Finding solutions for interspecific hybridization problems in chickpea (*Cicer arietinum* L.). In: *4th International Food Legumes Research Conference,* 18–22 October, New Delhi, India, p. 44.

Malhotra RS, Baum M, Udupa SM, Bayaa B, Kababbeh S, Khalaf G (2003) *Ascochyta* blight research in chickpea present status and future prospects. In: Sharma RN, Srivastava GK, Rathore AL, Sharma ML, Khan MA (eds) *Proceedings of the International Chickpea Conference Chickpea Research for the Millennium,* 20–22 January 2003, Raipur, Chhattisgarh, India, pp. 108–117.

Mallikarjuna N, Jadhav DR (2008) Techniques to produce hybrids between *Cicer arietinum* L. × *Cicer pinnatifidum* Jaub. *Indian J Genet* 68:398–405.

Mallikarjuna N, Sharma HC, Upadhyaya HD (2007) Exploitation of wild relatives of pigeonpea and chickpea for resistance to *Helicoverpa armigera. EJ SAT Agric Res Crop Improv* 3(1):4–7.

Mallikarjuna N, Coyne C, Cho S, Rynearson S, Rajesh PN, Jadhav DR, Muehlbauer FJ (2011) *Cicer*. In: Kole C (ed) *Wild Crop Relatives: Genomic and Breeding Resources, Legume Crops and Forages*. Springer, Berlin, pp. 63–82.

Pande S, Ramgopal D, Kishore GK, Mallikarjuna N, Sharma M, Pathak M, Narayana Rao J (2006) Evaluation of wild *Cicer* species for resistance to *Ascochyta* blight and Botrytis gray mold in controlled environment at ICRISAT, Patancheru, India. *Int Chickpea Pigeonpea Newsl* 13:25–26.

Pundir RPS, Mengesha MH (1995) Cross compatibility between chickpea and its wild relative *Cicer echinospermum* Davis. *Euphytica* 83:241–245.

Rajesh PN, Sant VJ, Gupta VS, Muehlbauer FJ, Ranjekar PK (2003) Genetic relationships among annual and perennial wild species of *Cicer* using inter simple sequence repeat (ISSR) polymorphism. *Euphytica* 129:15–23.

Rao NK, Reddy LJ, Bramel PJ (2003) Potential of wild species for genetic enhancement of some semi-arid food crops. *Genet Resour Crop Evol* 50:707–721.

Robertson LD, Ocampo B, Singh KB (1997) Morphological variation in wild annual *Cicer* species in comparison with the cultigen. *Euphytica* 95:309–319

Ryan JG (1997) A global perspective on pigeonpea and chickpea sustainable production systems: present status and future potential. In: Asthana AN, Ali M (eds) *Recent Advances in Pulses Research. Indian Society of Pulses Research and Development*. IIPR, Kanpur, India, pp. 1–31.

Sagi MS, Deokar AA, Tar'an B (2017) Genetic analysis of NBSLRR gene family in chickpea and their expression profiles in response to *ascochyta* blight infection. *Front Plant Sci* 8:838.

Sandhu JS, Gupta SK, Singh G, Sharma YR, Bains TS, Kaur L, Kaur A (2006) Interspecific hybridization between *Cicer arietinum* L. and *Cicer pinnatifidum* Jaub. et. Spach for improvement of yield and other traits. In: *4th International Food Legumes Research Conference,* October 18–22, 2005, Indian Society of Genetics and Plant Breeding, New Delhi, p. 192.

Santra DK, Tekeoglu M, Ratnaparkhe M, Kaiser WJ, Muehlbauer FJ (2000) Identification and mapping of QTLs conferring resistance to *Ascochyta* blight in chickpea. *Crop Sci* 40:1606–1612.

Saraf CS, Rupela OP, Hegde DM, Yadav RL, Shivkumar BG, Bhattarai S, Razzaque MA, Sattar MA (1998) Biological nitrogen fixation and residual effects of winter grain legumes in rice and wheat cropping systems of the IndoGangetic plain. In: Kumar JVDK, Johansen C, Rego TJ (eds) *Residual Effects of Legumes in Rice and Wheat Cropping Systems of the Indo- Gangetic Plain*. Oxford/IBH Publishing, New Delhi, pp. 14–30.

Saxena MS, Bajaj D, Kujur A, Das S, Badoni S, Kumar V, Singh M, Bansal KC, Tyagi AK, Parida SK (2014) Natural allelic diversity, genetic structure and linkage disequilibrium pattern in wild chickpea. *PLoS One* 9(9):e107484.

Sharma HC, Pampapathy G, Lanka SK, Ridsdill-Smith TJ (2005) Antibiosis mechanism of resistance to pod borer, *Helicoverpa armigera* in wild relatives of chickpea. *Euphytica* 142:107–117.

Siddique KHM, Brinsmead RB, Knight R, Knights EJ, Paull JG, Rose IA (2000) Adaptation of chickpea (*Cicer arietinum* L.) and faba bean (*Vicia faba* L.) to Australia. In: Knight R (ed) *Linking Research and Marketing Opportunities for Pulses in the 21st Century*. Springer, Berlin, pp. 289–303.

Singh KB (1997) Chickpea (*Cicer arietinum* L.). *Field Crops Res* 53:161–170.

Singh RP, Singh BD (1989) Recovery of rare interspecific hybrids of gram *Cicer arietinum* × *C. cuneatum* L. through tissue culture. *Curr Sci* 58:874–876.

Singh KB, Ocampo B (1993) Interspecific hybridization in annual *Cicer* species. *J Genet Breed* 47:199–204.

Singh KB, Ocampo B (1997) Exploitation of wild *Cicer* species for yield improvement in chickpea. *Theor Appl Genet* 95:418–423.

Singh NP, Pratap A (2016) Food legumes for nutritional security and health benefits. In: Singh U et al (eds) *Biofortification of Food Crops*. Springer, Berlin, pp. 41–50.

Singh KB, Malhotra RS, Saxena MC (1995) Additional sources of tolerance to cold in cultivated and wild *Cicer* species. *Crop Sci* 35:1491–1497.

Singh NP, Singh A, Asthana AN, Singh A (1999) Studies on inter-specific crossability barriers in chickpea. *Indian J Pulses Res* 12:13–19.

Singh S, Gumber RK, Joshi N, Singh K (2005) Introgression from wild *Cicer reticulatum* to cultivated chickpea for productivity and disease resistance. *Plant Breed* 124:477–480.

Singh R, Sharma P, Varshney RK, Sharma SK, Singh NK (2008) Chickpea improvement: role of wild species and genetic markers. *Biotechnol Genet Eng Rev* 25:267–314.

Singh S, Gumber R, Virdi K (2008) Inheritance of growth habit in *Cicer*: Evidences for epistatic effect of *C. reticulatum* gene over *C. arietinum*. In: Kharkwal MC (ed) *Proceedings of the 4th International Food Legumes Research Conference,* New Delhi, India, pp. 146–151.

Singh TP, Deshmukh PS, Kumar P (2008) Relationship between physiological traits in chickpea (*Cicer arietinum* L.) under rainfed condition. *Indian J Plant Physiol* 13:411–413.

Singh RP, Singh I, Singh S, Sandhu JS (2012) Assessment of genetic diversity among interspecific derivatives of *C. arietinum* with *C. pinnatifidum*. *J Food Legu* 25(2):150–152.

Singh M, Bisht IS, Dutta M, Kumar K, Basandrai AK, Kaur L, Sirari A, Khan Z et al (2014) Characterization and evaluation of wild annual *Cicer* species for agro-morphological traits and major biotic stresses under Northwestern Indian conditions. *Crop Sci* 54:229–239.

Singh S, Singh I, Kapoor K, Gaur PM, Chaturvedi SK, Singh NP, Sandhu JS (2014) Chickpea. In: Singh M et al (eds) *Broadening the Genetic Base of Grain Legumes*. Springer, Berlin, pp. 51–73.

Singh M, Kumar K, Bisht IS, Dutta M, Rana MK, Rana JC, Bansal KC, Sarker A (2015) Exploitation of wild annual *Cicer* species for widening the gene pool of chickpea cultivars. *Plant Breed* 134:186–192.

Srivastava R, Bajaj D, Malik A, Singh M, Parida SK (2016) Transcriptome landscape of perennial wild *Cicer microphyllum* uncovers functionally relevant molecular tags regulating agronomic traits in chickpea. *Sci Rep* 6:33616.

Stevenson PC, Veitch NC (1996) Isoflavones from the roots of *Cicer judaicum*. *Phytochemistry* 43:695–700.

Stevenson PC, Veitch NC (1998) A 2-arylbenzofuran from roots of *Cicer bijugum* associated with *Fusarium* wilt resistance. *Phytochemistry* 48:947–951.

Stevenson PC, Haware MP (1999) Maackiain in *Cicer bijugum Rech. F. associated* with resistance to Botrytis gray mold. *Biochem Syst Ecol* 27:761–767.

Sudupak MA, Akkaya MS, Kence A (2002) Analysis of genetic relationships among perennial and annual *Cicer* species growing in Turkey using RAPD markers. *Theor Appl Genet* 105:1220–1228.

Sudupak MA, Akkaya MS, Kence A (2004) Genetic relationships among perennial and annual *Cicer* species growing in Turkey assessed by AFLP fingerprinting. *Theor Appl Genet* 108:937–944.

Tayyar RI, Waines JG (1996) Genetic relationships among annual species of *Cicer* (*Fabaceae*) using isozyme variation. *Theor Appl Genet* 92:245–254.

Tekeoglu M, Rajesh PN, Muehlbauer FJ (2002) Integration of sequence tagged microsatellite sites to chickpea genetic map. *Theor Appl Genet* 105:847–854.

Tekeoglu M, Isuk M, Muehlbauer FJ (2004) QTL analysis of *ascochyta* blight resistance in chickpea. *Turk J Agric Forecast* 28:183–187.

Toker C (2005) Preliminary screening and selection for cold tolerance in annual wild *Cicer* species. *Genet Resour Crop Evol* 52:1–5.

Toker C, Canci H, Yildirim T (2007) Evaluation of perennial wild *Cicer* species for drought resistance. *Genet Resour Crop Evol* 54:1781–1786.

Upadhyaya HD (2008) Crop Germplasm and wild relatives: a source of novel variation for crop improvement. *Korean J Crop Sci* 53:12–17.

van der Maesen LJG (1987) Origin, history and taxonomy of chickpea. In: Saxena MC, Singh KB (eds) *The Chickpea*. CAB International, Cambridge, pp. 11–34.

van der Maesen LJG, Pundir RPS (1984) Availability and use of wild *Cicer* germplasm. *FAO/IBPGR Plant Genet Resour Newsl* 57:19–24.

van Dorrestein B, Baum M, Malhotra RS (1998) Interspecific hybridization between cultivated chickpea (*Cicer arietinum* L.) and the wild annual species *C. judaicum*, *C. pinnatifidum*. In: *Proceedings of 3rd European Conference on Grain Legumes*, 14–19 November 1998, Valladolid, Spain, pp. 362–363.

Varshney RK, Kudapa H, Roorkiwal M, Thudi M, Pandey MK, Saxena RK et al (2012) Advances in genetics and molecular breeding of three legume crops of semi-arid tropics using next-generation sequencing and high-throughput genotyping technologies. *J Biosci* 37:811–820.

Varshney RK, Song C, Saxena RK, Azam S, Yu S, Sharpe AG et al (2013) Draft genome sequence of chickpea (*Cicer arietinum*) provides a resource for trait improvement. *Nat Biotechnol* 31:240–246.

Vavilov NI (1951) The Origin, variation, immunity and breeding of cultivated plants. *Chron Bot* 13:1–366.

Velu G, Singh RP (2013) Phenotyping in wheat breeding. In: Panguluri SK, Kumar AA (eds) *Phenotyping for Plant Breeding: Applications of Phenotyping Methods for Crop*. Springer, Berlin, pp. 41–72.

Verma MM, Sandhu JS, Ravi K (1995) Characterization of the interspecific cross *Cicer arietinum* × *C. judaicum* (Boiss). *Plant Breed* 114:259–266.

2 Back to Wild

Designing Future Climate-Resilient Cultivars of Urdbean and Mungbean

Amandeep, Gurjeet Singh, Inderjit Singh, R. K. Gill, Parul Sharma, Aditya Pratap, Sanjeev Gupta, Harsh Nayyar, and Kadambot H. M. Siddique

INTRODUCTION

Today, global food security is seriously threatened by factors such as expanding human population, changing temperatures, depleting arable land, and decreasing productivity coupled with post-harvest losses. Additionally, surging diseases and new insect pests pose a new challenge to researchers across the globe for developing stress-resistant, high-yielding, nutrient-rich cultivars to reduce poverty and hunger (Singh et al., 2022; Bakala et al., 2020). Legumes are a member of the "Fabaceae" family, which has over 800 genera and 20,000 species. Among families of agriculturally important crops, it is ranked second after cereals and is the third biggest family of angiosperms (Smýkal and Konecná, 2014). They are a vital part of humans' daily diet since they are a source of vitamins, proteins and minerals, as well as cost-effective, helping to avoid chronic illnesses (Vaz Patto et al., 2015; Arnoldi et al., 2015). The largest amount of vegetable protein (around 33%) in the human diet is derived from grain legumes, which account for almost 27% of the global food supply (Young and Bharti, 2012).

In the leguminous family, mungbean (*Vigna radiata* L. *wilczek*) as well as urdbean *(Vigna mungo* (L.) *Hepper)* have become one of the most important crops as they provide a good source of protein. Both mungbean and urdbean are native to India where domestication began nearly 3,500 years ago (Vishnu-Mittre, 1974; Fuller and Harvey, 2006). These are members of the family Fabaceae, tribe Phaseoleae, genus *Vigna*, subgenus Ceratotropis. In the subgenus Ceratotropis, there were 23 species (Tomooka et al., 2010; Aitawade et al., 2012), which include widely cultivated Asian legume species *viz.*, moth bean (*V. aconitifolia*), adzuki bean (*V. angularis*), black gram (*V. mungo*), créole bean (*V. reflexo-pilosa var. glabra*), jungle bean (*V. trilobata*), Tooapee (*V. trinervia*), and rice bean (*V. umbellata*; Table 2.1). The genomic information of mungbean and urdbean has been summarized in Table 2.1. Mungbean

DOI: 10.1201/9781003434535-2

TABLE 2.1
Genomic Information of Mungbean and Urdbean

Genomic Information	Mungbean	Urdbean	References
Botanical name	*Vigna radiata*	*Vigna mungo*	Kang et al. (2014); Jegadeesan et al. (2021); Gupta et al. (2013); Nair et al. (2012)
2*n*	$2n = 2x = 22$	$2n = 2x = 22$	
Origin	South Asia	South Asia	
Major Cultivation area	South and Southeast Asia	South, East, Southeast Asia	
Genome size (mb)	579	475	
Number of genes	22,427	42,115	
Parents used for sequencing	VC1973A	Hepper	
Method used for sequencing	Whole genome sequencing	Whole genome sequencing	

and urdbean were domesticated and selected over several generations to become the cultivated varieties that are now grown all across southern and southeast Asia.

Mungbean and urdbean are mostly grown on small farms during the warm season in Asia (Kang et al., 2014). Mungbean has a short life cycle (approximately 60 days) as compared to urdbean which has a life cycle of around 100 days. It is cultivated on almost 7 million hectares of area worldwide. In India, mungbean was grown on approximately 5.55 m ha with a production of 3.17 m tonnes and productivity of 570 kg/ha whereas urdbean was cultivated on 4.63 m ha with a production of 2.78 m tonnes and average productivity of 600 kg/ha during 2021–22 (Anonymous, 2023). Mungbean as well as urdbean are high in easily digestible proteins, vitamins, minerals, carbohydrates, fibres and phytonutrients, making them an excellent source for improving human health. Due to their easy availability and digestibility, mungbean is sometimes considered a "poor man's" meat (Hall, 1992). Consuming these seeds along with other cereals tends to increase the quality of protein because of the high amount of sulphur in the legumes, which contain amino acids (Fang et al., 2017). Legumes are known to fix atmospheric nitrogen through symbiotic association with nitrogen-fixing bacteria. Therefore, these crops are beneficial for both economic and nutritional reasons because they tend to improve soil fertility and decrease the quantity of nitrogen fertilizer required by the soil. As a result, cereal grain number and straw yields will increase when legumes are grown in rotation with cereals (Yaqub et al., 2010).

In spite of their immense economic importance as well as health benefits, the frequency of genetic improvement in mungbean and urdbean over the past 20 years has been quite slow in comparison to that gained in cereal crops (Figure 2.1). The cause for this slow genetic gain in these legumes is the narrow genetic base coupled with the lack of new breeding techniques to introgress required genes associated with biotic and abiotic stresses from non-cultivated species to cultivated varieties (Araujo et al., 2015). The major primary factors that limit the genetic gain in mungbean and urdbean are: (a) accumulation of domestication syndrome traits forming a narrow genetic base, (b) use of the same breeding lines continuously in the legume breeding

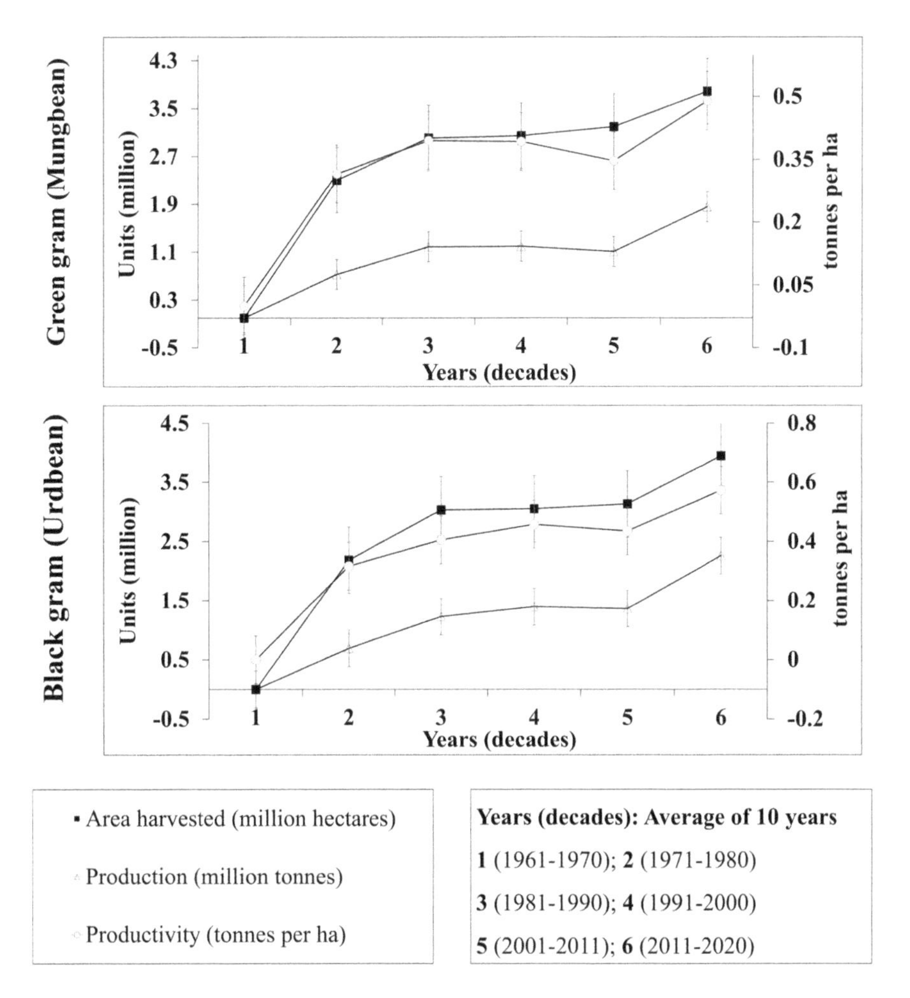

FIGURE 2.1 Area (in million hectare), production (in million tonnes), and productivity (in tonnes per hectare) of mungbean and urdbean.

programmes, (c) linkage drag of undesirable traits, (d) monophyletic evolution and (e) a lack of exploitation of wild relatives.

Along with these factors, the lack of phenotyping facilities and crossing barriers also reduces the potential of utilizing the genetic variability, which is considered as secondary and tertiary gene pools (Singh et al., 2013). Additionally, the improvement of legumes will also depend on sensitivity towards the environment, slow growth, a lack of appropriate management practices, and favourable government policies. These legume crops have a weak source-sink connection, slow accumulation of dry matter, lower canopy development and seedling vigour, less photosynthetic efficiency and premature leaf senescence. In addition to this, they have a lower leaf area index, *i.e.*, 0.7–0.2, which ultimately reduces gas exchange and photosynthetic capacity. The flowering capacity of mungbean and urdbean is very high as they produce many flowers, but most of the flowers fall before setting into fruits due to a lack of nutrient accumulation. In mungbean, nearly 70%–90% of produced flowers drop before setting into fruit (Alam Mondal et al., 2011). Moreover, these crops are vulnerable to photoperiod, high-temperature conditions and the relationship between environment and genotype. Additionally, during the early stages of crop growth and development, weed infestation is high, and they compete for nutrients, light, water and space, affecting the health of crop plants. Many insect pests also harbour on weeds, which act as their alternate hosts. Reduced productivity of these legume crops is also a result of poor management techniques, such as rainfed cultivation, low seed replacement and improper post-harvest handling. On the other hand, Government policies like lack of guaranteed marketing places, proper storage facilities, MSP (minimum support price), and processing industries are equally responsible for the present status of mungbean and urdbean.

In the last decade, the impact of global warming is evident as India has witnessed highly fluctuating weather conditions (Oldenborgh et al., 2018). Due to high temperatures, there is a change in the rainfall pattern as well as its distribution, which has led to water scarcity, and this water scarcity will increase drought-affected regions. Moreover, it will also affect those regions that have normal precipitation (McKersie, 2015). Thus, aberrant weather conditions will have a negative impact on crop production as the increasing temperatures will lead to the production of poor biomass; reduction in days to flowering, rate of fertilization, and seed production, which ultimately make the crops vulnerable to diseases and insect pests (Ali and Gupta, 2012). According to the Food and Agriculture Organization (FAO) report (FAOSTAT, 2016), climate change has put global food security more at risk; heightened the dangers of under-nutrition in resource-poor regions of the world due to heat, drought, salinity, and waterlogging; and increased the threat of newly emerging diseases and insect pests. Several researchers have conducted studies to assess the impact of drought on crop yields, and it has been estimated that drought can lead to yield reduction up to 20%–90% which can go up to 100% in the worst-case scenario.

Considering the global environmental changes, avoiding biotic and abiotic stress, and resisting diseases and pest outbreaks, improving the genetic makeup of these legume crops is of utmost importance, which can only be achieved through the introgression of traits like tolerance and escape through gene pooling, wherein crop wild races and landraces can be utilized (Dwivedi et al., 2016; Sharma et al., 2013). To broaden the narrow genetic base of leguminous crops, there is a requirement to fully exploit legume crops through the combination of phenotyping and functional genomics. Keeping in view the immense importance of mungbean and urdbean in food security and sustainable agricultural systems, this book chapter focuses on the challenges and strategies for developing climate-resilient cultivars of these legumes.

CONCEPT OF GENE POOL

In 1971, Harlan and de Wet introduced the concept of "gene pools" for the judicious and effective usage of germplasm. The total of all the alleles or genes existing in a crop species that could be harnessed by breeders for the improvement of the crop is called the gene pool (GP). GPs can be categorized into three categories based on the ease of hybridization or crossing with other species: primary gene pool (GP-1), secondary gene pool (GP-2) and tertiary gene pool (GP-3). GP-1 comprises all the germplasm lines such as cultivated varieties, elite germplasm lines and landraces, where fertilization among different individuals is possible without affecting normal seed set, segregation and recombination, enabling gene transfer from one individual to another through routine breeding. GP-2 includes species that have some degree of barriers to crossability with the crop plants. The crosses of this GP produce sterile hybrids and abnormal chromosome pairing, restricting the normal transfer of genes. However, this GP can be utilized with special efforts to overcome barriers of crossability and develop normal seeds. The transfer of genes from GP-2 to GP-1 is possible but with great difficulty. GP-3 is relatively more distant from GP-2, resulting in hybrids that, if successfully attempted, tend to be lethal or sterile due to abnormalities in embryo development. Normally, gene transfer is not possible, but successful hybrid embryos can be produced through special techniques such as embryo culture, tissue culture chromosome doubling or using bridging species. It was considered the outer limit of germplasm in terms of potential use, but the latest developments in cell and tissue culture technology have brought this GP within the reach of breeding programmes for most crop plants.

GENE POOL OF MUNGBEAN AND URDBEAN

Generally, the GP of a crop is mainly defined based on the crossability between domesticated forms and their close relatives, but cross-compatibility among *Vigna* species is not well-defined. As a result, the GP of mungbean and urdbean is not clearly defined. Based on hybridization studies, several researchers have classified the GP of mungbean and urdbean, as presented in Figure 2.2.

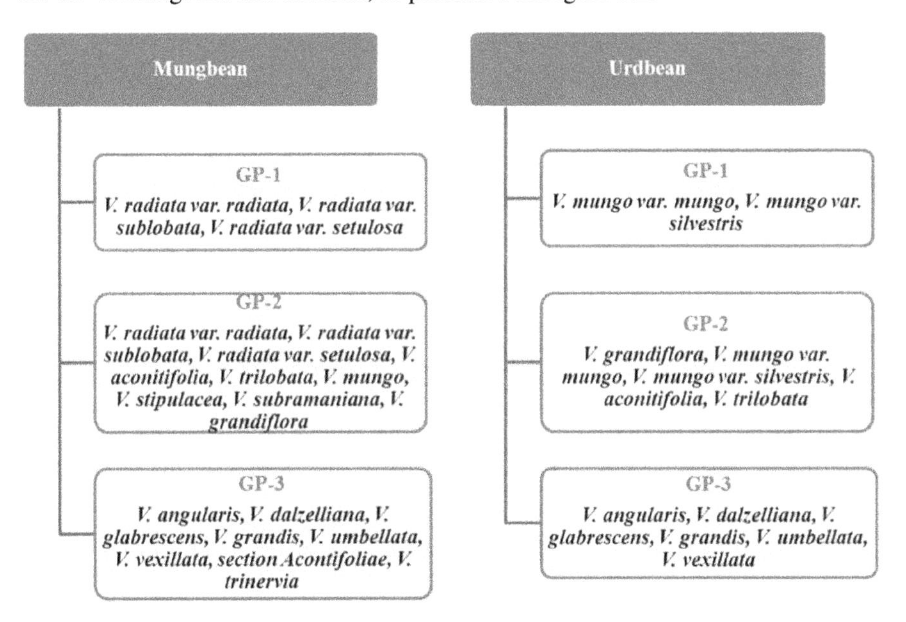

FIGURE 2.2 Different gene pools of mungbean and urdbean. (Kumar et al., 2011.)

GERMPLASM COLLECTIONS OF MUNGBEAN AND URDBEAN

Ex-situ conservation of plant genetic resources (PGRs), which contain a wide range of diversity, is crucial for the development of new varieties to meet the food security needs of a growing global population and to address the problems posed by the adverse effects of climate change. Grain legumes are among the most important crops that can help solve both of these problems. Since humans first began practicing farming, selecting seeds and storing seeds in the Neolithic era, the need to conserve plant genetic resources for food and agriculture (PGRFA) has been recognized. However, systematic explorations, collections and conservation of crop germplasm began in 1916 by the renowned Russian geneticist N. I. Vavilov. Today, over 7.4 million accessions are conserved ex-situ worldwide.

Many institutes and centers around the world are involved in the maintenance of mungbean genetic resources. These include the National Bureau of Plant Genetic Resources (NBPGR), New Delhi; the University of the Philippines; the Institute of Crop Germplasm Resources of the Chinese Academy of Agricultural Sciences; AVRDC-The World Vegetable Center, Taiwan; and the Plant Genetic Resources Conservation Unit of the University of Georgia, the United States of America (Ebert, 2012). Core collections of mungbean have been established in countries such as the United States of America, India, China and Korea to facilitate the effective use of genetic resources and improve access for breeders. The World Vegetable Centre AVRDC recently created a hub of collections consisting of 1,481 accessions and a core consisting of 296 accessions (Schafleitner et al., 2015). The mini-core was generated through molecular analysis employing 20 SSR markers. With 12,153 accessions, AVRDC-The World Vegetable Centre currently has the largest collection of *Vigna* germplasm in the world, serving as a valuable source for interspecific hybridization. In addition, a significant number of mungbean germplasm accessions (~4,325) are preserved at the NBPGR, New Delhi.

In the case of urdbean, the Bangladesh Agricultural Research Institute, Dhaka, has a collection of 339 accessions, while NBPGR, New Delhi, has a collection of 1,429 accessions. AVRDC-The World Vegetable Centre presently maintains a collection of 478 accessions of urdbean. Additionally, the Regional Plant Introduction Station in Georgia, the United States of America, holds a collection of 277 accessions of urdbean.

CONCEPT OF INTROGRESSION AND PRE-BREEDING

New gene combinations resulting from spontaneous hybridization between crop plants and their wild relatives, as well as the trait introgression through spontaneous hybridization are considered major drivers of gene flow to expand the genetic diversity of crop plants (Arnold, 1992; Ellstrand et al., 1999). However, determining the extent and significance of such introgression is very difficult (Jarvis and Hodgkin, 1999). Therefore, planned introgression of desirable traits has become an essential part of plant breeding since 1949, when conventional breeding techniques such as backcrossing for trait introgression were introduced by Dr. Edgar Anderson (Anderson, 1949). Pre-breeding and distant hybridization are considered useful combinations for introducing certain traits from wild species into elite lines to increase their genetic diversity (Simmonds, 1993; Gill et al., 2011).

Pre-breeding refers to all the procedures involved in identifying desirable traits or genes in unadapted germplasm (exotic or wild donor parents that cannot be used directly in breeding programmes) and transferring those traits or genes into the backgrounds of cultivars that are well-adapted to their environment (recipients). Pre-breeding provides a great opportunity to increase the genetic diversity of the GP-1 by harnessing the genetic diversity present in wild species. This will provide a steady flow of new and useful genetic diversity into the breeding pipelines for the creation of novel cultivars with high resistance and broader adaptability (Shimelis and Laing, 2012). In order to develop stable introgression lines pre-breeding activities must be initiated to transfer desired genes and quantitative trait loci (QTLs) found in wild relatives and landraces (Jarvis and Hodgkin, 1999). These ILs can be crossed directly with working collections to create high-yielding, superior-quality varieties that are also very resilient to changing climates, new pests and diseases (Sharma, 2017).

Like other crops, the wild relatives of legumes exhibit significant genetic variation in several important traits. Legume breeders are reluctant to use wild species in their breeding programmes, despite the potential benefits for crop improvement. This reluctance results from inadequate germplasm characterization, cross-incompatibility and associated linkage drags. Linkage drag is a term used to describe the decrease in fitness of an individual caused by the alien introgression of both undesirable and desirable genes. Along with the desired trait, linkage drag may also introduce undesired traits such as delayed maturity, pod cracking, an unattractive seed coat color and texture, anti-nutritional factors, etc. The most time and resource-intensive stage of the IL development process is the need for numerous backcrosses to eliminate such linkage drags (Sharma et al., 2013; Gudi et al., 2020). Breeders can, however, use modern breeding techniques, such as marker-assisted backcrossing (MABC), to eliminate linkage drag and hasten the recovery of the recurrent parent genome.

WILD RELATIVES AS POTENTIAL DONORS FOR IMPORTANT TRAITS

Modern agricultural practices have domesticated several crop plants from their wild relatives into cultivars with altered phenotypes as well as genotypes (Gros-Balthazard and Flowers, 2021). Plants that were chosen for cultivation during the early stages of domestication were indistinguishable from their wild relatives. The accumulation of domestication syndrome traits, however, caused cultivated plants to diverge from their wild species throughout the course of artificial selection. The genetic diversity of crops decreased during domestication due to artificial selection for economically important traits (such as enhanced grain yield, self-fertility, etc.). Additionally, domestication significantly reduced effective population sizes, which changed the population's genotypic frequencies. It also resulted in a population domestication bottleneck, further lowering the genetic variability of crop plants (Gros-Balthazard and Flowers, 2021). Agricultural productivity has suffered from lowered genetic diversity, changing climatic conditions and a rise in insect populations. It became necessary for plant breeders to search for valuable genetic resources present in crop

TABLE 2.2
Wild Species of Mungbean and Urdbean Identified as Promising Donors for Rendering Economic Traits

Traits	Important Traits	Wild Relatives		References
		Mungbean	Urdbean	
Resistance to pest	Bruchid beetles	*V. radiata var. sublobata*	*V. umbellata*	Miyagi et al. (2004); Schafleitner et al. (2016); Somta et al. (2006)
	Pod bug	–	*V. unguiculata subsp. dekindtiana*	Koona et al. (2002)
	Nematode	–	*V. angularis*	Kushida et al. (2013)
Resistance to disease	Yellow mosaic virus	*V. radiata var. sublobata*	*V. umbellat*	Pal and Inderjit (2000); Pandiyan et al. (2008)
Yield traits	Long pods and profuse poding	*V. radiata var. sublobata*	–	Tripathy et al. (2016)
	Flower colour, pod pubescence, germination habit	*V. racemosa, V. reticulata, V. vexillata*	*V. ambacensis*	Harouna et al. (2020)
Abiotic stresses	Heat and warm weather		*V. angularis*	Tomooka et al. (2001)
	Photo and thermo insensitivity	*V. glabrescens*	*V. umbellata*	Pratap et al. (2014)
	Salt tolerance	*V. luteola, V. marina*	*V. vexillata, V. trilobata*	Yoshida et al. (2020); Iseki et al. (2016)
	Salinity, heat and drought	*V. aconitifolia, V. unguiculata,*	*V. monantha, V. exilis*	Van Zonneveld et al. (2020)
Quality Traits	Low trypsin inhibitor activity	*V. tenuicaulis*	–	Konarev et al. (2002)
	Chymotrypsin absent	*V. subramaniana*	*V. grandiflora*	

wild relatives (CWRs) to maximize genetic gain and meet specific dietary requirements. To breach the yield plateau, favourable alleles preserved in natural reservoirs might be investigated. Numerous crops have shown that their wild relatives are potential sources of economic traits, such as productivity and resistance to biotic and abiotic stresses (Dempewolf et al., 2017). To develop high-yielding, climate-resilient legumes, potential donors from wild species can also be used in legume breeding programmes. Some examples of the introgression of beneficial genes from wild species into legumes are presented in Table 2.2.

CROSSABILITY STUDIES IN MUNGBEAN AND URDBEAN

Mungbean and urdbean are the two major short-duration legume crops, and they are also intercropped with cereals to improve soil fertility (Mehandi et al., 2019). It is believed that the wild progenitor of urdbean, i.e., *V. mungo* var. *silvestris*, was domesticated around 4,500 years ago in India (Chandel et al., 1984). Mungbean is an excellent source of alimentary proteins, vitamin B9, and minerals, and it is an extensively grown crop. In these crops, a number of biotic factors (bruchids, Cercospora leaf spot, powdery mildew and YMV) and abiotic (heat, drought, waterlogging and photoperiod) will significantly reduce yields. By broadening the genetic base through distant hybridization, these problems can be overcome. Wild relatives of *Vigna* have useful genes for vigorous and erect growth, sturdy stems, broad leaves and long and profuse pods with biotic and abiotic stress tolerance (Tripathy et al., 2016; van Zonneveld et al., 2020). Wild accessions showing tolerance to bruchids (viz., *V. radiata* var. *sublobata*) (Tomooka et al., 1992), pod bug (*V. unguiculata* subsp. Dekindtiana (Harms) Verdc.), nematode (*V. angularis (*Wild.) Ohwi & H. Ohashi) (Kushida et al., 2013), MYMD (mungbean yellow mosaic disease) (viz., *V. sublobata*) (Pal and Singh, 2000) and heat and salt stress (viz., *V. angularis*, *V. luteola* (Jacq.) Benth., *V. marina* (Burm.) Merr. and *V. vexillate*) were identified (Yoshida et al., 2020). Using high-throughput genotyping and phenotyping methods, the bruchid and MYMD resistance genes present in *V. sublobata* accessions were subsequently introgressed successfully (Miyagi et al., 2004; Schafleitner et al., 2016). Additionally, efforts were attempted to introduce genes for yield and resistance from ricebean into the mungbean and urdbean plants (Basavaraja et al., 2019). Furthermore, through distant hybridization between black gram and ricebean, the Punjab Agricultural University, Ludhiana, has developed a superior yielding black gram variety (Mash-114) with resistance to mungbean yellow mosaic virus (MYMV), Cercospora leaf spot and bacterial leaf spot.

APPROACHES FOR DEVELOPING FUTURE CLIMATE-RESILIENT CULTIVARS OF MUNGBEAN AND URDBEAN

Conventional Breeding

To create suitable plant varieties that can adapt to various environmental niches and cropping systems, conventional breeding methods have been applied. Breeders primarily focused on agro-morphological traits, which are visually adaptable and highly heritable qualities. Climate change may make the disease and pest scenario a severe issue in the upcoming years. Pulses that are climate-smart must therefore possess disease and insect pest resistance. Programmes for breeding pulses with resistance to various diseases and insect pests routinely screen germplasm under both natural and artificial circumstances to find sources of resistance (Choudhary et al., 2013). Knowledge of the genetics of resistance traits and racial composition of pathogens has accelerated the development of cultivars having adaptability under epidemic conditions. Several varieties suitable for production in various agro-climatic zones of India have been developed as a consequence of conventional phenotype-based breeding programmes. Kumar et al. (2017) used F_4 and F_5 generations of recombinant inbred lines RILs (146 and 155 lines) from two crosses, Chinamung × BL-849 and Chinamung × LM-1668, which showed different responses for powdery mildew resistance. One of the 146 F_4 RILs examined in the cross between Chinamung and BL-849, C1-34-23, was shown to be highly resistant (R0). Out of the 155 F_4 RILs screened, 39 RILs were found to be moderately resistant (R2) in the cross Chinamung × LM-1668. Similar resistance responses were seen in the lines C1-34-23 of the cross between Chinamung and BL-849 and in the RILs C2-14-11 and C2-16-13 of the cross between Chinamung and LM-1668, which could be used in future research to screen for yield-related traits to develop highly resistant breeding lines with high-yielding ability in mungbean.

Although there was little MYMD resistance observed in the mungbean GP, it was not enough to develop resistant cultivars. Several lines with high levels of resistance against MYMD were developed through mutation breeding using moderately resistant accessions and hybrids produced from them. The genotype generated from mutation breeding and the high-yielding cultivars were crossed to create the line NM94, which is now recognized in many nations as a MYMD-resistant line. NM94 has been found to be susceptible in areas where the MYMVD-urd bean strain is prevalent. Punjab Agricultural University, Ludhiana, India, developed ML1628, which is resistant to many species and strains of the virus that cause MYMVD (Nair et al., 2017). Mungbean has been determined to have genetic resistance to bruchid infestation, and this resistance has been used to produce resistant varieties in China, Korea, and at the World Veg.

The serious efforts for varietal development in urdbean started in the early twentieth century with an aim to improve locally adapted but genetically variable populations through mass selection and the pure line selection method of breeding. KM1 was the first variety developed through hybridization. Several other varieties like Pant U19 and U30 were developed from a cross between UP1 and UP2. With the inception of hybridization followed by selection, several other varieties that were MYMV resistant were developed.

The first powdery mildew-resistant urdbean variety LBG17 was developed in 1983. A total of nine varieties have been developed to date through interspecific hybridization. Consequently, more than 100 varieties have been developed in India to date.

Despite the systematic and continuous breeding efforts through conventional methods, substantial genetic gain in mungbean and urdbean could not be achieved. This is because most of the economically important traits, including biotic and abiotic constraints, are complexly inherited with large genotype × environment (G × E) interactions (Kumar and Ali, 2006). Hence, a paradigm shift is needed in the breeding strategies to strengthen our traditional crop improvement programmes. Utilization of genomic tools in conventional breeding programmes such as DNA markers, genetic engineering and genome editing is the way forward.

OMICS-BASED APPROACHES

Several attempts have been made to map the important QTLs controlling economically important traits using various types of DNA markers through both conventional linkage and association mapping methods as a prerequisite to implementing marker-assisted selection in breeding mungbean and urdbean. The availability of the complete genome sequence and low-cost, high-performance genotyping technologies, including genotyping by sequencing, make it easier to map breeders' desired traits. Several genetic linkage maps have been constructed in green gram cultivars using Restriction Fragment Length Polymorphism (RFLP), Random Amplified Polymorphic DNA (RAPD) and simple sequence repeat (SSR) markers (Fatokun et al., 1993). The genetic maps developed in mungbean are mainly based on populations derived from F_{2s} or RILs from interspecific crosses (Chaitieng et al., 2002). These maps have been used to map genes for azuki bean weevil resistance (Young et al., 1992) and seed colour (Lambrides et al., 2004) and to identify QTLs for seed weight, hard seed, powdery mildew resistance (Young et al., 1992; Chaitieng et al., 2002; Kasettranan et al., 2010) and *Cercospora* leaf spot resistance. For the purpose of mapping bruchid resistance genes, genotyping by sequencing (GBS) produced more than 6000 single nucleotide polymorphism markers. One highly significant QTL associated with bruchid resistance was mapped to chromosome 5, suggesting that TC1966 and V2802 (donor parents) contain the same resistance locus.

Singh et al. (2018) developed a mapping population from a cross between a susceptible cultivar, Sonali, and a resistant wild relative of greengram (*Vigna radiata* var. *sublobata*) to map molecular markers linked with mungbean yellow mosaic Indian virus (MYMIV) resistance and yield-attributing traits in mungbean. A total of 224 markers were used for the detection of polymorphism between the parents, and out of these 224 markers, 46 markers showed polymorphism. Twenty-two polymorphic markers were employed for the construction of linkage maps, which consist of 11 linkage groups. QTL analysis identified molecular markers linked with MYMIV resistance and agronomic traits, viz., number of pods per plant, number of seeds per pod and 100-seed weight.

The major focus on the use of molecular markers in urdbean breeding has been on resistance to MYMV. Several QTLs linked to MYMV have been reported by researchers in mungbean (Alam et al., 2011). ICAR-Indian Institute of Pulses Research (Kanpur) developed a recombinant inbred line (RIL) population through a cross between the resistant parent DPU88-and the susceptible parent AKU9904 to develop and validate markers related to yellow mosaic disease resistance loci (Gupta et al., 2013).

Though various genomic resources have been developed for mungbean and urdbean to enhancing breeding efficiency, the limited marker polymorphism within these species makes it difficult to use marker-assisted selection for their improvement. The utility of most mapped loci for breeding has not yet been confirmed. Another important research direction for adaptation strategies in mungbean to stressful environments is wide crosses that aim to introgress traits from related wild species. One example is the introgression of MYMD immunity from *V. mungo* (Lekhi et al., 2018). Crossing barriers affect this approach, but several *Vigna* species are cross-fertile (Kaur et al., 2017).

Genome-Wide Association Mapping

In mungbean, Nobel et al. (2018) carried out a pilot genome-wide association study of seed coat colour to characterize the genetic diversity, population structure and linkage disequilibrium and to indicate its utility. To identify 22,230 polymorphic genome-wide single nucleotide polymorphism (SNPs), a diversity panel of 466 farmed accessions and 16 wild accessions was genotyped. Of these, 16,462 SNPs were physically mapped across the 11 mungbean chromosomes. With an average marker density of 57.81 SNPs/Mb, 1,497 SNPs were discovered on each chromosome, ranging from 903 on chromosome 3–2,306 on chromosome 7. Comparing the farmed accessions to the wild accessions, it was found that the level of polymorphism was significantly lower in the cultivated accessions. Five genomic regions linked to mungbean seed coat color were identified, two of which were also in close proximity to seed coat color genes in other species. This mungbean diversity panel is a useful tool for genetically analyzing key agronomic traits to hasten mungbean breeding. This study represents the first high-resolution quantification of linkage disequilibrium (LD) decay in mungbean, defining the extent of LD within and between cultivated and wild mungbean. The genome-wide association study provides an example of how the data can be used to identify genomic regions responsible for phenotypic traits.

Many of these investigations may currently be ongoing in numerous laboratories around the world. The near-future completion of crop species' genome sequencing projects, powered by more affordable sequencing technologies, will undoubtedly lay the groundwork for whole genome association studies (Kim et al., 2015), accounting for uncommon and common copy number variants (CNVs) (Estivill and Armengol, 2007) and epigenomic details of traits of interest in plants, which are successfully used in human genetics. This will provide crop breeding and genomics programmes with more effective association mapping tools for tagging true functional linkages conditioning genetic diversity and, subsequently, its effective utilization. The integration of genomic tools and conventional breeding triggers new breeding strategies like gene pyramiding, marker-assisted recurrent selection, marker-assisted pedigree selection, and genome selection (GS), which greatly accelerate the breeding process. Genomic-assisted breeding (GAB) has emerged as a potent method for plant breeding in recent years. GAB facilitates the integration of genomic tools with high-throughput phenotyping to support breeding practices by using molecular markers to predict phenotypes from genotypes. Breeders can utilize GAB to begin with a sizable population of single genotypically characterized offspring and then only use a chosen subset for more expensive phenotypic analysis (Cooper et al., 2014). Additionally, genotypic evaluation can be done out of season, for example, in winter nurseries, where yield trials are typically not conducted, which also aids in quickening the breeding process. Because of its benefits of high accuracy, direct improvement, quick breeding cycles and high selection efficiency, GAB is particularly helpful for the improvement of complex traits. The ultimate objective of GAB is to identify the optimum allele (or haplotype) combinations, ideal gene networks and particular genomic locations to support crop development. Thus, according to Leng et al. (2017), GAB promises to hasten the creation of novel plant types and further the advancement of modern agriculture. Therefore, using these cutting-edge methods of plant breeding will significantly enhance the crop improvement in mungbean.

Genome Editing Approaches

New molecular genetics technologies such as genome editing can efficiently transfer desirable genes from other species or create novel genetic variations that are not found in nature (Somers et al., 2003). Other techniques, such as mutagenesis, can also be employed to create novel variations that are not naturally occurring. Mutations are sudden heritable changes that occur spontaneously. Techniques such as TILLING and ECO-TILLING enable the detection of directed mutations existing in certain genes and population-wide spontaneous mutations, respectively (Till et al., 2006). Site-specific recombination, whole genome editing (both major and minor genes), and the elimination of harmful mutations and linkage drag are all accomplished with CRISPR-Cas9. This technique can be utilized in homologous recombination or remote hybridization to knockout genes such as sterility genes, lethal genes and necrotic and chlorotic genes linked to fertilization barriers (de Maagd et al., 2020). Additionally, it can be applied to verify potential genes linked to traits of interest. For example, Talakayala et al. (2022) identified two gRNAs encoding AC1 (replication-associated protein) and AV1 (coat protein) of MYMV and cloned them into a pMDC100 plant expression vector with *npt* II as the plant selection marker gene. Transgenic greengram plants were generated through *Agrobacterium* using cotyledonary nodes as explants. The putative transformed plants were confirmed by polymerase chain reaction (PCR) and dot blot detection assays. The T_2 generation transgenic plants were agro-infiltrated to evaluate the mosaic symptoms in the leaf region. The transcripts of AC1 and AV1 were checked using quantitative real-time PCR. The viral genome editing was further screened and evaluated using a T_7EI genome editing assay. This is the first report on genome editing in greengram, which would help to curb viral pathogens in pulse crops. In the case of urdbean, several research projects are ongoing worldwide to alter the genome of this crop to cope with various stresses. However, to date, there is no report of genome editing in urdbean.

CONCLUSION AND FUTURE PERSPECTIVES

As climate conditions have been changing and arable land has been depleting over the past decade, it has become necessary to change cropping patterns to achieve sustainability in farming. Mungbean and urdbean can fulfil these requirements due to their inherent capability of nitrogen fixation, which improves soil fertility, and their role in ensuring food security. Despite their significant health benefits and economic importance, progress in legume improvement has been slow due to their narrow genetic base. The huge variation present in the genetic resources of these legumes can help broaden this narrow genetic base. Considering the abundant wild species present in gene banks, targeted donors should be identified. High-throughput phenotyping will play a major role in identifying the targeted donors considering the lack of resources to characterize such a vast collection of germplasm. Moreover, the pan-genome information will provide profiles of haplotypes and allelic variations for genes of interest. After completing germplasm characterization, these germplasm lines can be directly used as parents in varietal development or in the development of inbred lines. Another major problem in the legume improvement is the pre-fertilization and post-fertilization barriers that hinder alien gene introgression. To overcome these incompatibility barriers, advanced molecular breeding techniques can be employed to produce pre-breeding lines and interspecific hybrids. These pre-breeding inbred lines can be evaluated under multi-location trails, and promising, stable lines can be utilized in GAB to develop superior cultivars of these climate-resilient legumes. One important aspect of CWRs is that valuable alleles can be identified only when they are incorporated into a cultivar background. The introgression of several genes has provided several breakthrough results in mungbean and urdbean.

REFERENCES

Ahmar, S., Gill, R. A., Jung, K. H., Faheem, A., Qasim, M. U., Mubeen, M., et al. (2020). Conventional and molecular techniques from simple breeding to speed breeding in crop plants: recent advances and future outlook. *International Journal of Molecular Sciences*, 21, 2590. https://doi.org/10.3390/ijms21072590

Aitawade, M. M., Sutar, S. P., Rao, S. R., Malik, S. K., Yadav, S. R., & Bhat, K. V. (2012). Section ceratotropis of subgenus ceratotropis of Vigna (leguminosae-papilionoideae) in India with a new species from northern Western Ghats. *Rheedea*, 22(1), 20–27.

Alam Mondal, M. M., Ali Fakir, M. S., Prodhan, A. A. U. D., Ismail, M. R., & Ashrafuzzaman, M. (2011). Deflowering effect on vasculature and yield attributes in raceme of mungbean [*Vigna radiata* (L.) Wilczek]. *Australian Journal of Crop Science*, 5(11), 1339–1344.

Ali, M., Gupta, S. (2012). Carrying capacity of Indian agriculture: Pulse crops. *Current Science, 25*, 874–881.

Anderson, E. (1949). *Introgressive Hybridization*. John Wiley and Sons, Inc/Chapman and Hall Ltd, New York/London.

Anonymous. (2023). Project coordinator's report. All India Coordinated Research Project on *kharif* crops, p. 57.

Arnold, M. L. (1992). Natural hybridization as an evolutionary process. *Annual review of Ecology and Systematics*, 23(1), 237–261.

Arnoldi, A., Zanoni, C., Lammi, C., & Boschin, G. (2015). The role of grain legumes in the prevention of hypercholesterolemia and hypertension. *Critical Reviews in Plant Sciences*, 34(1–3), 144–168.

Bakala, H. S., Singh, G., & Srivastava, P. (2020). Smart breeding for climate resilient agriculture. In *Plant Breeding-Current and Future Views*. IntechOpen, London, UK.

Basavaraja, T., Murthy, N., Kumar, L. V., & Mallikarjun, K. (2019). Studies on cross compatibility in interspecific crosses of *Vigna radiata* × *Vigna umbellata* species. *Legume Research*, 42, 699–704.

Chaitieng, B., Kaga, A., Han, O. K., Wang, X. W., Wongkaew, S., Laosuwan, P., & Vaughan, D. A. (2002). Mapping a new source of resistance to powdery mildew in mungbean. *Plant Breeding*, 121(6), 521–525.

Chandel, K. P. S., Lester, R. N., & Starling, R. J. (1984). The wild ancestors of urid and mung beans (*Vigna mungo* (L.) Hepper and *V. radiata* (L.) Wilczek). *Botanical Journal of the Linnean Society*, 89, 85–96.

Choudhary, A. K., Kumar, S., Patil, B. S., Sharma, M., Kemal, S., Ontagodi, T. P., & Vijayakumar, A. G. (2013). Narrowing yield gaps through genetic improvement for Fusarium wilt resistance in three pulse crops of the semi-arid tropics. *Sabrao Journal of Breeding and Genetics*, 45(03), 341–370.

Cooper, M., Messina, C. D., Podlich, D., Totir, L. R., Baumgarten, A., Hausmann, N. J., & Graham, G. (2014). Predicting the future of plant breeding: complementing empirical evaluation with genetic prediction. *Crop and Pasture Science*, 65(4), 311–336.

de Maagd, R. A., Loonen, A., Chouaref, J., Pelé, A., Meijer-Dekens, F., Fransz, P.,et al. (2020). CRISPR/Cas inactivation of RECQ 4 increases homeologous crossovers in an interspecific tomato hybrid. *Plant Biotechnology Journal*, 18, 805–813. https://doi.org/10.1111/pbi.13248

Dempewolf, H., Baute, G., Anderson, J., Kilian, B., Smith, C., & Guarino, L. (2017). Past and future use of wild relatives in crop breeding. *Crop Science*, 57(3), 1070–1082.

Dwivedi, S. L., Ceccarelli, S., Blair, M. W., Upadhyaya, H. D., Are, A. K., & Ortiz, R. (2016). Landrace germplasm for improving yield and abiotic stress adaptation. *Trends in Plant Science*, 21(1), 31–42.

Ebert, A. W. (2012). Ex situ conservation of plant genetic resources of major vegetables. In *Conservation of Tropical Plant Species* (pp. 373–417). Springer New York, New York, NY.

Ellstrand, N. C., Prentice, H. C., & Hancock, J. F. (1999). Gene flow and introgression from domesticated plants into their wild relatives. *Annual review of Ecology and Systematics*, 30(1), 539–563.

Estivill, X., & Armengol, L. (2007). Copy number variants and common disorders: filling the gaps and exploring complexity in genome-wide association studies. *PLoS Genetics*, 3(10), e190.

Fang, Y., Du, Y., Wang, J., Wu, A., Qiao, S., Xu, B., & Chen, Y. (2017). Moderate drought stress affected root growth and grain yield in old, modern and newly released cultivars of winter wheat. *Frontiers in Plant Science*, 8, 672.

FAO, & FAOSTAT. (2016). *Food and Agriculture Organization of the United Nations*. FAO, Rome, Italy.

Fatokun, C. A., Danesh, D., Young, N. D., & Stewart, E. L. (1993). Molecular taxonomic relationships in the genus Vigna based on RFLP analysis. *Theoretical and Applied Genetics*, 86, 97–104.

Fuller, D. Q., & Harvey, E. L. (2006). The archaeobotany of Indian pulses: identification, processing and evidence for cultivation. *Environmental Archaeology*, 11(2), 219–246.

Gill, B. S., Friebe, B. R., & White, F. F. (2011). Alien introgressions represent a rich source of genes for crop improvement. *Proceedings of the National Academy of Sciences*, 108(19), 7657–7658.

Gros-Balthazard, M., & Flowers, J. M. (2021). A brief history of the origin of domesticated date palms. In *The Date Palm Genome, Vol. 1: Phylogeny, Biodiversity and Mapping* (pp. 55–74). Springer International Publishing, Cham.

Gudi, S., Atri, C., Goyal, A., Kaur, N., Akhtar, J., Mittal, M., & Banga, S. S. (2020). Physical mapping of introgressed chromosome fragment carrying the fertility restoring (Rfo) gene for Ogura CMS in *Brassica juncea* L. Czern & Coss. *Theoretical and Applied Genetics*, 133, 2949–2959.

Gupta, S., Gupta, D. S., Anjum, K. T., Pratap, A., & Kumar, J. (2013). Transferability of simple sequence repeat markers in blackgram (*Vigna mungo* L. Hepper). *Australian Journal of Crop Science*, 7(3), 345–353.

Hall, A. E. (1992). Breeding for heat tolerance. *Plant Breed Reviews*, 10, 129–168.

Harouna, D. V., Venkataramana, P. B., Matemu, A. O., & Ndakidemi, P. A.(2020). Agro-morphological exploration of some unexplored wild *Vigna* legumes for domestication. *Agronomy*, 10, 111. https://doi.org/10.3390/agronomy10010111

https://doi.org/10.2134/agronj1945.00021962003700020006x

Iseki, K., Takahashi, Y., Muto, C., Naito, K., & Tomooka, N. (2016). Diversity and evolution of salt tolerance in the genus *Vigna*. *PLoS One*, 11(10), e0164711.

Jarvis, D. I., & Hodgkin, T. (1999). Wild relatives and crop cultivars: detecting natural introgression and farmer selection of new genetic combinations in agroecosystems. *Molecular Ecology*, 8, S159–S173.

Jegadeesan, S., Raizada, A., Dhanasekar, P., & Suprasanna, P. (2021). Draft genome sequence of the pulse crop blackgram [*Vigna mungo* (L.) Hepper] reveals potential R-genes. *Scientific Reports*, 11(1), 11247.

Kang, Y. J., Kim, S. K., Kim, M. Y., Lestari, P., Kim, K. H., Ha, B. K., & Lee, S. H. (2014). Genome sequence of mungbean and insights into evolution within Vigna species. *Nature Communications*, 5(1), 5443.

Kasettranan, W., Somta, P., & Srinives, P. (2010). Mapping of quantitative trait loci controlling powdery mildew resistance in mungbean (*Vigna radiata* (L.) Wilczek). *Journal of Crop Science and Biotechnology*, 13, 155–161.

Kaur, S., Bains, T. S., & Singh, P. (2017). Creating variability through interspecific hybridization and its utilization for genetic improvement in mungbean [*Vigna radiata* (L.) Wilczek]. *Journal of Applied and Natural Science*, 9(2), 1101–1106.

Kim, S. K., Nair, R. M., Lee, J., & Lee, S. H. (2015). Genomic resources in mungbean for future breeding programs. *Frontiers in Plant Science*, 6, 626.

Konarev, A.V., Tomooka, N., & Vaughan, D. A. (2002). Proteinase inhibitor polymorphism in the genus *Vigna* subgenus Ceratotropis and its biosystematic implications. *Euphytica*, 123(2), 165–177.

Koona, P., Osisanya, E., Jackai, L., Tamo, M., & Markham, R. (2002). Resistance in accessions of cowpea to the coreid pod-bug *Clavigralla tomentosicollis* (Hemiptera: Coreidae). *Journal of Economic Entomology*, 95(6), 1281–1288.

Kumar, S., & Ali, M. (2006). GE interaction and its breeding implications in pulses. *The Botanica*, 56, 31–36.

Kumar, S., Imtiaz, M., Gupta, S., & Pratap, A. (2011). Distant hybridization and alien gene introgression. *Biology and Breeding of Food Legumes*, 2011, 81–110.

Kumar, A., Adarsha, H. S., Shanthala, J. A., & Savithramma, D. L. (2017). Differential response of F4 and F5 green gram [*Vigna radiata* (L.) Wilczek] recombinant inbred lines (RILs) to powdery mildew infection. *Journal of Pharmacognosy and Phytochemistry*, 6(5), 1147–1153.

Kushida, A., Tazawa, A., Aoyama, S., & Tomooka, N. (2013). Novel sources of resistance to the soybean cyst nematode (*Heterodera glycines*) found in wild relatives of azuki bean (*Vigna angularis*) and their characteristics of resistance. *Genetic Resources and Crop Evolution*, 60(3), 985–994.

Lambrides, C. J., Godwin, I. D., Lawn, R. J., & Imrie, B. C. (2004). Segregation distortion for seed testa color in mungbean (*Vigna radiata* L. Wilcek). *Journal of Heredity*, 95(6), 532–535.

Lekhi, P., Gill, R. K., Kaur, S., & Bains, T. S. (2018). Generation of interspecific hybrids for introgression of mungbean yellow mosaic virus resistance in [*Vigna radiata* (L.) Wilczek]. *Legume Research-An International Journal*, 41(4), 526–531.

Leng, P. F., Lübberstedt, T., & Xu, M. L. (2017). Genomics-assisted breeding-a revolutionary strategy for crop improvement. *Journal of Integrative Agriculture*, 16(12), 2674–2685.

McKersie, B. (2015). Planning for food security in a changing climate. *Journal of Experimental Botany, 66,* 3435–3450.

Mehandi, S., Quatadah, S., Mishra, S. P., Singh, I., Praveen, N., & Dwivedi, N. (2019). "Mungbean (Vigna radiata l. wilczek): retrospect and prospects," in *Legume Crops-Characterization and Breeding for Improved Food Security* (pp. 49–66). IntechOpen, London.

Miyagi, M., Humphry, M., Ma, Z. Y., Lambrides, C. J., Bateson, M., & Liu, C. J. (2004). Construction of bacterial artificial chromosome libraries and their application in developing PCR-based markers closely linked to a major locus conditioning bruchid resistance in mungbean (*Vigna radiata* L. Wilczek). *Theoretical and Applied Genetics*, 110, 151–156.

Nair, R. M., Schafleitner, R., Kenyon, L., Srinivasan, R., Easdown, W., Ebert, A. W., & Hanson, P. (2012). Genetic improvement of mungbean. *Sabrao Journal of Breeding and Genetics*, 44(2), 177–190.

Nair, R. M., Götz, M., Winter, S., Giri, R. R., Boddepalli, V. N., Sirari, A., & Kenyon, L. (2017). Identification of mungbean lines with tolerance or resistance to yellow mosaic in fields in India where different begomovirus species and different *Bemisia tabaci* cryptic species predominate. *European Journal of Plant Pathology*, 149, 349–365.

Noble, T. J., Tao, Y., Mace, E. S., Williams, B., Jordan, D. R., Douglas, C. A., & Mundree, S. G. (2018). Characterization of linkage disequilibrium and population structure in a mungbean diversity panel. *Frontiers in Plant Science*, 8, 2102.

O'Connor, D. J., Wright, G. C., Dieters, M. J., George, D. L., Hunter, M. N., Tatnell, J. R., et al. (2013). Development and application of speed breeding technologies in a commercial peanut breeding program. *Peanut Science*, 40, 107–114. https://doi.org/10.3146/PS12-12.1

Ohnishi, T., Yoshino, M., Yamakawa, H., & Kinoshita, T. (2011). The biotron breeding system: a rapid and reliable procedure for genetic studies and breeding in rice. *Plant and Cell Physiology*, 52, 1249–1257. https://doi.org/10.1093/pcp/pcr066

Pal, S.S., & Singh, I. (2000). Transfer of YMV resistance in cultivar SML 32 of *Vigna radiata* from other related *Vigna* species. *Plant Disease Research*, 15(1), 67–69.

Pandiyan, M., Ramamoorthi, N., Ganesh, S., Jebaraj, S., Pagarajan, P., & Balasubramanian, P. (2008). Broadening the genetic base and introgression of MYMV resistance and yield improvement through unexplored genes from wild relatives in mungbean. *Plant Mutation Reports*, 2(2), 33–43.

Pratap, A., Basu, P. S., Gupta, S., Malviya, N., Rajan, N., Tomar, R., Madhavan, L., Nadarajan, N., & Singh, N. P. (2014). Identification and characterization of sources for photo-and thermo-insensitivity in *Vigna* species. *Plant Breeding*, 133(6), 756–764.

Schafleitner, R., Nair, R. M., Rathore, A., Wang, Y. W., Lin, C. Y., Chu, S. H., & Ebert, A. W. (2015). The AVRDC-the world vegetable center mungbean (*Vigna radiata*) core and mini core collections. *BMC Genomics*, 16(1), 1–11.

Schafleitner, R., Huang, S. M., Chu, S. H., Yen, J. Y., Lin, C. Y., Yan, M. R., & Nair, R. (2016). Identification of single nucleotide polymorphism markers associated with resistance to bruchids (Callosobruchus spp.) in wild mungbean (*Vigna radiata* var. *sublobata*) and cultivated *V. radiata* through genotyping by sequencing and quantitative trait locus analysis. *BMC Plant Biology*, 16(1), 1–15.

Sharma, S. (2017). Pre-breeding using wild species for genetic enhancement of grain legumes at ICRISAT. *Crop Science*, 57(3), 1132–1144.

Sharma, S., Upadhyaya, H. D., Varshney, R. K., & Gowda, C. L. L. (2013). Pre-breeding for diversification of primary gene pool and genetic enhancement of grain legumes. *Frontiers in Plant Science*, 4, 309.

Shimelis, H., & Laing, M. (2012). Timelines in conventional crop improvement: pre-breeding and breeding procedures. *Australian Journal of Crop Science*, 6, 1542–1549.

Simmonds, N. W. (1993). Introgression and incorporation. Strategies for the use of crop genetic resources. *Biological Reviews*, 68(4), 539–562.

Singh, I., Sandhu, J. S., Gupta, S. K., & Singh, S. (2013). Introgression of productivity and other desirable traits from ricebean (*Vigna umbellata*) into black gram (*Vigna mungo*). *Plant Breeding*, 132(4), 401–406.

Singh, N., Mallick, J., Sagolsem, D., Mandal, N., & Bhattacharyya, S. (2018). Mapping of molecular markers linked with MYMIV and yield attributing traits in mungbean. *Indian Journal of Genetics and Plant Breeding*, 78(1), 118–126.

Singh, G., Gudi, S. A., Upadhyay, P., Shekhawat, P. K., Nayak, G., Goyal, L., Kumar, D., Kumar, P., Kamboj, A., Thada, A., Shekhar, S., Koli, G. K., Meghana, D. P., Halladakeri, P., Kaur, R., Kumar, S., Saini, P., Singh, I., & Ayoubi, H. (2022). Unlocking the hidden variation from wild repository for accelerating genetic gain in legumes. *Frontiers in Plant Science*, 13, 1035878. https://doi.org/10.3389/fpls.2022.1035878

Singh, G., Kaur, N., Khanna, R., Kaur, R., Gudi, S., Kaur, R., et al. (2022). 2Gs and plant architecture: breaking grain yield ceiling through breeding approaches for next wave of revolution in rice (*Oryza sativa* L.). *Critical Reviews in Biotechnology*, 44, 1–24. https://doi.org/10.1080/07388551.2022.2112648

Sinha, P., Singh, V. K., Saxena, R. K., Khan, A. W., Abbai, R., Chitikineni, A., et al. (2020). Superior haplotypes for haplotype-based breeding for drought tolerance in pigeonpea (*Cajanus cajan* L.). *Plant Biotechnology Journal*, 18, 2482–2490. https://doi.org/10.1111/pbi.13422

Smartt, J. (1985). Evolution of grain legumes. III. Pulses in the genus *Vigna*. *Experimental Agriculture*, 21(2), 87–100.

Smýkal, P., & Konečná, E. (2014). Advances in pea genomics. *Legumes in the Omic Era*, 2014, 301–337.

Somers, D. A., Samac, D. A., & Olhoft, P. M. (2003). Recent advances in legume transformation. *Plant Physiology*, 131, 892–899. https://doi.org/10.1104/pp.102.017681

Somta, P, Kaga, A., Tomooka, N., Kashiwaba, K., Isemura, T., Chaitieng, B., Srinives, P., & Vaughan, D. A.(2006). Development of an interspecifi c Vigna linkage map between *Vigna umbellata* (Thunb.) Ohwi & Ohashi and *V. nakashimae* (Ohwi) Ohwi & Ohashi and its use in analysis of bruchid resistance and comparative genomics. *Plant Breeding*, 125, 77–84.

Talakayala, A., Mekala, G. K., Reddy, M. K., Ankanagari, S., & Garladinne, M. (2022). Manipulating resistance to mungbean yellow mosaic virus in greengram (*Vigna radiata* L): through CRISPR/Cas9 mediated editing of the viral genome. *Frontiers in Sustainable Food Systems*, 6, 911574.

Till, B. J., Zerr, T., Comai, L., & Henikoff, S. (2006). A protocol for TILLING and ecotilling in plants and animals. *Nature Protocols*, 1, 2465–2477. https://doi.org/10.1038/nprot.2006.329

Tomooka, N., Lairungreang, C., Nakeeraks, P., Egawa, Y., & Thavarasook, C.(1992). Development of bruchid-resistant mungbean line using wild mungbean germplasm in Thailand. *Plant Breeding*, 109, 60–66.

Tomooka, N., Vaughan, D., Xu, R. Q., Kashiwaba, K., & Kaga, A. (2001). Japanese native *Vigna* genetic resources. *Japan Agricultural Research Quarterly*, 35(1), 1–9.

Tomooka, N., Kaga, A., Isemura, T., & Vaughan, D. (2010). Vigna. In *Wild Crop Relatives: Genomic and Breeding Resources: Legume Crops and Forages* (pp. 291–311). Springer, Berlin, Heidelberg.

Tripathy, S. K., Nayak, P., Lenka, D., Swain, D., Baisakh, B., Mohanty, P. N., Senapati, P., Dash, G., Dash, S., & Mohapatra, P. (2016). Morphological diversity of local land races and wild forms of mungbean. *Legume Research-An International Journal*, 39(4), 485–493.

Van Zonneveld, M., Rakha, M., Tan, S. Y., Chou, Y. Y., Chang, C. H., Yen, J. Y., Schafleitner, R., Nair, R., Naito, K., & Solberg, S. O. (2020). Mapping patterns of abiotic and biotic stress resilience uncovers conservation gaps and breeding potential of *Vigna* wild relatives. *Scientific Reports* 10(1), 1–11.

Van Oldenborgh, G. J., Philip, S., Kew, S., van Weele, M., Uhe, P., Otto, F., Singh, R., Pai, I., Cullen, H., Achuta Rao, K. (2018). Extreme heat in India and anthropogenic climate change. *Natural Hazards and Earth System Sciences*, *18*, 365.

Vaz Patto, M. C., Amarowicz, R., Aryee, A. N., Boye, J. I., Chung, H. J., Martin-Cabrejas, M. A., & Domoney, C. (2015). Achievements and challenges in improving the nutritional quality of food legumes. *Critical Reviews in Plant Sciences*, 34(1–3), 105–143.

Vishnu-Mittre, B. (1974). "Palaeobotanical evidence in India," in *Evolutionary Studies in World Crops: Diversity and Change in the Indian Sub-continent*, ed. J. Hutchinson (pp. 3–30). Cambridge: Cambridge University Press.

Yaqub, M., Mahmood, T., Akhtar, M., Iqbal, M. M., & Ali, S. (2010). Induction of mungbean [*Vigna radiata* (L.) Wilczek] as a grain legume in the annual rice-wheat double cropping system. *Pakistan Journal of Botany*, 42(5), 3125–3135.

Yoshida, J., Tomooka, N., Khaing, T. Y., Shantha, P. S., Naito, H., Matsuda, Y., & Ehara, H. (2020). Unique responses of three highly salt-tolerant wild Vigna species against salt stress. *Plant Production Science*, 23(1), 114–128.

Young, N. D., Kumar, L., Menancio-Hautea, D., Danesh, D., Talekar, N. S., Shanmugasundaram, S., et al (1992) RFLP mapping of major bruchid resistance gene in mungbean (*Vigna radiata*). *Theoretical and Applied Genetics*, 84, 839–844.

3 Exploiting *Arachis* Wild Relatives for Increasing Genetic Diversity and Resilience in Groundnut

Deekshitha Bomireddy, Vinay Sharma, Soraya Leal-Bertioli, Daniel Fonceka, Ramesh S. Bhat, Boshou Liao, Madhavilatha Kommana, Huifang Jiang, Jiaping Wang, Sandip K. Bera, Li Huang, Xingjun Wang, and Manish K. Pandey

INTRODUCTION

Groundnut or peanut (*Arachis hypogaea* L.), as a prominent oilseed and grain legume, is an excellent cash crop cultivated in over 108 countries representing warm temperate, tropical, and subtropical climatic zones in the world. It is a self-pollinated allotetraploid (AABB; $2n = 4x = 40$) with a genome size of ~2.7 Gb (Bertioli et al., 2019). Groundnut is an annual herbaceous legume belonging to the Leguminosae family and Papilionaceae sub-family (Stalker and Wilson, 2016). Groundnut is popularized because of its diversified and multipurpose consumption in human nutrition as cooking oil, confectionery, dietary food, and even as livestock fodder (Pandey et al., 2012). Globally, groundnut ranks fourth after soybean, rapeseed, and sunflower in terms of oilseed production, with production of 66.3 million tonnes covering 34.1 million hectares of cultivated area and an average productivity of 19.8 quintals per hectare in 2020 (Food and Agriculture Organization, 2020).

Groundnut is mainly valued and cultivated for its kernels because of their high energy content provided by protein (16%–34%) and oil (31%–56%) (Jivani et al., 2012). Furthermore, groundnut kernels are rich in monounsaturated fatty acids along with a wide variety of health-promoting nutrients such as vitamins, minerals, and antioxidants. Two-thirds of global groundnut production is used for vegetable oil extraction and the remaining part is used for making edible food products. Groundnut and its products are being promoted as nutritional foods to combat protein and micronutrient malnutrition among resource-poor communities. Groundnut cake, after extracting oil, is used as a protein-rich meal for livestock and also as a soil amendment. As with other leguminous crops, groundnut can improve soil

 DOI: 10.1201/9781003434535-3

fertility through the process of biological nitrogen fixation. Owing to its beneficial and unique characteristics, groundnut plays a pivotal role in sustainable agriculture systems (Sharma et al., 2017).

The evolution, domestication, and subsequent genetic improvement have led to the development of high-yielding cultivars, many with resistance to biotic and abiotic stresses. The emphasis on developing a genetically uniform, steady stream of familiar cultivars has made the breeding efforts focus on limited species and varieties at the cost of genetically diverse, locally adapted traditional lines. Eventually, this has resulted in the loss of adaptive alleles and the fixation of deleterious alleles making the crops more vulnerable to diseases and pests, thus creating a gap between actual and potential yields (Warschefsky et al., 2014). Furthermore, due to ever-changing climatic constraints, new pests and diseases are emerging as a menace to crop production and productivity (Chakraborty and Newton, 2011).

Drought, salinity, heat, and chilling are the predominant abiotic stresses adversely affecting groundnut production. Biotic stresses such as aflatoxin contamination, rust, early and late leaf spots, peanut bud necrosis virus, rosette disease, bacterial wilt, stem rot, pod rot, and insect pests like leaf miners and tobacco caterpillars are widespread constraints for yield and seed quality. Even though these pests and diseases can be controlled to some extent by chemical measures, such methods are not economically feasible as they tend to be costly for smallholder farmers and harmful to the environment. Additionally, pests and diseases can develop resistance during long-term chemical control. A narrow genetic base coupled with low productivity due to biotic and abiotic stresses necessitates identifying and utilizing diverse germplasm sources with a broad genetic base to develop novel high-yielding cultivars. Hence, using wild species for developing high-yielding cultivars with host-plant resistance and other acceptable market traits is fundamental for groundnut improvement.

The genus *Arachis* has 84 species assembled under nine sections possessing vast genetic variability. Several were already identified as potent sources of valuable traits, including abiotic and biotic stress resistance (Michelotto et al., 2015). Most of the diversity in the available germplasm remains unutilized as breeders are reluctant to use them due to their ploidy differences, hybridization barriers, infertility of hybrids and linkage drag (Stalker, 2017; Stalker et al., 2013). Combining advanced, efficient hybridization schemes with marker-assisted technologies provides new avenues for overcoming several of these problems. Pre-breeding provides an opportunity to introduce wild diploid genes into cultivated groundnuts by developing adaptable synthetic amphiploids that act as bridging sources. Marker-assisted backcrossing (MABC) is essential for the introgression of wild genes as it enables rapid parent genome recovery and the pyramiding of multiple useful genes into a single cultivar, thus saving considerable time. Several groundnut varieties are resistant to early and late leaf spots, root-knot nematode, and tomato-spotted wilt virus (TSWV) have been developed and released by deploying MABC (Holbrook et al., 2017; Bertioli et al., 2021b). In this chapter, we intend to review and discuss the importance of wild relatives of *Arachis*, gene introgressions from wild into cultivated groundnuts, challenges of gene introgression, opportunities for wild germplasm utilization, and prospects pertaining to groundnut improvement.

ORIGIN, DISTRIBUTION, TAXONOMY, AND GENE POOLS OF GENUS *ARACHIS*

Groundnut is a new world crop indigenous to South America and likely to have originated in the tropical wetlands of Brazil, where most of the wild species were traced (Stalker, 2017). The wild species of groundnut range widely from northeastern Brazil to southern Uruguay, and from the eastern Atlantic coast to the Andean lowlands of the west (Valls et al., 1985). Groundnut was likely spread to Africa, Asia, Europe, and the Pacific Islands during the 16th and 17th centuries by the discovery voyages of the Spanish, Portuguese, British, and Dutch (Krapovickas, 1969, 1973; Gregory et al., 1980). Currently, groundnut is primarily grown in tropical and subtropical regions of more than 100 countries located between 40°N and 40°S of the equator, where average rainfall ranges from 500 to 1,200 mm, and mean daily temperatures exceed 20°C.

Groundnut belongs to the *Fabaceae/Leguminosae* family, *Papilionoideae* sub-family, tribe *Aeschynomeneae*, subtribe *Stylosanthinae*, and the genus *Arachis* (Krapovickas et al., 2007). Based on geographical distributions, sexual compatibilities, cytogenetic, and morphological features, the genus *Arachis* has been classified into nine taxonomic sections comprising 84 species with a large diversity of genomes (Figure 3.1) (Krapovickas and Gregory, 1994; Valls and Simpson, 2005; Stalker et al., 2017). Of these, 72 species are diploids ($2n=2x=20$), four species are aneuploids ($2n=2x=18$), and five species are tetraploid ($2n=2x=40$) (Friend et al., 2010). From the section *Arachis*, *A. hypogaea* is the widely cultivated allotetraploid groundnut, which is the result of hybridization between two wild diploid species, thought to be *A. duranensis* (female parent with A genome) and *A. ipaensis* (male parent with B genome) (Kochert et al., 1996; Bertioli et al., 2016).

In the cultivated groundnut, two subspecies (*A. hypogaea* ssp. *hypogaea* and *A. hypogaea* ssp. *fastigiata*) have been recognized primarily based on morphological differences (Krapovickas and Gregory, 1994). *A. hypogaea* ssp. *hypogaea* is characterized by a spreading growth habit, alternate branching pattern, absence of flowers on the central stem, long growth cycle (120–160 days), and presence of seed dormancy. In contrast, *A. hypogaea* ssp. *fastigiata* is characterized by an erect growth habit, sequential branching, presence of flowers on the central stem, shorter growth cycle (85–130 days), and lack of seed dormancy. These two subspecies are subdivided into botanical varieties based on the morphological variation. *A. hypogaea* ssp. *hypogaea* is classified into *var. hypogaea* (Virginia runner/bunch) and *var. hirsuta* (Peruvian runner) with the latter being differentiated by hirsute leaflets and even longer growth duration. *A. hypogaea* ssp. *fastigiata* is divided into *var. fastigiata* (Valencia), *var. vulgaris* (Spanish bunch), *var. peruviana,* and *var. aequatoriana*, with the former two being erect in plant habit and relatively shorter in growth period (Krapovickas, 1969, 1973).

The genetic diversity of the genus *Arachis* has been divided into primary, secondary, and tertiary gene pools based on the amount of gene flow among all the species (Singh and Simpson, 1994; Rami et al., 2013). The primary gene pool includes *A. hypogaea* and *A. monticola*, the two tetraploid *Arachis* species that are

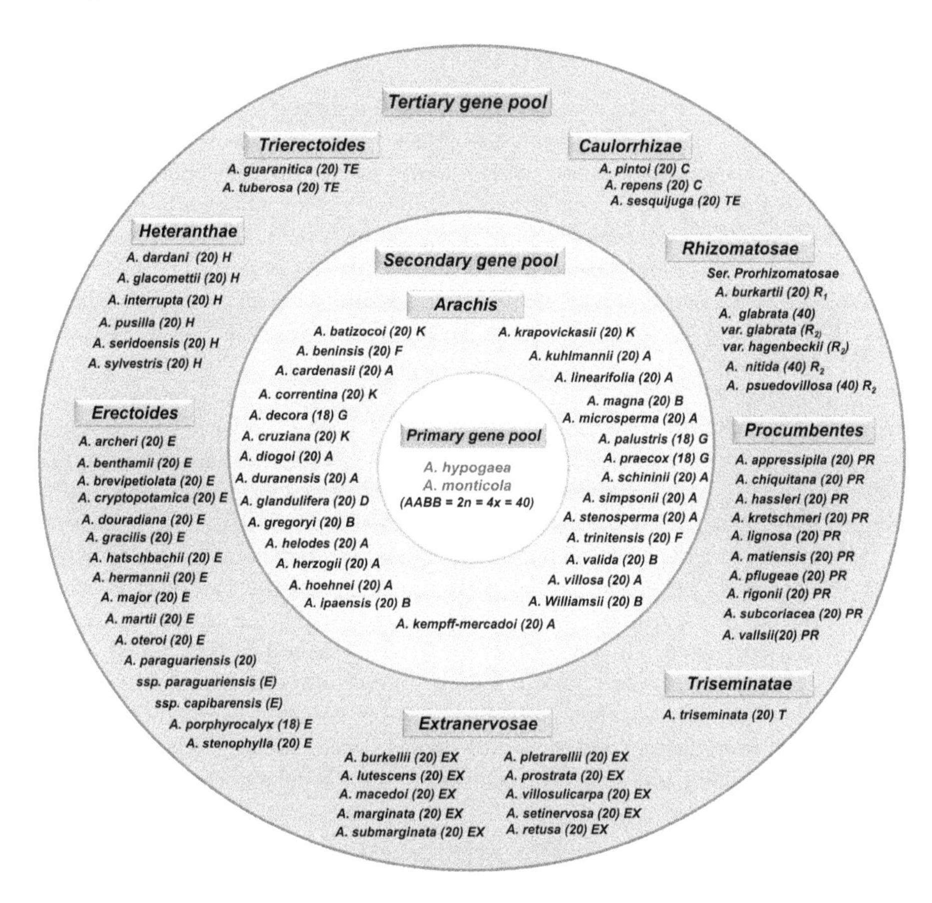

FIGURE 3.1 Species, sections diversity, and gene pools of the genus *Arachis*: 81 species of genus *Arachis* belonging to 9 sections with two ploidy levels (diploid and tetraploid) and 16 genome types (A, B, AB, D, F, G, K, EX, T, PR, H, C, T, E, R_1, and R_2). Primary gene pool includes cross-compatible *Arachis hypogaea* and *A. monticola*. Secondary gene pool includes all other diploid species of section *Arachis*. The sections other than *Arachis* that are weakly compatible and cross-incompatible with cultivated groundnuts are classified under the tertiary gene pool.

cross-compatible. *A. monticola* is very closely related to *A. hypogaea* and is considered to be *A. hypogaea*'s immediate tetraploid ancestor and is probably reported to share the same origin (Lu and Pickersgill, 1993; Grabiele et al., 2012; Leal-Bertioli et al., 2021). Whereas, all other diploid species of section *Arachis* closely related to *A. hypogaea* and can be useful for groundnut improvement are classified under a secondary gene pool (Rami et al., 2013). The sections other than *Arachis* that are weakly compatible and cross-incompatible with cultivated groundnuts are classified as tertiary gene pools. Generally, the gene flow between gene pools and different sections within the tertiary gene pool is highly limited. The ploidy barrier between cultivated tetraploid groundnut and most of its wild diploid relatives, except for *A. monticola* (with $2n=40$), has created a strong genetic bottleneck and severely prevented the introgression of wild genes into cultivated groundnuts.

DESIRABLE TRAITS IDENTIFIED IN WILD RELATIVES OF *ARACHIS*

Crop wild relatives (CWRs) are rich and valuable genetic resources for breeders, containing several "game-changing" traits or genes that can improve crop resilience and global agricultural production. Wild *Arachis* species have been identified with extremely high levels of resistance or tolerance to most of the pathogens, insects, and abiotic stresses affecting groundnuts (Stalker and Moss, 1987; Dwivedi et al., 2007; Chen et al., 2008). Several accessions of *A. cardenesii* were earlier reported with varied levels of resistance to early leaf spot (ELS) and late leaf spot (LLS) (Nigam et al., 1991; Stalker, 1992), rust (Subrahmanyam et al., 1983, 1985c), rosette disease, peanut bud necrosis, and TSWV (Subrahmanyam et al., 1985a, 2001; Reddy et al., 2000). Similarly, accessions of *A. kempff mercadoi*, *A. duranensis*, and *A. diogoi* also demonstrated resistance to several diseases like LLS, TSWV, corn earworm, groundnut aphids, leaf hoppers, etc., (Subrahmanyam et al., 1980; Amin and Mohammad, 1980; Lyerly et al., 2002; Stevenson et al., 1993; Stalker and Campbell, 1983; Amin, 1985; Nelson et al., 1989). Accessions of *A. cardenesii, A. cruziana, A. batizocoi, A. duranensis, A. helodes, A. hoehnei, A. kuhlmannii* exhibited resistance to nematodes (Nelson et al., 1989; Sharma et al., 1999). Also, accessions of wild *Arachis* superior for both nutritional quality and agronomic traits, including days to 50% flowering, high oil, protein, and sugar content have been widely identified (Upadhyaya et al., 2011). As 29 diploid ($2n=2x=20$) species of section *Arachis* have various degrees of compatibility with the cultivated groundnut, they are considered important for crop improvement. Fortunately, species resistant to most severe insects and pests affecting groundnut production have been identified within this diploid group of wild species.

STATUS OF GROUNDNUT GENETIC RESOURCES IN GLOBAL GENE BANKS

Plant genetic resources are reservoirs of natural genetic diversity and furnish the raw materials required for crop improvement to strengthen the global efforts for ensuring future food security. The RS Paroda genebank at International Crops Research Institute for the Semi-Arid Tropics (ICRISAT). Hyderabad, holds the largest active collection of 15,896 accessions of groundnut that provide ample natural genetic variation for use in breeding programmes. Of these, 15,416 are advanced/improved cultivars, traditional cultivars/landraces, breeding/research materials, and the remaining 480 are wild relatives of groundnuts belonging to 48 species. Besides ICRISAT, the major genebanks housing groundnut germplasm are the United States Department of Agriculture (USDA), the United States of America, with 9,964 accessions, Oil Crops Research Institute - Chinese Academy of Agricultural Sciences (OCRI-CAAS), China with 9,103 accessions (Zhou et al., 2022), the National Bureau of Plant Genetic Resources (NBPGR), New Delhi, India, with 14,593 accessions, and the Directorate of Groundnut Research (DGR), Junagadh, India, with 8,934 accessions. The germplasm banks that contain the most wild species are EMBRAPA (Empresa Brasileira de Pesquisa Agropecuária) – Brasilia, Brazil, with 1,559 accessions of 84 species, and IBONE (Instituto de Botánica del Nordeste, UNNE-CONICET) – Corrientes, Argentina, with 259 accessions comprising 57 wild species (52 described, plus 6–8 new but still-undescribed species) and 213 accessions of wild species hybrids. These germplasm collections have been characterized for several important agronomic and economic traits, and from them, different diversity panels have been developed for use in breeding programmes. To accelerate the use of available groundnut germplasm in breeding programmes, ICRISAT, the United States of America, and China have developed core and mini-core collections of *A. hypogaea* (Holbrook et al., 1993; Upadhyaya et al., 2002; Jiang et al., 2013). Accessions to these collections have contributed significantly to a better understanding of overall groundnut genetics and to the development of appropriate breeding strategies for target traits.

The low levels of resistance/tolerance to biotic and abiotic stresses and the narrow genetic base of the cultivated gene pools made it important to exploit new and diverse sources of variation. As CWRs are reservoirs of many useful genes and alleles, they are genetically more diverse, have natural defense mechanisms to withstand climate extremities, and are of immense importance for introgressing useful traits in crop breeding programmes. Presently, emphasis has been made on conserving the accessions safely with duplication both within and outside the Consultative Group on International Agricultural Research (CGIAR) system. To store seed samples of a wide variety of plant species, Svalbard Global Seed Vault has been established at the global level on the Norwegian island of Spitsbergen in an underground cavern (*Svalbard Global Seed Vault*). This seed vault will act as a refuge for seeds at times of large-scale global or regional crises and provide insurance against the loss of seeds in genebanks.

GENE INTROGRESSIONS FROM WILD TO CULTIVATED GENEPOOL

The limited genetic spectrum available in the cultivated groundnut germplasm calls for the exploration and exploitation of wild *Arachis* species for genetic improvement. Despite realizing the potential of wild *Arachis* species in improving biotic and abiotic stress resistance in commercial cultivars, they are not adequately utilized in breeding programmes (Sharma et al., 2017). Major limitations include hybridization barriers between wild and cultivated species, poor viability and sterility of F_1 hybrids, ploidy/genome level differences, poor agronomic performance, and linkage drag, thus making breeders reluctant to use wild relatives in commercial breeding programmes (Bohra et al., 2022).

The low percentage of pollination, genome incompatibilities, and sterility of progenies improve cultivated groundnuts by using diploid species extremely difficult and time-consuming. Introgressions from wild *Arachis* species into *A. hypogaea* appear to be in large blocks, restricting the use of progenies because of association or linkage with undesirable traits (Nagy et al., 2010). As *A. monticola* is the second tetraploid species of the section *Arachis* and is considered the weedy and close relative of *A. hypogaea*, the first interspecific hybridization was attempted between them. Spancross, an early maturing bunch cultivar, and Tamnut 74 were released by Hammons (1970) and Simpson and Smith (1975), respectively, from interspecific hybridization. *A. monticola* has susceptibility to several diseases and insects similar to that of *A. hypogaea*; therefore, not much effort has been made for the introgression of genomic regions. Diploid wild species, on the other hand, have significantly greater resistance to biotic and abiotic stresses, with greater potential for crop improvement than the tetraploid *A. monticola*.

In earlier studies, several hybrid combinations generated within sections failed due to species incompatibilities, particularly when the wild species were used as female parents. An extensive hybridization programme using 91 exotic *Arachis* germplasm collections generated 1,075 cross-combinations and reported cross-compatibility relationships among species of the genus *Arachis* (Gregory and Gregory, 1979). As highly sterile F_1s were observed from intersectional crosses, their study demonstrated that hybridization between species belonging to the same section is more successful than crosses between the species of different sections. Gregory and Gregory (1979) and Stalker (1981) attempted to develop complex hybrids to overcome the crossing barriers, but fertility could not be restored in any of the progenies. As a result, the introgression of genes from wild relatives into cultivated *A. hypogaea* is believed to be restricted to the species present within the section *Arachis*. Even within the section *Arachis*, hybrid sterility may be imposed because of ploidy and/or genomic differences or due to irregular meiosis in colchicine-treated hybrids (Stalker, 2017). The problem of sterile progeny and the difficulties of backcrossing hexaploids can be overcome through ploidy manipulations as well as the generation of tetraploid derivatives that can then be crossed with *A. hypogaea*.

To introgress genes from diploid species to *A. hypogaea*, several approaches were used with varying degrees of success. At first, crosses were made by directly hybridizing tetraploid *A. hypogaea* with diploids to produce sterile triploid hybrids ($3x=30$). The F_1 cuttings were then treated with colchicine to restore fertility at the hexaploid level ($6x=60$) (Stalker et al., 1979). Backcrossing these hexaploids with *A. hypogaea* resulted in vigorous pentaploids ($5x=50$) with partial fertility and sporadic flowering, making them difficult to use in crossing programmes. However, when selfed, they sometimes produced a few seeds where the ploidy level of the progenies stabilizes at the tetraploid level. However, during selfing generations, progenies lost the chromosomes of wild species apparently and obtained no tetraploid lines with preferential traits (Stalker et al., 1995). Similarly, backcrossing hexaploids with diploid species theoretically tends to reduce chromosome numbers to the tetraploid level in one generation, but thousands of crosses made between hexaploids and diploids did not result in any viable progenies (Stalker et al., 1995). Allowing hexaploids to self-pollinate is another alternative method, where plants spontaneously lose chromosomes (though random and infrequent) and stabilize at the tetraploid level. The benefit of this approach is the chromosomal association of wild and *A. hypogaea* species for many generations at a high ploidy level, which increases the chances of recombination. Tetraploid progenies highly variable for seed colour, size, and other morphological characters were derived from *A. hypogaea* × *A. cardenasii* hexaploids after 12 generations of selfing (Company et al., 1982).

Arachis cardenasii has served as a major source of resistance to multiple diseases of groundnut. A total of 18 lines with resistance to ELS, nematode, and several other insect pests have been released in the United States from selections of *A. cardenasii* introgression lines (Stalker et al., 2002a, 2002b; Stalker and Lynch, 2002; Isleib et al., 2006; Tallury et al., 2011). The early releases were hybridized with the most leaf spot-resistant cultivated accessions available to pyramid genes for disease resistance, and significantly improved lines for resistance were selected. The introgression line GP-NC WS 5 derived, from a cross between *A. hypogaea* (PI 261942) and *A. cardenasii* was subsequently crossed with AgraTech 108, resulting in two improved root-knot nematode-resistant lines namely, NR 0812 and NR 0817 (Anderson et al., 2006). N96076L, generated from the *A. hypogaea* × *A. cardenasii* germplasm line (NC GP WS 13), was released as a multiple disease-resistant line for *Sclerotinia* blight, ELS, TSWV, and *Cylindrocladium* black rot. Subsequently, N96076 was used as a multiple disease-resistant source for the most widely grown cultivar in the Virginia–Carolina production region, Bailey (Isleib et al., 2011). High-throughput genotyping revealed that 251 groundnut lines and cultivars from 30 countries on all continents had previously unnoticed genetic segments from the wild relatives, revealing a large-scale contribution of *A. cardenasii* to groundnut improvement worldwide (Bertioli et al., 2021a).

During the mid-1980s, a set of interspecific hybrid lines (coded CS lines) selected for pest and disease resistance at North Carolina State University were forwarded to the ICRISAT germplasm system, where they were renumbered as ICGV lines. These ICGV lines have been distributed to groundnut scientists across India, Australia, Brazil, China, and several other countries, where they are being used as sources of resistance for rust, LLS, and other disease pests (Bertioli et al., 2021a). GPBD-4, a Spanish-type cultivar in India (Gowda et al., 2002), is highly resistant to rust and LLS and was selected from the KRG 1 × CS 16 (ICGV 86855) cross, where CS 16 is a disease-resistant line derived from the *A. hypogaea* × *A. cardenasii* populations. From the selection of CS 9, ICGV 87165 was released as a multiple disease-resistant line for LLS and rust, with moderate resistance to tobacco caterpillar, leaf miner, and bacterial wilt (Moss et al., 1997). Similarly, the ICGV-SM 86715 germplasm line, originating from the *A. hypogaea* × *A. diogoi* (GKP 10602) interspecific cross was released in Malawi as "Veronica," possessing high levels of resistance to ELS, LLS, rust, and pepper spot (Moss et al., 1998). Yuanza-9102, an early maturing high-yielding cultivar with resistance to bacterial wilt, was developed from an interspecific cross with *A. chacoense* (later renamed *A. diogoi*) as a parent in Henan, China. Additional resistant germplasm lines with LLS and rust resistance were also identified by scientists of ICRISAT (Singh et al., 2003). This is the base of several cultivars used in Uganda (Okello et al., 2018).

Another route for introgression, is the tetraploid route, first used to create a wild-derived tetraploid. Firstly, an A genome hybrid was made by crossing *A. cardenasii* with *A. diogoi*. Then, the B genome species *A. batizocoi* was crossed with the A genome hybrid to create a sterile AB hybrid. This hybrid was treated with colchicine to double the chromosome number and restore fertility. This tetraploid [*A. batizocoi* × (*A. cardenasii* × *A. diogoi*)]^{4x} was registered as TxAG-6 (Simpson et al., 1993)]. Using this route, several lines with root-knot nematode resistance have been produced using *Arachis stenosperma* as the resistance donor (Ballen-Taborda et al., 2021, 2022).

CONSTRAINTS TO UTILIZATION OF WILD SPECIES FOR GROUNDNUT IMPROVEMENT

The other key deterrent to using wild species in crop improvement is the presence of strong linkage drag that negatively influences the phenotypic assessment of progenies derived from interspecific crosses (Bohra et al., 2022). To break these undesirable associations while using wild or unknown germplasm, more time and effort are required during the developmental process, discouraging their use in genetic improvement programmes (Sharma et al., 2013). Nevertheless, the over-cautiousness towards linkage drag has resulted in the continuous narrowing down of genetic variability among existing cultivars.

In general, elite cultivars are bred to perform well across a range of environments in response to managerial inputs, whereas, wild relatives exhibit poor adaptability beyond their natural zones of adaptation (Cowling, 2013). When wild members are grown outside their natural distribution range, the expression of beneficial alleles may be masked, making the performance of wild accessions and their derived interspecific progenies less promising during standard breeding trials (Wang et al., 2017). All these factors make the breeding programme comparatively lengthier and more cumbersome, leading the breeders to rely on their existing collection, resulting in the re-circulation of the same genotypes and a narrow genetic base for the modern cultivars (Sharma et al., 2013).

Despite the challenges, increasingly efficient hybridization schemes and the ease of marker technologies that enable efficient hybrid confirmation and tracking of wild genetics in breeding programmes allow the use of wild diploid species for improving cultivated groundnut on a previously unheard-of scale (Stalker, 2017; Stalker et al., 2013; Bertioli et al., 2021b; Chu et al., 2021). Introgression can be accomplished by utilizing appropriate pre-breeding pathways combined with various parasexual techniques. Finding suitable hotspots for improved natural seed set in interspecific hybrids is essential. In addition to the parents, suitable bridging species that combine better for fertility and seed set should be identified to introduce more wild gene sources into the cultivated background.

PRE-BREEDING FOR HARNESSING NOVEL GENES FROM WILD *ARACHIS* SPECIES

Pre-breeding offers a unique platform for introgressing genes from diploid species to *A. hypogaea* enhancing the utilization of wild germplasm and ensuring a continuous supply of genetic variability in a conventionally usable form into the breeding pipeline. Pre-breeding is comparatively much faster than going through the triploid-hexaploid procedure and has the advantage of avoiding the uncertainty of recovering tetraploid progenies and several generations of sterile hybrids (Stalker, 2017). The prerequisite for pre-breeding involves the identification of desirable genes and/or traits from unadapted exotic/wild germplasm and using them as donors in a crossing programme with well-adapted cultivars as recipients to generate a new set of intermediate pre-breeding populations. These pre-breeding populations are further evaluated to identify introgression lines (ILs) with a high frequency of desirable genes/traits from wild species and good agronomic performance, which can be readily used in specific crop breeding programmes to develop new cultivars with a broad genetic base (Sharma et al., 2013).

Wild species of the section *Arachis* are mostly diploids ($2n = 2x = 18$ or 20), with A, B, D, F, G or K genomes (Robledo and Seijo, 2010). As discussed earlier, the frequent utilization of these wild diploid *Arachis* species is hindered in groundnut improvement programmes because of cross-incompatible or hybridization barriers arising mostly from ploidy-level differences between wild and cultivated tetraploid groundnut (Halward and Stalker, 1987). New sources of 15 tetraploid groundnut synthetics were developed at ICRISAT by crossing diploid wild *Arachis* accessions (having A, B, and K genomes) in different combinations followed by chromosomal doubling of inter- and intra-genomic diploid F_1 hybrids using colchicine treatment (Mallikarjuna et al., 2011). These synthetics were extensively screened for rust and LLS resistance and are being used for developing pre-breeding populations using popular groundnut cultivars (Mallikarjuna et al., 2012; Sharma et al., 2017). For instance, using the synthetics ISATGR 40, ISATGR 278-18, ISATGR 265-5, and ISATGR 121250 as donors and TMV 2, Tifrunner, ICGV 91114, and ICGV 87846 as recipients, five advanced backcross populations exhibiting significant variability for biotic stresses as well as for morpho-agronomic traits have been developed. Phenotyping two of the pre-breeding populations developed from ICGV 87846 × ISATGR 265-5 and ICGV 91114 × ISATGR 121250 resulted in the identification of ILs having high levels of rust and LLS resistance (Sharma et al., 2017). These advanced backcross populations have been shared with various partners for multilocation screening for LLS and rust resistance and identified 29 best-performing ILs for use in groundnut improvement programmes. Twenty-nine ILs selected from these pre-breeding populations were screened for *Spodoptera litura* along with resistant checks, and ICGIL 17101 was the potential donor for incorporating *Spodoptera* resistance in groundnut cultivars (Dange and Naidu, 2021). Similarly, the pre-breeding lines ICGIL 17101 and ICGIL 17124 showed resistance to *Aspergillus flavus*, while ICGIL 17124 recorded higher pod yield as well (Naidu et al., 2022). The ILs, ICGIL 17105, ICGIL 17110, and ICGIL 17112 identified as resistant sources to rust and leaf spots are being used in breeding programmes of ICRISAT and UAS-Dharwad, India. A list of tetraploid groundnut synthetics developed worldwide is presented in Table 3.1.

Many elite germplasm lines and cultivars were developed by utilizing synthetic amphidiploids, resulting from the interspecific hybridization of wild *Arachis* species. For example, COAN and NemaTAM are root-knot nematode-resistant groundnut cultivars derived from complex interspecific crosses involving *A. cardenasii* Krapov. and W. C. Gregory (Simpson and Starr, 2001; Simpson et al., 2003). Tifguard, a runner-type groundnut cultivar having resistance to root-knot nematode and TSWV, was released by the Georgia Agricultural Experiment Station and United States Department of Agriculture - Agricultural Research Service (USDA-ARS) (Holbrook et al., 2008). Webb is a nematode-resistant, high-yielding, and high-oleate cultivar with moderate resistance to Sclerotinia blight, resulting from wild introgressions (Simpson et al., 2013). Georgia-14N and TifNV-High O/L are other examples of high oleate, TSWV, and root-knot nematode-resistant groundnut cultivars derived from wild introgressions (Branch and Brenneman, 2015; Holbrook et al., 2017). The synthetic developed by Favero et al. (2006) was used in Senegal to develop several introgression lines with Fleur11, a popular Spanish variety, as a recurrent parent. Among the population developed, six ILs that showed consistently bigger pods and seeds (Tossim et al., 2020) were registered as varieties. For making the availability of alleles from three wild species accessions in breeding, GA-MagSten1 and GA-BatSten1 allotetraploids derived from crosses between *A. magna* × *A. stenosperma* and *A. batizocoi* × *A. stenosperma,* respectively. These artificially induced

TABLE 3.1
Groundnut Synthetic Tetraploids Developed Worldwide

S No	Name of the Synthetic	Parents Involved in Developing Synthetic	Species of the Involved Parents	Genome Status	Source	References
1	ISATGR 5B	ICG 8960 × ICG 8209	*A. magna* × *A. batizocoi*	BBKK	ICRISAT	Mallikarjuna et al. (2011)
2	ISATGR 40A	ICG 8206 × ICG 8123	*A. ipaensis* × *A. duranensis*	BBAA	ICRISAT	Mallikarjuna et al. (2011)
3	ISATGR 47A	ICG 13256 × ICG 8123	*A. valida* × *A. duranensis*	BBAA	ICRISAT	Mallikarjuna et al. (2011)
4	ISATGR 48A	ICG 13256 × ICG 8123	*A. valida* × *A. duranensis*	BBAA	ICRISAT	Mallikarjuna et al. (2011)
5	ISATGR 65B	ICG 11548 × ICG 8123	*A. valida* × *A. duranensis*	BBAA	ICRISAT	Mallikarjuna et al. (2011)
6	ISATGR 72B	ICG 8138 × ICG 8216	*A. duranensis* × *A. cardenasii*	AAAA	ICRISAT	Mallikarjuna et al. (2011)
7	ISATGR 154	ICG 11548 × ICG 8123	*A. valida* × *A. duranensis*	BBAA	ICRISAT	Mallikarjuna et al. (2011)
8	ISATGR 168B	ICG 11548 × ICG 8123	*A. valida* × *A. duranensis*	BBAA	ICRISAT	Mallikarjuna et al. (2011)
9	ISATGR 265-5A	ICG 8164 × ICG 8190	*A. kempff mercadoi* × *A. hoehnei*	AAAA	ICRISAT	Mallikarjuna et al. (2011)
10	ISATGR 278-18	ICG 8138 × ICG 13160	*A. duranensis* × *A. batizocoi*	AAKK	ICRISAT	Mallikarjuna et al. (2011)
11	ISATGR 160	ICG 4983 × ICG 8216	*A. diogoi* × *A. cardenasii*	AAAA	ICRISAT	Mallikarjuna et al. (2011)
12	ISATGR 163B	ICG 8164 × ICG 15160	*A. kempff mercadoi* × *A. stenosperma*	AAAA	ICRISAT	Mallikarjuna et al. (2011)
13	ISATGR 173A	ICG 8139 × ICG 13160	*A. duranensis* × *A. batizocoi*	AAKK	ICRISAT	Mallikarjuna et al. (2011)
14	ISATGR 206B	ICG 8123 × ICG 11548	*A. duranensis* × *A. valida*	AABB	ICRISAT	Mallikarjuna et al. (2011)
15	ISATGR 1212	ICG 8123 × ICG 8206	*A. duranensis* × *A. ipaensis*	AABB	ICRISAT	Mallikarjuna et al. (2011)
16	AiAd	K30076 × V14167	*A. ipaensis* × *A. duranensis*	BBAA	ESALQ	Favero et al. (2006)
17	GA-MagSten1 (PI 695417)	K30097 × V10309	*A. magna* × *A. stenosperma*	BBAA	UGA	Bertioli et al. (2021b)
18	GA-BatSten1 (PI 695418)	K9484 × V10309 PI	*A. batizocoi* × *A. stenosperma*	KKAA	UGA	Bertioli et al. (2021b)

(*Continued*)

TABLE 3.1 (*Continued*)
Groundnut Synthetic Tetraploids Developed Worldwide

S No	Name of the Synthetic	Parents Involved in Developing Synthetic	Species of the Involved Parents	Genome Status	Source	References
19	IpaCor2-GA-NC (PI 695391)	KG 30076 × GKP9530	*A. ipaënsis* × *A. correntina*	BBAA	UGA	Chu et al. (2021)
20	IpaDur3-GA-NC (PI 695392)	KG30076 × KGBSPSc 30060	*A. ipaënsis* × *A. duranensis*	BBAA	UGA	Chu et al. (2021)
21	ValSten1-GA-NC (PI 695393)	KG30011 × V10309	*A. valida* × *A. stenosperma*	BBAA	UGA	Chu et al. (2021)
22	BatDur	K9484 × SeSN2848	*A. batizocoi* × *A. duranensis*	KKAA	UFPB	Dutra et al. (2018)
23	An 2	V 6389 × V 9401	*A. gregoryi* × *A. linearifolia*	AABB	EGRB	Michelotto et al. (2016)
24	An 4	KG 30076 × V 14167	*A.ipaënsis* × *A. duranensis*	BBAA	EGRB	Michelotto et al. (2016)
25	MagSten	K 30097 × V 15076	*A. magna* × *A. stenosperma*	BBAA	ESALQ	Favero et al. (2015b)

EGRB, Embrapa Genetic Resources and Biotechnology, Brazil; ESALQ, Escola Superior de Agricultura Luiz de Queiroz, Brazil; ICRISAT, International Crops Research Institute for the Semi-Arid Tropics, Patancheru, India; UFPB, The Federal University of Paraíba, Brazil; UGA, University of Georgia, USA.

tetraploids carrying resistance to root-knot nematode and early and late leaf spots are cross-compatible with cultivated groundnuts and are used in US breeding programmes (Bertioli et al., 2021b). Similarly, three more allotetraploids PI 695391, PI 695392, and PI 695393 with high levels of resistance to leaf spots and TSWV were developed from (*A. ipaënsis* KG 30076 × *A. correntina* GKP9530)^{4x}, (*A. ipaënsis* KG30076 × *A. duranensis* KGBSPSc 30060)^{4x} and (*A. valida* KG30011 × *A. stenosperma* V10309)^{4x} respectively (Chu et al., 2021). These new sources of disease resistance will assist the groundnut breeding community in improving the performance of cultivated groundnuts. By crossing an elite commercial cultivar BR1 with an allotetraploid derived from the wild species *A. duranensis* and *A. batizocoi,* two BC_1F_6 lines (53 P4 and 96 P9) with earliness, drought-tolerant traits, and good pod yield were identified (Dutra et al., 2018). With the aim of generating induced mutants, IL-3, and IL-4 introgression lines (derived from ABK genomes) were subjected to sodium azide and gamma mutagenesis, identifying 12 superior mutants (Joshi et al., 2019). These mutants were significantly more resistant to leaf spots and rust over the respective parents and GPBD-4 (best check) and are under initial trials for variety development for commercial release. Similarly, induced mutagenesis of three interspecific derivatives (ICGIL 17105, ICGIL 17110, and ICGIL 17112) identified superior mutants with foliar disease resistance and high pod yield, and they are being evaluated for development as commercial varieties (Tilak and Bhat, 2021).

MARKER OR GENOMICS-ASSISTED PRE-BREEDING TO FASTEN UP TRAIT DISCOVERY AND INTROGRESSION

As wild diploid *Arachis* species possess significantly more genetic variation than the cultivated tetraploid groundnut, developing molecular markers for them is considered important as they aid in the efficient introgression of wild genes. The genome complexity and low genetic variation of tetraploid groundnuts led to the construction of linkage maps of wild *Arachis* species initially (Stalker, 2017). Among various available marker systems, the first genetic map was constructed with restriction fragment length polymorphisms (RFLP), utilizing the population derived from the cross between *A. stenosperma* and *A. cardenasii*, where 117 RFLPs were mapped onto 11 linkage groups (Halward et al., 1993). Subsequently, an amplified fragment length polymorphisms (AFLPs)-based A-genome map was constructed for an F_2 population derived from *A. kuhlmannii* Krapov. and W. C. Gregory × *A. diogoi* (Milla, 2003). Using the A. *stenosperma* × A. *cardenasii* population, the first random amplified polymorphic DNA (RAPD)-based map was published by Garcia et al. (2005). A more saturated linkage map for A-genome was generated for *A. duranensis* × *A. stenosperma* derived F_2 population using 170 simple sequence repeat (SSR) markers. Nagy et al. (2012) published a more saturated map with 1,724 single-stranded DNA conformation polymorphism (SSCP) markers, SSRs, and single-nucleotide polymorphic markers (SNP) using a population derived from a cross between two *A. duranensis* (A genome) accessions.

Similarly, using an F_2 population derived from *A. ipaënsis* × *A. magna*, the B genome was mapped with 149 SSRs onto 10 linkage groups (Moretzsohn et al., 2009). The first tetraploid map with RFLPs was constructed using Florunner × TxAG-6 (synthetic amphidiploid) cross progenies (Burow et al., 2001). An SSR-based genetic map with 298 SSR markers was constructed using 88 BC_1F_1 tetraploid individuals from a cross Fleur11 × synthetic amphidiploid (*A. ipaensis* × *A. duranensis*) (Fonceka et al., 2009). A tetraploid SSR-based genetic map with 357 markers was also developed using 90 F_2 individuals from a cross Fleur11 × synthetic amphidiploid ISATGR 278-18 (A. *batizocoï* × A. *duranensis*). The inheritance of the SSR loci and the genetic map allowed providing evidence of homeologous recombinations between the A and K genomes, assessing their frequencies, and locating the chromosomes and regions within chromosomes where they occurred (Nguepjop et al., 2016). Given that homeologous recombination can change allele frequency and speed up the accumulation of rare but favourable alleles, it can be exploited in breeding programmes to improve important traits. A large amount of synteny between A and B genomes and several inversions and translocations were observed by comparing high-density A- and B-genome maps (Guo et al., 2012).

Molecular mapping aids in identifying quantitative trait loci (QTLs) and molecular markers closely associated with agronomically important traits, such as disease resistance and drought tolerance-related traits (Nagy et al., 2012; Stalker et al., 2017). Molecular markers help eliminate linkage drag or adverse genes and ensure that useful genes are being incorporated into the cultivated genome. Markers linked to root-knot nematode resistance were first derived from *A. cardenasii* species, where two sequence-characterized amplified region (SCAR) markers were developed for reduced galling and egg number (Garcia et al., 1996). Three RAPD markers in the interspecific hybrid TxAG-6 associated with nematode resistance, originated from *A. cardenasii*, were identified (Burow et al., 1996). Subsequently, a high-yielding, nematode-resistant cultivar NemaTAM was developed through marker-assisted selection (Simpson et al., 2003). As a single dominant gene is believed to contribute to the highest level of resistance, resistance may break down under sufficient selection pressure. Therefore, other species like *A. stenosperma* are being explored as they were reported as potent sources for nematode resistance (Singsit et al., 1995; Leal-Bertioli et al., 2010). Eight genes associated with *M. arenaria* resistance in the roots of *A. stenosperma* have been identified (Guimarães et al., 2010). A breeding scheme outlined to pyramid the high oleic acid trait and nematode resistance using marker-assisted selection has greatly increased the efficiency of developing groundnut breeding lines and cultivars (Chu et al., 2011).

The resistance to the two most devastating diseases of groundnut, early and late leaf spots caused by *Cercospora arachidicola* and *Cercosporidium personatum,* respectively, is a complex trait including several components like sporulation, incubation period, latent period, lesion number and diameter, and per cent defoliation (Aquino et al., 1995). Whereas the cultivated gene pool has moderate levels of resistance to these diseases, several wild diploid *Arachis* species are reported to be highly resistant to either or both of these diseases. To enhance the selection efficiency, three RAPD markers associated with the lesion diameter of ELS were identified from a population derived from *A. hypogaea* × *A. cardenasii* introgression line with "NC 7" (Stalker and Mozingo, 2001). Five QTLs and 34 resistance gene analogues (RGAs) for LLS resistance were mapped in a mapping population developed from the cross *A. duranensis* × A. *stenosperma* (Leal-Bertioli et al., 2009). Eleven LLS QTLs explaining 2%–7% phenotypic variance explained (PVE) were reported in a population derived from GPBD-4 (derived from *A. cardenasii*) (Khedikar et al., 2010). Similarly, using GPBD-4 as a parent in a mapping population, another major QTL for LLS was found, and the associated marker has been used in groundnut breeding programmes (Sujay et al., 2012). Two candidate genomic regions (A02 and A03) conferring resistance to LLS and rust identified from previous studies have been validated and used for developing foliar disease-resistant lines by marker-assisted selection (Kolekar et al., 2017; Pandey et al., 2017). QTLs conferring resistance to LLS and Groundnut Rosette Disease (GRD) were recently reported on chromosomes B04 and B5 using the AB-QTL population derived from the cross between Fleur11 and ISATGR 278-18 (Essandoh et al., 2022), confirming *A. batizocoi* as a valuable source of disease resistance that can be used in breeding programmes. Wild genomic segments conferring resistance to root-knot nematode and LLS were identified in *A. stenosperma* Krapov. & W. C. Greg (Leal-Bertioli et al., 2009, 2016; Ballén-Taborda et al., 2019), and a robust rust-resistant QTL was identified in *A. magna* Krapov., W. C. Greg & C. E. Simpson (Leal-Bertioli et al., 2015a). A synthetic allotetraploid developed using two wild species was found to be resistant to both LLS and rust (Fávero et al., 2015a; Michelotto et al., 2016).

Rust is the major foliar fungal disease caused by *Puccinia arachidis*, affecting most tropical production areas of Asia, Africa, and South America. Moderate levels of rust resistance in the cultivated germplasm have led to the utilization of wild

Arachis species that are highly resistant. Rust resistance is mainly believed to have originated from a major gene of *A. cardenasii*, and GPBD-4 has been used as a resistant parent in several mapping populations. Khedikar et al. (2010) identified a candidate SSR marker, IPAHM103, associated with a major rust-resistant QTL. TSWV is one of the most devastating diseases across all groundnut-producing areas of the United States. The evaluation of a segregating F_2 population from the cross *A. kuhlmannii* × *A. diogoi* identified five AFLP markers linked with TSWV resistance (Milla, 2003; Milla et al., 2004). Other wild species like *A. kuhlmannii*, *A. stenosperma*, *A. villosa*, and interspecific populations are also promising for introgression of thrips resistance (*Enneothrips flavens*) (Janini et al., 2010).

Wild species harbour alleles for several agro-morphological traits, yield component traits, and traits conferring adaptation to environmental stresses. Twenty-nine RFLP markers were mapped on BC_3F_2 lines derived from the cross *A. hypogaea* × *A. cardenasii* associated with lateral branches, main stem length, pod, and seed size (Burow et al., 2011). Fonceka et al. (2012a, 2012b) produced AB-QTL and Chromosome Segment Substitution Line (CSSL) populations derived from crosses of cultivar Fleur 11 and an amphidiploid (*A. ipaënsis* × *A. duranensis*)^{4x}. In the AB-QTL study, 95 QTLs were mapped for date to flowering, plant architecture, pod and seed morphology and size, yield component, and drought tolerance (Fonceka et al., 2012a). About half of the QTL positive effects were associated with alleles of the wild parents. In the CSSL population, several QTLs were mapped for plant architecture (Fonceka et al., 2012b), pod and seed weight (Tossim et al., 2020), and biological nitrogen fixation (Nzepang et al., 2023). The fatty acid profile of the CSSL population was analyzed and showed three lines (CSSL 84, CSSL 100, and CSSL 111) with elevated oleic acid content and reduced linoleic and palmitic acid relative to the recurrent parent Fleur 11 (Gimode et al., 2020), attesting to the potential of wild species for improving quality-related traits. Moreover, one CSSL was recently used to develop near isogenic lines (NILs) that allowed fine-mapping of a QTL region involved in seed size (Alyr et al., 2020). Using SSR markers, significant variation in oil content was found among the *Arachis* species and identified three associated alleles for higher oil content (Huang et al., 2012). Markers for pod and seed number and drought tolerance attributed to the wild parents were discovered by comparing QTLs obtained under water-stressed and well-irrigated conditions. As SNP markers can be analyzed with high-throughput genotyping methods and are the most abundant molecular markers in the genome, Nagy et al. (2012) developed an SNP-based map of diploid *Arachis*. A large-scale SNP genotyping array comprising polymorphic SNPs between diploid and tetraploid species was developed to genotype both cultivated and interspecific populations (Clevenger et al., 2017). Genotyping by sequencing of a subset of the population derived from an interspecific cross between *A. diogoi* (resistance to spotted wilt) and "Gregory" displayed SNP loci of *A. diogoi* genome on every chromosome of the A sub-genome of the introgression lines suggesting successful genomic introgression of *A. diogoi* contributing to the resistance (Hanson et al., 2020). Similarly, evaluating a synthesized interspecific RIL population (IAC-886 × *IpaDur1* (allotetraploid)) demonstrated a wide range of variation for agro-morphological traits and leaf spot resistance in some introgression lines. In addition, QTLs identified for agronomic and resistance traits will be useful for using marker-assisted selection in groundnut breeding programmes

(Suassuna et al., 2020). More recently, an induced allotetraploid (*A. stenosperma* × *A. magna*)^{4x} was used as a donor in a successive backcrossing scheme with IAC OL 4, a high-yielding Brazilian cultivar, where high-throughput SNP genotyping was used for background selection and SSR markers associated with LLS and rust resistance were used for foreground selection (Moretzsohn et al., 2023). They have successfully identified agronomically adapted lines with high-yield potential, high cultivated genome recovery, and retention of wild chromosome segments from both the wild parents (having resistance to rust and LLS). These introgressed segments include four previously identified QTLs (Leal-Bertioli et al., 2019) with resistance to both diseases and four additional QTLs. With the advancing sequencing technology, it now becomes a routine practice and an efficient aid in developing useful markers to facilitate gene introgressions through MABC.

CHALLENGES, OPPORTUNITIES AND FUTURE PROSPECTS

Wild relatives of several crops are conserved in genebanks worldwide and offer a novel source of germplasm into the breeding pipeline. In the face of ever-changing climatic conditions and emerging abiotic stresses (such as extreme temperatures, salinity, erratic rainfall patterns) and biotic stresses (insect pests and diseases), these wild relatives serve as repositories of novel variation and provide favourable alleles for addressing existing stresses. Pre-breeding has greater potential to generate genetic variability using wild species; however, it faces several technical challenges such as lack of knowledge on crossability relationships, characterization, and linkage drag, hindering progress in pre-breeding. Large-sized advanced backcross (AB) populations and utilizing molecular markers for precise introgression of desirable genes with minimum linkage drag have been employed to overcome these challenges. The AB-QTL approach allows for the simultaneous identification and transfer of exotic QTLs into elite breeding lines. Using AB populations, leveraging genomic tools enables efficient tracking of desired and genome-specific introgressions among breeding lines. Phenotypic and genotypic data from these AB populations have been used to identify genomic regions for several agronomically important traits.

Genotyping and genetic analysis of two AB populations developed at ICRISAT from crosses, ICGV 87846 × ISATGR 265-5A and ICGV 91114 × ISATGR 1212, provided a novel source of diversity and identified several wild segments (Khera et al., 2019). From these populations, several introgression lines with disease resistance and agronomic traits are identified. QTL analysis revealed that cultivated genomic segments primarily contributed favourable alleles for yield and oil quality traits, while wild genomic segments contributed favourable alleles for foliar disease resistance. Additionally, the development of introgression lines facilitates the rapid selection of QTL-NILs for use in hybridization programmes. Therefore, genomics-assisted pre-breeding harnesses unadapted exotic germplasm for quantitative trait improvement by integrating genomic tools with conventional breeding approaches. However, biases towards recurrent parent alleles, reduced recombination opportunity, extensive linkage disequilibrium (LD), and deployment of exotic QTLs without extensive validation limit the applications of this approach. To overcome these bottlenecks, the

BC-NAM has been designed, combining nested association mapping (NAM) with backcrossing. This approach involves crossing tetraploid synthetic genotypes with a common elite parent(s). Furthermore, the combination of next-generation sequencing technologies, genome-wide marker data generation, and the integration of genomic selection (GS) with optimum contribution selection (OCS) in pre-breeding programmes will broaden the genetic base of elite breeding gene pools and optimize the rate of genetic gain by recurrently introducing valuable sources of exotic diversity (Bohra et al., 2022).

Next-generation sequencing technologies and analysis tools have enabled the availability of high-quality reference genomes for many crops, including groundnut. However, single reference genome-based analysis may overlook critical structural variations (SVs) that distinguish one accession from another. There are implications that introgressed genomic segments from exotic/wild donors may contain genes not present in the reference genome. With the decreasing cost of sequencing and increased availability of high-quality genome sequencing data for diverse accessions, relying solely on a single reference genome to describe the entire landscape of genetic variation within a species is insufficient. *De novo* genome assembly and resequencing of various landraces and wild relatives can reveal the entire gene repertoire within a given species. Pangenome studies strongly support the importance of SVs in domestication and crop breeding (Zhao et al., 2018). Therefore, developing pangenomes helps capture the genetic diversity that may have been overlooked or lost during the domestication process. In the case of *Arachis*, the Axiom_*Arachis* 58K SNP array has been a landmark for marker-assisted breeding (Pandey et al., 2017) aiding in the identification of genomic regions and genes associated with biotic and abiotic stress response (Stein et al., 2018). Similarly, constructing "super-pangenomes," which extend the current pangenome strategy to a genus-level approach, provides a more comprehensive illustration of the wide range of diversity present among CWR genomes (Khan et al., 2020).

Technological advancements have opened up opportunities for the *de novo* domestication of wild plants as a viable solution for designing ideal crops capable of withstanding environmental stresses. *De novo* domestication involves introducing desirable/domesticated genes with known function either directly into undomesticated wild species or manipulating wild species at the gene level to achieve domestication. Advances in the "omics" technologies have facilitated the identification and understanding of multiple key traits, enabling the *de novo* domestication of CWR (Figure 3.2) (Kumar et al., 2021). Gene-editing technologies, such as CRISPR/Cas9, have become increasingly popular for domesticating CWR and neglected crops within a short time frame (Fernie and Yan, 2019). Identification of major domestication genes and CRISPR/Cas9 editing techniques, along with other scientific breakthroughs, have integrated various genomic changes in crop plants that allowed for *de novo* domestication of wild species and re-domestication of existing crops within a single generation without any associated drag on their inherent traits (Schindele et al., 2020). All these efforts contribute to harnessing the rich and potential diversity of wild species, leading to improvements in yield, nutrition, and resilience in modern crop varieties.

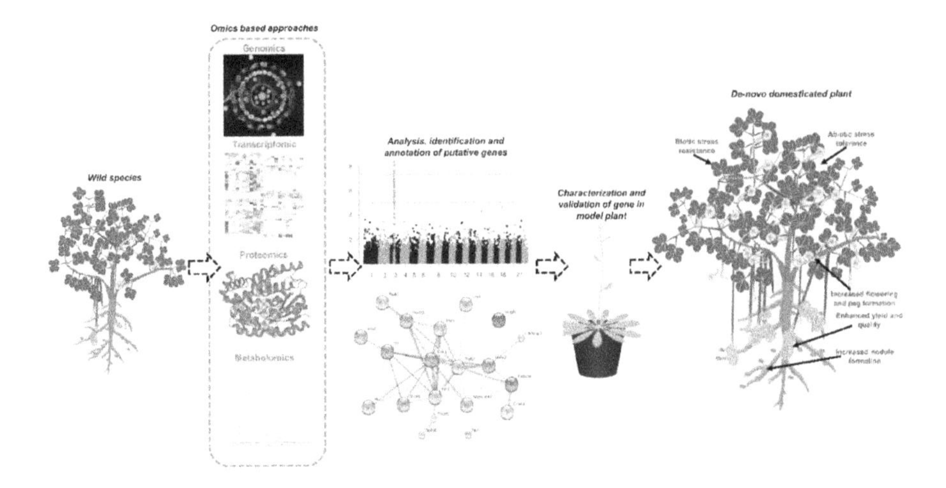

FIGURE 3.2 Schematic diagram depicting the relevance of OMICS approaches in gene characterization and development of designer crops utilizing *de novo* domestication strategy.

CONCLUSION

The availability of sufficient genetic variability in wild relatives and landraces, carrying favourable alleles/genes, is the prerequisite for the success of the crop improvement programme. In the case of groundnuts, wild relatives and traditional landraces possess a vast array of beneficial traits essential for improving crop resilience in challenging climatic conditions. However, incorporating wild species alleles in groundnuts improvement programmes is limited by various challenges, including hybridization barriers, difficulties in producing fertile hybrids, and linkage drag, among others. To address these challenges, initiating systematic and targeted pre-breeding activities is imperative for the continuous transfer of genetic variability to the breeding pipeline for cultivar improvement. While pre-breeding offers significant advantages for cultivar improvement, it is a time-consuming process, and the presence of linkage drag when utilizing wild species makes it complex and challenging.

Fortunately, advancements in genomics platforms provide promising solutions to overcome these challenges and accelerate the utilization of natural variation for cultivar improvement. By leveraging genomics tools, breeders can efficiently identify and incorporate desirable traits (alleles/genes) from wild *Arachis* relatives into elite cultivars.

Furthermore, *de novo* domestication and pangenome studies hold great potential in unleashing the utility of wild *Arachis* relatives. By utilizing such approaches, researchers can harness the rich and untapped diversity of wild *Arachis* relatives to develop groundnut cultivars that are more resilient, productive, and better equipped to meet the challenges of pest pressure, diseases, and environmental stresses. Ultimately, these efforts contribute to the larger goal of achieving global food security.

REFERENCES

Alyr, M.H., Pallu, J., Sambou, A., Nguepjop, J.R., Seye, M., Tossim, H.-A., Djiboune, Y.R., Sane, D., Rami, J.-F., & Fonceka, D. (2020). Fine-mapping of a wild genomic region involved in pod and seed size reduction on chromosome A07 in peanut (*Arachis hypogaea* L.). *Genes*, 11, 1402.

Amin, P.W. (1985). Resistance of wild species of groundnut to insect and mite pests. In: B. J. Moss and S. D. Feakin, editors, *Proceedings of an International Workshop on Cytogenetics of Arachis*, ICRISAT, Patancheru, pp. 57–60.

Amin, P.W., & Mohammad, A.B. (1980). Groundnut pest research at ICRISAT. In: *Proceedings of an International Workshop on Groundnuts*, ICRISAT, Patancheru. pp. 158–166.

Anderson, W.F., Holbrook, C.C., & Timper, P. (2006). Registration of root-knot nematode resistant peanut germplasm lines NR 0812 and NR 0817. *Crop Science*, 46, 481.

Aquino, V.M., Shokes, F.M., Gorbet, D.W., & Nutter, F.W. (1995). Late leaf spot progression on peanut as affected by components of partial resistance. *Plant Disease*, 79, 74–78. https://doi.org/10.1094/PD-79-0074

Ballén-Taborda, C., Chu, Y., Ozias-Akins, P., Timper, P., Holbrook, C.C., Jackson, S.A., Bertioli, D.J., & Leal-Bertioli, S.C. (2019). A new source of root-knot nematode resistance from Arachis stenosperma incorporated into allotetraploid peanut (*Arachis hypogaea*). *Scientific Reports*, 9(1), 17702.

Ballén-Taborda, C., Chu, Y., Ozias-Akins, P., Timper, P., Jackson, S.A., Bertioli, D.J., & Leal-Bertioli, S.C. (2021). Validation of resistance to root-knot nematode incorporated in peanut from the wild relative *Arachis stenosperma*. *Agronomy Journal*, 113(3), 2293–2302.

Ballén-Taborda, C., Chu, Y., Ozias-Akins, P., Holbrook, C.C., Timper, P., Jackson, S.A., Bertioli, D.J., & Leal-Bertioli, S. (2022). Development and genetic characterization of peanut advanced backcross lines that incorporate root-knot nematode resistance from *Arachis stenosperma*. *Frontiers in Plant Science*, 12, 3078.

Bertioli, D.J., Cannon, S.B., Froenicke, L., Huang, G., Farmer, A.D., Cannon, E.K., Liu, X., Gao, D., Clevenger, J., Dash, S. and Ren, L. (2016). The genome sequences of *Arachis duranensis* and *Arachis ipaensis*, the diploid ancestors of cultivated peanut. *Nature genetics*, 48(4): 438–446.

Bertioli, D.J., Jenkins, J., Clevenger, J., Dudchenko, O., Gao, D., Seijo, G., Leal-Bertioli, S.C., Ren, L., Farmer, A.D., Pandey, M.K., & Samoluk, S.S. (2019). The genome sequence of segmental allotetraploid peanut *Arachis hypogaea*. *Nature Genetics*, 51(5), 877–884

Bertioli, D.J., Clevenger, J., Godoy, I.J., Stalker, H.T., Wood, S., Santos, J.F., Ballén-Taborda, C., Abernathy, B., Azevedo, V., Campbell, J., & Chavarro, C. (2021a). Legacy genetics of *Arachis cardenasii* in the peanut crop shows the profound benefits of international seed exchange. *Proceedings of the National Academy of Sciences of the United States of America*, 118, 1–9.

Bertioli, D.J., Gao, D., Ballen-Taborda, C., Chu, Y., Ozias-Akins, P., Jackson, S.A., Holbrook, C.C., & Leal-Bertioli, S.C.M. (2021b). Registration of *GA-BatSten1* and *GA-MagSten1*, two induced allotetraploids derived from peanut wild relatives with superior resistance to leaf spots, rust, and root-knot nematode. *Journal of Plant Registrations*, 15(2), 372–378.

Bohra, A., Kilian, B., Sivasankar, S., Caccamo, M., Mba, C., McCouch, S.R., & Varshney, R.K. (2022). Reap the crop wild relatives for breeding future crops. *Trends in Biotechnology*, 40(4), 412–431.

Branch, W., & Brenneman, T.B. (2015). Registration of 'Georgia-14N' peanut. *Journal of Plant Registrations*, 9, 159–161.

Burow, M.D., Simpson, C.E., Paterson, A.H., & Starr, J.L. (1996). Identification of peanut (*Arachis hypogaea*) RAPD markers diagnostic of root-knot nematode (*Meloidogyne arenaria* (Neal) Chitwood) resistance. *Molecular Breeding*, 2, 369–379.

Burow, M., Simpson, C., Starr, J., & Paterson, A. (2001). Transmission genetics of chromatin from a synthetic amphidiploid to cultivated peanut (*Arachis hypogaea* L.). Broadening the gene pool of a monophyletic polyploid species. *Genetics*, 159, 823–837.

Burow, M.D., Simpson, C.E., Starr, J.L., Park, C.H., & Paterson, A.H. (2011). QTL analysis of early leaf spot resistance and agronomic traits in an introgression population of peanut. In: *Abstracts of the 5th International Conference of the Peanut Research Community: Advances in Arachis through genomics and biotechnology*, Brasilia, Brazil. EMBRAPA. American Peanut Council, Alexandria, VA.

Chakraborty, S., & Newton, A.C. (2011). Climate change, plant diseases and food security: an overview. *Plant Pathology*, 60, 2–14.

Chen, B.Y., Jiang, H.F., Ren, X.P., Liao, B.S., & Huang, J.Q. (2008). Germplasm identification and molecular characterization of resistance to bacterial wilt in wild *Arachis* species. *Acta Agriculturae Boreali-Sinica*, 23(3), 6

Chu, Y., Wu, C.L., Holbrook, C.C., Tillman, B.L., Pearson, G., & Oziakas-Akins, P. (2011). Marker-assisted selection to pyramid nematode resistance and the high oleic trait in peanut. *Plant Genome*, 4, 110–117.

Chu, Y., Stalker, H.T., Marasigan, K., Levinson, C.M., Gao, D., Bertioli, D.J., Leal-Bertioli, S.C., Holbrook, C.C., Jackson, S.A., & Ozias-Akins, P. (2021). Registration of three peanut allotetraploid interspecific hybrids resistant to late leaf spot disease and tomato spotted wilt. *Journal of Plant Registrations*, 15(3):562–572.

Clevenger, J., Chu, Y., Chavarro, C., Agarwal, G., Bertioli, D.J., Leal-Bertioli, S.C., Pandey, M.K., Vaughn, J., Abernathy, B., Barkley, N.A., & Hovav, R. (2017). Genome-wide SNP genotyping resolves signatures of selection and tetrasomic recombination in peanut. *Molecular Plant*, 10(2), 309–322.

Company, M., Stalker, H.T., & Wynne, J.C. (1982). Cytology and leafspot resistance in *Arachis hypogaea* wild species hybrids. *Euphytica*, 31, 885–893

Cowling, W.A. (2013). Sustainable plant breeding. *Plant Breeding*, 132(1), 1–9.

Dange, S.B., & Naidu, G.K. (2021). Assessment of groundnut pre-breeding material for resistance to *Spodoptera litura* and productivity traits in northern transitional zone of Karnataka. *Journal of Farm Sciences*, 34(1), 5–9.

Dutra, W.F., Guerra, Y.L., Ramos, J.P., Fernandes, P.D., Silva, C.R., Bertioli, D.J., Leal-Bertioli, S.C., & Santos, R.C. (2018). Introgression of wild alleles into the tetraploid peanut crop to improve water use efficiency, earliness and yield. *PLoS One*, 13(6), e0198776.

Dwivedi, S.L., Bertioli, D.J., Crouch, J.H., Valls, J.M.F., Upadhyaya, H.D., & Favero, A. (2007). Peanut. In: C. Kole, editor, *Genome Mapping and Molecular Breeding in Plants. Oilseeds*, Springer-Verlab, Berlin, Heidelberg, pp. 115–151.

Essandoh, D.A., Odong, T., Okello, D.K., Fonceka, D., Nguepjop, J., Sambou, A., Ballén-Taborda, C., Chavarro, C., Bertioli, D.J., & Leal-Bertioli, S.C.M. (2022). Quantitative trait analysis shows the potential for alleles from the wild species arachis batizocoi and *A. duranensis* to improve groundnut disease resistance and yield in East Africa. *Agronomy*, 12, 2202.

Favero, A.P., Simpson, C.E., Valls, F.M.J., & Velo, N.A. (2006). Study of evolution of cultivated peanut through crossability studies among *Arachis ipaënsis*, *A. duranensis* and *A. hypogaea*. *Crop Science*, 46, 1546–1552.

Fávero, A.P., Santos, R.F., Simpson, C.E., Valls, J.F.M., & Vello, N.A. (2015a). Successful crosses between fungal-resistant wild species of *Arachis* (Section *Arachis*) and *Arachis hypogaea*. *Genetics and Molecular Biology*, 38, 353–365.

Fávero, A.P., Pádua, J.G., Costa, T., Gimenes, M.A., Godoy, I.J., Moretzsohn, M.D.C. and Michelotto, M.D. (2015b). New hybrids from peanut (Arachis hypogaea L.) and synthetic amphidiploid crosses show promise in increasing pest and disease tolerance. *Genetics and Molecular Research*, 14, 16694–16703

Fernie, A.R., & Yan, J.B. (2019). De novo domestication: an alternative route toward new crops for the future. *Molecular Plant*, 12, 615–631.

Foncéka, D., Hodo-Abalo, T., Rivallan, R. et al. (2009). Genetic mapping of wild introgressions into cultivated peanut: a way toward enlarging the genetic basis of a recent allotetraploid. *BMC Plant Biology*, 9, 103.

Fonceka, D., Tossim, H.A., Rivallan, R., Vignes, H., Faye, I., Ndoye, O., Moretzsohn, M.C., Bertioli, D.J., Glaszmann, J.C., Courtois, B., & Rami, J.F. (2012a). Fostered and left behind alleles in peanut: interspecific QTL mapping reveals footprints of domestication and useful natural variation for breeding. *BMC Plant Biology*, 12, 1–16.

Fonceka, D., Tossim, H.A., Rivallan, R., Vignes, H., Lacut, E., De Bellis, F., Faye, I., Ndoye, O., Leal-Bertioli, S.C., Valls, J.F., & Bertioli, D.J. (2012b). Construction of chromosome segment substitution lines in peanut (*Arachis hypogaea* L.) using a wild synthetic and QTL mapping for plant morphology. *PLoS ONE*, 7(11), e48642.

Food and Agriculture Organization. (2020). *Statistical Database*. FAO, Rome, Italy. https://faostat.fao.org/

Friend, S., Quandt, D., Tallury, S., Stalker, H., & Hilu, K. (2010). Species, genomes, and section relationships in the genus *Arachis* (Fabaceae): a molecular phylogeny. *Plant Systematics and Evolution*, 290(1–4), 185–199.

Garcia, G.M., Stalker, H.T., Shroeder, E., & Kochert, G. (1996). Identification of RAPD, SCAR, and RFLP markers tightly linked to nematode resistance genes introgressed from *Arachis cardenasii* into *Arachis hypogaea*. *Genome*, 39, 836–845.

Garcia, G.M., Stalker, H.T., Schroeder, E., Lyerly, J.H., & Kochert, G. (2005). A RAPD-based linkage map of peanut based on a backcross population between the two diploid species *Arachis stenosperma* and *Arachis cardenasii*. *Peanut Science*, 32, 1–8.

Gimode, D., Chu, Y., Dean, L., Holbrook, C., Fonceka, D., & Ozias-Akins, P. (2020). Seed composition survey of a peanut CSSL population reveals introgression lines with elevated oleic/linoleic profiles. *Peanut Science*, 47(3), 139–149.

Gowda, M.V.C., Motagi, B.N., Naidu, G.K., Diddimani, S.B., & Sheshagiri, R. (2002). GPBD 4: a Spanish bunch groundnut genotype resistant to rust and late leaf spot. *International Arachis Newsletter*, 22, 29–32.

Grabiele, M., Chalup, L., Robledo, G., & Seijo, G. (2012). Genetic and geographic origin of domesticated peanut as evidenced by 5S rDNA and chloroplast DNA sequences. *Plant Systematics and Evolution*, 298, 1151–1165.

Gregory, M.P., & Gregory, W.C. (1979). Exotic germplasm of *Arachis* L. interspecific hybrids. *Journal of Heredity*, 70, 185–193.

Gregory, W.C., Krapovickas, A., & Gregory, M.P. (1980). Structures, variation, evolution and classification in *Arachis*. In: R. J. Summerfield and A. H. Bunting, editors, *Advances in Legume Science*, Royal Botanic Gardens, London, pp. 469–481.

Guimaraes, P.M., Brasileiro, A.C.M., Araujo, A.C.G., Leal-Bertioli, S.C.M., da Silva, F.R., & Morgante, C.V. (2010). A study of gene expression in the nematode resistant wild peanut relative, *Arachis stenosperma*, in response to challenge with *Meloidogyne arenaria*. *Tropical Plant Biology*, 3, 183–192. https://doi.org/10.1007/s12042-010-9056-z

Guo, Y., Khanal, S., Tang, S., Bowers, J.E., Heesacker, A.F., Khalilian, N., et al. (2012). Comparative mapping in intraspecific populations uncovers a high degree of macrosynteny between A- and B- genome diploid species of peanut. *BMC Genomics*, 13, 608.

Halward, T.M., & Stalker, H.T. (1987). Incompatibility Mechanism in Interspecific Peanut Hybrids. *Crop science*, 27, 456–460.

Halward, T.M., Stalker, H.T., & Kochert, G. (1993). Development of an RFLP linkage map in diploid peanut species. *Theoretical and Applied Genetics*, 87, 379–384.

Hammons, R.O. (1970). Registration of Spancross peanuts (Reg. No. 3). *Crop Science*, 10, 459–460.

Hanson, E., Zhou, H., Tallury, S.P., Yang, X., Paudel, D., Tillman, B., & Wang, J. (2020). Identifying chromosomal introgressions from a wild species *Arachis diogoi* into interspecific peanut hybrids. *Plant Breeding*, 139(5), 969–976.

Holbrook, C.C., Anderson, W.F., & Pittman, R.N. (1993). Selection of a core collection from the U.S. Germplasm collection of peanut. *Crop Science*, *33*, 859–861.

Holbrook, C.C., Timper, P., Culbreath, A.K., & Kvien, C.K. (2008). Registration of 'Tifguard' peanut. *Journal of Plant Registrations*, 2, 2.

Holbrook, C., Ozias-Akins, P., Chu, Y., Culbreath, A., Kvien, C., & Brenneman, T. (2017). Registration of 'TifNV-High O/L' peanut. *Journal of Plant Registrations*, 11, 228–230.

Huang, L., Jiang, H., Ren, X., Chen, Y., Xiao, Y., Zhao, X. et al. (2012). Abundant microsatellite diversity and oil content in wild *Arachis* species. *PLoS One*, 7, e50002. https://doi.org/10.1371/journal.pone.0050

Isleib, T.G., Rice, P.W., Mozingo, R.W., Copeland, S.C., Graeber, J.B., & Stalker, H.T. (2006). Registration of N96076L peanut germplasm. *Crop Science*, 46, 2329–2330.

Isleib, T.G., Milla-Lewis, S.R., Pattee, H.E., Copeland, S.C., Zuleta, M.C., Shew, B.B. et al. (2011). Registration of 'Bailey' peanut. *Journal of Plant Registrations*, 5, 27–39.

Janini, J.C., Boiça Júnior, A.L., Godoy, I.J., Michelotto, M.D., & Favero, A.P. (2010). Avaliaçao de especies silvestres e cultivares de amendoim para resistencia a Enneothrips flavens Moulton. *(In Portuguese.) Bragantia*, 69, 891–898.

Jiang, H.F., Ren, X.P., Chen, Y.N., Huang, L., Zhou, X.J., Huang, J.Q., Froenicke, L., Yu, J.J., Guo, B.Z., Liao, & B.S. (2013). Phenotypic evaluation of the Chinese mini-mini core collection of peanut (*Arachis hypogaea* L.) and assessment for resistance to bacterial wilt disease caused by *Ralstonia solanacearum. Plant Genetic Resources*, 11, 77–83.

Jivani, L.L., Vachhani, J.H., Vaghasia, P.M., Kachhadia, V.H., Patel, M.B., & Shekhat, H.G. (2012). Genetics and inter-relationship of oil and protein content in crosses involving bunch genotypes of groundnut (*Arachis hypogaea* L.). *International Journal of Agricultural Sciences*, 8(2), 338–340.

Joshi, P., Jadhav, M.P., Yadawad, A., Shirasawa, K., & Bhat, R.S. (2019). Foliar disease resistant and productive mutants from the introgression lines of peanut (*Arachis hypogaea* L.). *Plant Breed*, 139 (1), 148–155.

Khan, A.W. et al. (2020). Super-pangenome by integrating the wild-side of a species for accelerated crop improvement. *Trends in Plant Science*, 25, 148–158.

Khedikar, Y.P., Gowda, M.V.C., Sarvamangala, C., Patgar, K.V., Upadhyaya, H.D., & Varshney, R.K. (2010). A QTL study on late leaf spot and rust revealed one major QTL for molecular breeding for rust resistance in groundnut (*Arachis hypogaea* L.). *Theoretical and Applied Genetics*, 121, 971–984.

Khera, P., Pandey, M.K., Mallikarjuna, N., Sriswathi, M., Roorkiwal, M., Janila, P., Sharma, S., Shilpa, K., Sudini, H., Guo, B., & Varshney, R.K. (2019). Genetic imprints of domestication for disease resistance, oil quality, and yield component traits in groundnut (*Arachis hypogaea* L.). *Molecular Genetics and Genomics*, 294, 365–378.

Kochert, G., Stalker, H.T., Gimenes, M., Galgaro, L., Lopes, C.R., & Moore, K. (1996). RFLP and cytogenetic evidence on the origin and evolution of allotetraploid domesticated peanut, *Arachis hypogaea* (Leguminosae). *American Journal of Botany*, 83(10), 1282–1291.

Kolekar, R.M., Sukruth, M., Shirasawa, K., Nadaf, H.L., Motagi, B.N., Lingaraju, S., Patil, P.V., & Bhat, R.S. (2017). Marker-assisted backcrossing to develop foliar disease-resistant genotypes in TMV 2 variety of peanut (*Arachis hypogaea* L.). *Plant Breeding*, 136(6), 948–953.

Krapovickas, A. (1969). The origin, variability and spread of the groundnut (*Arachis hypogaea*). In: R. J. Ucko and W. C. Dimbledy, editors. *The Domestication and Exploitation of Plant and Animals*, Greald Duckworth Co. Ltd, London, pp. 427–441.

Krapovickas, A. (1973). Evolution of the genus *Arachis*. In: R. Moav, editor. *Agricultural Genetics-Selected Topics*, Wiley, New York, pp. 135–151.

Krapovickas, A., & Gregory, W.C. (I994). Taxonomy of the genus *Arachis*. *Bonplandia*, 8, 1–186.

Krapovickas, A., Gregory, W.C., Williams, D.E., & Simpson, C.E. (2007). Taxonomy of the genus *Arachis* (Leguminosae). *Bonplandia*, 16, 7–205.

Kumar, R., Sharma, V., Suresh, S., Ramrao, D.P., Veershetty, A., Kumar, S., Priscilla, K., Hangargi, B., Narasanna, R., Pandey, M.K., & Naik, G.R. (2021). Understanding omics driven plant improvement and de novo crop domestication: some examples. *Frontiers in Genetics*, 12, 637141.

Leal-Bertioli, S.C.M., Jose, A.C.V.F., Alves-Freitas, D.M.T., Moretzsohn, M.C., Guimaraes, P.M., Nielen, S. et al. (2009). Identification of candidate genome regions controlling disease resistance in *Arachis*. *BMC Plant Biology*, 9, 112.

Leal-Bertioli, S.C.M., Farias, M.P., Silva, P.T., Guimaraes, P.M., Brasileiro, A.C.M., Bertioli, D.J., & Araujo, A.C.G. (2010). Ultrastructure of the initial interaction of *Puccinia arachidis* and *Cercosporidium personatum* with leaves of *Arachis hypogaea* and *Arachis stenosperma*. *Journal of Phytopathology*, 158, 792–796.

Leal-Bertioli, S.C., Cavalcante, U., Gouvea, E.G., Ballén-Taborda, C., Shirasawa, K., Guimarães, P.M., Jackson, S.A., Bertioli, D.J., & Moretzsohn, M.C. (2015a). Identification of QTLs for rust resistance in the peanut wild species *Arachis magna* and the development of KASP markers for marker-assisted selection. *G3; Genes|Genomes|Genetics*, 5, 1403–1413.

Leal-Bertioli, S.C., Moretzsohn, M.C., Roberts, P.A., Ballén-Taborda, C., Borba, T.C., Valdisser, P.A., Vianello, R.P., Araújo, A.C.G., Guimarães, P.M., & Bertioli, D.J. (2016). Genetic mapping of resistance to *Meloidogyne arenaria* in *Arachis stenosperma*: a new source of nematode resistance for peanut. *G3; Genes|Genomes|Genetics*, 6, 377–390.

Leal-Bertioli, S.C., Nascimento, E.F., Chavarro, M.C.F., Custódio, A.R., Hopkins, M.S., Moretzsohn, M.C., Bertioli, D.J., & Araújo, A.C.G. (2021). Spontaneous generation of diversity in Arachis neopolyploids (*Arachis ipaënsis* × *Arachis duranensis*) 4*x* replays the early stages of peanut evolution. *G3 Genes|Genomes|Genetics*, 11(11), jkab289.

Lu, J., & Pickersgill, B. (1993). Isozyme variation and species relationships in peanut and its wild relatives *Arachis* L. - Leguminosae. *Theoretical and Applied Genetics*, 85, 550–560.

Lyerly, J.H., Stalker, H.T., Moyer, J.W., & Hoffman, K. (2002). Evaluation of the wild species of peanut for resistance to tomato spotted wilt virus. *Peanut Science*, 29, 79–84.

Mallikarjuna, N., Senthilvel, S., & Hoisington, D. (2011). Development of new sources of tetraploid Arachis to broaden the genetic base of cultivated groundnut (*Arachis hypogaea* L.). *Genetic Resources and Crop Evolution*, 58, 889–907.

Mallikarjuna, N., Jadhav, D.R., Reddy, K., Husain, F., & Das, K. (2012). Screening new *Arachis* amphidiploids, autotetraploids for resistance to late leaf spot by detached leaf technique. *European Journal of Plant Pathology*, 132, 17–21.

Michelotto, M.D., Barioni, W. Jr., de Resende, M.D.V., de Godoy, I.J., Leonardecz, E., & Favero, A.P. (2015). Identification of fungus resistant wild accessions and interspecific hybrids of the genus *Arachis*. *PLoS One*, 10(6), e0128811.

Michelotto, M.D., Godoy, I.J., Santos, J.F., Martins, A.L.M., Leonardecz, E., & Fávero, A.P. (2016). Identifying *Arachis* amphidiploids resistant to foliar fungal diseases. *Crop Science*, 56, 1792–1798.

Milla, S.R. (2003). Relationship and utilization of Arachis germplasm in peanut improvement, Ph.D. diss., North Carolina State Univ., Raleigh.

Milla, S.R., Tallury, S.P., Stalker, H.T., & Isleib, T.G. (2004). Identification of molecular markers associated with tomato spotted wilt virus in a genetic linkage map *of Arachis kuhlmannii* × A. *diogoi*. *Proceedings of American Peanut Research and Education Society*, 36, 27.

Moretzsohn, M., Barbosa, A., Alves-Freitas, D., Teixeira, C., Leal-Bertioli, S., Guimaraes, P. et al. (2009). A linkage map for the B-genome of *Arachis* (Fabaceae) and its synteny to the A-genome. *BMC Plant Biology*, 9, 40.

Moretzsohn, M.C., Santos, J.F., Moraes, A.R.A., Custódio, A.R., Michelotto, M.D., Maharjan, N., Leal-Bertioli, S.C.M., Godoy, I.J., & Bertioli, D.J. (2023). Marker-assisted introgression of wild chromosome segments conferring resistance to fungal foliar diseases into peanut (*Arachis hypogaea* L.). *Frontiers of Plant Science*, 14, 764.

Moss, J.P., Singh, A.K., Reddy, L.J., Nigam, S.N., Subrahmanyam, P., McDonald, D., & Reddy, A.G.S. (1997). Registration of ICGV 87165 peanut germplasm line with multiple resistance. *Crop Science*, 37, 1028.

Moss, J.P., Singh, A.K., Nigam, S.N., Hilderbrand, G.L., Goviden, N., & Ismael, F.M. (1998). Registration of ICGV-SM 87165 peanut germplasm. *Crop Science*, 38, 572.

Nagy, E.D., Chu, Y., Guo, Y., Khanal, S., Tang, S., Li, Y. et al. (2010). Recombination is suppressed in an alien introgression in peanut harboring *Rma*, a dominant root-knot nematode resistance gene. *Molecular Plant Breeding*, 26, 357–370.

Nagy, E.D., Guo, Y., Tang, S., Bowers, J.E., Okashah, R.A., Taylor, C.A. et al. (2012). A high-density genetic map of *Arachis duranensis*, a diploid ancestor of cultivated peanut. *BMC Genomics*, 13, 469.

Naidu, G.K., Dange, S.B., & Nadaf, H.L. (2022). Potential groundnut pre-breeding genotypes with resistance to *Aspergillus flavus*. *Electronic Journal of Plant Breeding*, 13(1), 243–248.

Nelson, S.C., Simpson, C.E., & Starr, J.L. (1989). Resistance to Meloidogyne arenaria in *Arachis* sp. germplasm. *Journal of Nematology*, 21, 654–660.

Nguepjop, J.R., Hodo-Abalo, T., Bell, J.M., Rami, J-F., Sharma, S., Courtois, B., Mallikarjuna, M., Sane, D., & Fonceka, D. (2016). Evidence of genomic exchanges between homeologous chromosomes in a cross of peanut with newly synthetizes allotetraploid hybrids. *Frontiers of Plant Science*, 7, 1635. https://doi.org/10.3389/fpls.2016.01635

Nigam, S.N., Dwivedi, S.L., & Gibbons, R.W. (1991). Groundnut breeding: constraints, achievements, and future possibilities. *Plant Breeding*, 61, 1127–1136.

Nzepang, D.T., Gully, D., Nguepjop, J.R., Zaiya Zazou, A., Tossim, H.-A., Sambou, A., Rami, J.-F., Hocher, V., Fall, S., Svistoonoff, S., & Fonceka, D. (2003). Mapping of QTLs associated with biological nitrogen fixation traits in peanuts (*Arachis hypogaea* L.) using an interspecific population derived from the cross between the cultivated species and its wild ancestors. *Genes*, 14, 797.

Okello, D.K., Deom, C.M., Puppala, N., Monyo, E., & Bravo-Ureta, B. (2018). Registration of 'serenut 6T' groundnut. *Journal of Plant Registrations*, 12(1), 43–47.

Pandey, M.K., Monyo, E., Ozias-Akins, P., Liang, X., Guimaraes, P., Nigam, S.N., et al. (2012). Advances in *Arachis* genomics for peanut improvement. *Biotechnology Advances*, 30, 639–651.

Pandey, M.K., Khan, A.W., Singh, V.K., Vishwakarma, M.K., Shasidhar, Y., Kumar, V., Garg, V., Bhat, R.S., Chitikineni, A., Janila, P., & Guo, B. (2017). QTL-seq approach identified genomic regions and diagnostic markers for rust and late leaf spot resistance in groundnut (*Arachis hypogaea* L.). *Plant Biotechnology Journal*, 15(8), 927–941.

Rami, J.F., Leal-Bertioli, S., Fonceka, D., Marcio, M., & David, B. (2013). Groundnut. In: A. Pratap and J. Kumar, editors, *Alien Gene Transfer in Crop Plants. Volume 2: Achievements and Impacts*, Springer, New York, pp. 253–279.

Reddy, A.S., Reddy, L.J., Mallikarjuna, N., Abdurahman, M.D., Reddy, Y.V., Bramel, P.J., & Reddy, D.V.R. (2000). Identification of resistance to peanut bud necrosis virus (PBNV) in wild *Arachis* germplasm. *Annals of Applied Biology*, 137, 135–139.

Robledo, G., & Seijo, G. (2010). Species relationships among the wild B genome of *Arachis* species section Arachis based on FISH mapping of rDNA loci and heterochromatin detection: a new proposal for genome arrangement. *Theoretical and Applied Genetics*, 121, 1033–1046.

Schindele, A., Dorn, A., & Puchta, H. (2020). CRISPR/Cas brings plant biology and breeding into the fast lane. *Current Opinion in Biotechnology*, 61, 7–14.

Sharma, S.B., Ansari, M.A., Varaprasad, K.S., Singh, A.K., & Reddy, L.J. (1999). Resistance to *Meloidogyne javanica* in wild *Arachis* species. *Genetic Resources and Crop Evolution*, 46, 557–568.

Sharma, S., Upadhyaya, H.D., Varshney, R.K., & Gowda, C.L.L. (2013). Pre-breeding for diversification of primary gene pool and genetic enhancement of grain legumes. *Frontiers in Plant Science*, 4, 309.

Sharma, S., Pandey, M.K., Sudini, H.K., Upadhyaya, H.D., & Varshney, R.K. (2017). Harnessing genetic diversity of wild *Arachis* species for genetic enhancement of cultivated peanut. *Crop Science*, 57(3), 1121–1131. Simpson, C.E., & Smith, O.D. (1975). Registration of Tamnut 74 peanut (Reg. No. 19). *Crop Science*, 15, 603–604.

Simpson, C.E., & Starr, J.L. (2001). Registration of COAN peanut. *Crop Science*, 41, 918.

Simpson, C.E., Nelson, S.C., Starr, J., Woodward, K.E., & Smith, O.D. (1993). Registration of TxAG-6 and TxAG-7 peanut germplasm lines. *Crop Science*, 33, 1418.

Simpson, C.E., Starr, J.L., Baring, M.R., Burow, M.D., Carson, J.M., & Wilson, J.N. (2013). Registration of 'Webb' peanut. *Journal of Plant Registrations*, 7, 265–268.

Simpson, C.E., Starr, J.L., Church, G.T., Burow, M.D., & Paterson, A.H. (2003). Registration of "NemaTAM" peanut. *Crop Science*, 43, 1561.

Singh, A.K., & Simpson, C.E. (1994). Biosystematics and genetic resources. In: J. Smartt, editor, *The Groundnut Crop: A Scientific Basis for Improvement*, Chapman & Hall, London, pp. 96–137.

Singh, A.K., Dwivedi, S.L., Pande, S., Moss, J.P., Nigam, S.N., & Sastri, D.C. (2003). Registration of rust and late leaf spot resistant peanut germplasm lines. *Crop Science*, 43, 440–441.

Singsit, C., Holbrook, C.C., Culbreath, A.K., & Ozias-Akins, P. (1995). Progenies of an interspecific hybrid between *Arachis hypogaea* and *A. stenosperma* pest resistance and molecular homogeneity. *Euphytica*, 83, 9–14.

Stalker, H.T. (1981). Hybrids in the genus *Arachis* between sections *Erectoides* and *Arachis*. *Crop Science*, 21, 359–362.

Stalker, H.T. (1992). Utilizing Arachis germplasm resources. In: S. N. Nigam, editor, *Groundnut, a Global Perspective: Proceedings of an International Workshop*, ICRISAT, Patancheru, pp. 281–295.

Stalker, H.T. (2017). Utilizing wild species for peanut improvement. *Crop Science*, 57(3), 1102–1120.

Stalker, H.T., & Campbell, W.V. (1983). Resistance of wild species of peanuts to an insect complex. *Peanut Science*, 10, 30–33.

Stalker, H.T., & Lynch, R.E. (2002). Registration of four insect resistant peanut germplasm lines. *Crop Science*, 42, 313–314.

Stalker, H.T., & Moss, J.P. (1987). Speciation, cytogenetics, and utilization of *Arachis* species. *Advances in Agronomy*, 41, 1–40.

Stalker, H.T., & Mozingo, L.J. (2001). Molecular markers of *Arachis* and marker-assisted selection. *Peanut Science*, 28, 117–123.

Stalker, H.T., & Wilson, R.F. (2016). *Peanuts Genetics, Processing, and Utilization*. AOCS Press. Published by Elsevier Inc, Cambridge, MA. ISBN 978-1-63067-038-2.

Stalker, H.T., Beute, M.K., Shew, B.B., & Barker, K.R. (2002a). Registration of two root-knot nematode-resistant peanut germplasm lines. *Crop Science*, 42, 312–313.

Stalker, H.T., Beute, M.K., Shew, B.B., & Isleib, T.G. (2002b). Registration of five leaf spot-resistant peanut germplasm lines. *Crop Science*, 42, 314–316.

Stalker, H.T., Dhesi, J.S., & Kochert, G.D. (1995). Variation within the species *A. duranensis*, a possible progenitor of the cultivated peanut. *Genome*, 38, 1201–1212. https://doi.org/10.1139/g95-158

Stalker, H.T., Wynne, J.C., & Company, M. (1979). Variation in progenies of an *Arachis hypogaea* × diploid wild species hybrid. *Euphytica*, 28(3), 675–684.

Stalker, H.T., Tallury, S., Ozias-Akins, P., Bertioli, D.J., & Leal-Bertioli, S. (2013). The value of diploid peanut relatives for breeding and genomics. *Peanut Science*, 40, 70–88.

Stein, J.C. et al. (2018). Genomes of 13 domesticated and wild rice relatives highlight genetic conservation, turnover and innovation across the genus Oryza. *Nature Genetics*, 50, 285–296.

Stevenson, P.C., Blaney, W.M., Simmonds, M.J.S., & Wightman, J.A. (1993). The identification and characterization of resistance in wild species of *Arachis* to *Spodoptera litura* (Lepidoptera: Noctuidae). *Bulletin of Entomological Research*, 83, 421–429.

Suassuna, T., Suassuna, N., Martins, K., Matos, R., Heuert, J., Bertioli, D., Leal-Bertioli, S., & Moretzsohn, M. (2020). Broadening the variability for peanut breeding with a wild species-derived induced allotetraploid. *Agronomy*, 10(12), 1917.

Subrahmanyam, P., Mehan, V.K., Nevill, D.J., & McDonald, D. (1980). Research on fungal diseases of groundnut at ICRISAT. *Proceedings of the International Workshop on Groundnuts*, ICRISAT, Patancheru, pp. 193–198.

Subrahmanyam, P., Moss, J.P., & Rao, V.R. (1983). Resistance to peanut rust in wild *Arachis* species. *Plant Disease*, 67, 209–212.

Subrahmanyam, P., Moss, J.P., McDonald, D., Rao, P.V.S., & Rao, V.R. (1985c). Resistance to leaf spot caused by Cercosporidium personatum in wild *Arachis* species. *Plant Disease*, 69, 951–954.

Subrahmanyam, P., Nolt, A.M., Reddy, B.L., Reddy, D.V.R., & McDonald, D. (1985a). Resistance to groundnut diseases in wild *Arachis* species. *Proceedings of an International Workshop on Cytogenetics of Arachis*, ICRISAT, Patancheru, pp. 49–55.

Subrahmanyam, P., Naidu, R.A., Reddy, L.J., Kumar, P.L., & Ferguson, M.E. (2001). Resistance to groundnut rosette disease in wild *Arachis* species. *Annals of Applied Biology*, 139, 45–50.

Sujay, V., Gowda, M.V.C., Pandey, M. K., Bhat, R.S., Khedikar, Y.P., Nadaf, H.L., et al. (2012). Quantitative trait locus analysis and construction of consensus genetic map for foliar disease resistance based on two recombinant inbred line populations in cultivated groundnut (*Arachis hypogaea* L.). *Molecular Breeding*, 30, 773–788.

Tallury, S.P., Hollowell, J., & Isleib, T.G. (2011). Greenhouse evaluation of section Arachis wild species for Sclerotinia blight and CBR resistance. *Proceedings of American Peanut Research and Education Society*, 43, 69.

Tilak, I.S., & Bhat, R.S. (2021). High yielding mutants from the introgression lines of groundnut (*Arachis hypogaea* L.). *Journal of Oilseeds Research*, 38(2), 132–136.

Tossim, H.-A., Nguepjop, J.R., Diatta, C., Sambou, A., Seye, M., Sane, D., Rami, J.-F., & Fonceka, D. (2020). Assessment of 16 peanut (*Arachis hypogaea* L.) CSSLs derived from an interspecific cross for yield and yield component traits: QTL validation. *Agronomy*, 10, 583.

Upadhyaya, H.D., Bramel, P.J., Ortiz, R., & Singh, S. (2002). Developing a mini core of peanut for utilization of genetic resources. *Crop Science*, 42, 2150–2156.

Upadhyaya, H.D., Dwivedi, S.L., Nadaf, H.L. and Singh, S. (2011). Phenotypic diversity and identification of wild Arachis accessions with useful agronomic and nutritional traits. *Euphytica*, 182, 103–115.

Valls, J.F.M., & Simpson, C.E. (2005). New species of *Arachis* from Brazil, Paraguay and Bolivia. *Bonplandin*, 14, 35–64.

Valls, J.F.M., Ramanatha Rao, V., Simpson, C.E., & Krapovickas, A. (1985). Current status of collection and conservation of South American groundnut germplasm with emphasis on wild species of *Arachis*. In: J. P. Moss, editor, *Proceedings of the International Workshop on Cytogenetics of Arachis*, ICRISAT, Patancheru. pp. 15–35.

Wang, C., Hu, S., Gardner, C., & Lübberstedt, T. (2017). Emerging avenues for utilization of exotic germplasm. *Trends in Plant Science*, 22(7), 624–637.

Warschefsky, E., Penmetsa, V.R., Cook, D.R., & Von Wettberg, E.J.B. (2014). Back to the wilds: Tapping evolutionary adaptations for resilient crops through systematic hybridization with crop wild relatives. *American Journal of Botany*, 101, 1791–1800.

Zhao, Q. et al. (2018). Pan-genome analysis highlights the extent of genomic variation in cultivated and wild rice. *Nature Genetics*, 50, 278–284.

Zhou, X.J., Ren, X.P., Luo, H.Y., Huang, L., Liu, N., Chen, W.G., Lei, Y., Liao, B.S., Jiang, H.F. (2022). Safe conservation and utilization of peanut germplasm resources in the oil crops middle term genebank of China. *Oil Crops Sciences*, 7, 9–13.

4 Crop Wild Relatives of Lentil (*Lens culinaris* Medik.)

Uttarayan Dasgupta and Anamika Barman

INTRODUCTION

Lentil (*Lens culinaris* Medik.) is the oldest pulse crop known to man and one of the earliest domesticated crops. There is evidence that the Egyptians, Romans, and Hebrews consumed this bean, and lentil artefacts from 8000 BC have been discovered on archaeological digs on the banks of the Euphrates River. The *lens*-shaped lentil seeds are termed *Adas* (Middle East), *Mercimek* (Turkey), *Masoor* (Pakistan and India), *Messer* (Ethiopia), and *Heramame* (Japan). Lentils are grown in over 40 countries, with Canada, India, Australia, Turkey, and the United States being the leading producers (Kaale et al., 2023). 5.61 million tonnes of lentils are produced globally on an area of 5.58 million hectares, with a realised yield of 1.004 t/ha (FAOSTAT, 2023). In India, lentil is the second-most important pulse after chickpeas, with a production of 1.49 million tonnes, which is second only to Canada worldwide. The production potential of improved lentil varieties in Ethiopian farmer fields has been estimated to be up to 3 t/ha and nearly 5 t/ha in research fields. From several producing zones in India, yield differences of 30%–105% with an average of 42% have also been recorded. A multitude of reasons inhibit the ability of the genotype to live up to its true potential. Because lentils are only grown in South Asia as a post-rainy season crop that relies on residual moisture, an early end to the rainy season has a negative impact on their establishment. At the time of grain filling, the crop must also deal with a rapid increase in temperature and dwindling soil moisture, which results in forced maturity and a reduction in yield (Kumar et al., 2013). It is also troubled by diseases like *Fusarium* wilt (*Fusarium oxysporum* f.sp. *lentis*), rusts (*Uromyces viciae-fabae*), and *Ascochyta* blight (*Ascochyta lentis*), which also hinder its productivity (Garkoti et al., 2013).

To overcome these challenges through breeding efforts, it becomes necessary to exploit the available genetic diversity. The gradual loss of alleles and genetic bottlenecks caused by past plant breeding efforts, domestication, and the extinction of plant species present challenges, though. The point being, domestication and plant breeding both rely on human selection, and all forms of selection reduce genetic diversity by only advancing genotypes that are better for specific features. To recover the lost genetic diversity, it is quintessential to find sources from which new genes and alleles can be mined to tailor an improved variety. One of the main sources is

DOI: 10.1201/9781003434535-4

crop wild relatives (CWRs), which are ancestors or progenitors of domesticated crop species as well as other close relatives throughout evolutionary history (Kapazaglou et al., 2023; Choudhary et al., 2017). Like other pulses, wild *Lens* species have been found to harbour valuable genes that can be utilised in breeding improved varieties. Recent advances in genomics-assisted breeding will help breeders accumulate useful alleles from the secondary and tertiary gene pools into the elite background of existing successful varieties, making it possible to push beyond the current yield plateau.

ORIGIN, DISTRIBUTION, GENE POOLS, AND PHYLOGENETIC RELATIONSHIPS OF CWRS TO CULTIVATED LENTIL

Lentil is one of the oldest domesticated crops and is a self-pollinated diploid legume ($2n=2x=14$) comprising the *Lens* genus, falling under the Papilionaceae subfamily of the Leguminosae family (Laskar et al., 2019). The cultivated lentil is *Lens culinaris* subsp. *Culinaris*, which has two different races, macrosperma and microsperma, the former having larger seeds than the latter (Cubero et al., 2009). Its scientific name, *Lens culinaris* was given by a German botanist and physician Medikus way back in 1787 (Cokkizgin and Shtaya, 2013). Recent advances in sequencing technologies have enabled scientists to properly classify the *Lens* genus into seven taxa falling under four main species distributed through four gene pools. The *L. culinaris* has four subspecies (subsp. *culinaris*, subsp. *orientalis*, subsp. *odemensis*, and subsp. *tomentosus*). The primary genepool contains *L. culinaris*, *L. orientalis*, and *L. tomentosus*; the secondary genepool has *L. odemensis* and *L. lamottei*; the tertiary and quaternary gene pools each contains one species, *L. ervoides* and *L. nigricans*, respectively. *Orientalis* was found to be the only wild species that gave rise to the domesticated lentil (*culinaris*), with contributions from other wild species to the domesticated gene pool being negligible (Zohary, 1972; Liber et al., 2021; Wong et al., 2015). Due to their strong resemblance to *L. culinaris* and their ease in producing seeds after interspecific mating, *L. tomentosus* and *L. orientalis* are considered members of the primary gene pool. *L. culinaris* ssp. *odemensis* has been found to have a distinct lineage from the subspecies of *Lens culinaris* and to be in a sister clade to *L. lamottei*. Thus, it was classified under the secondary gene pool with *L. lamottei* (Koul et al., 2017; Wong et al., 2015). The crossability of *Lens ervoides* and *L. nigricans* with cultivated lentil is very difficult and generally results in hybrid embryo breakdown (Gupta and Sharma, 2006). However, in some cases, it has been possible to develop populations like interspecific recombinant inbred lines and advanced backcross populations by crossing *L. ervoides* and *L. culinaris* and overcoming the crossing barrier using the F_1 embryo rescue technique (Tullu et al., 2013; Gela et al., 2021). For these reasons, *L. ervoides* has been classified under the tertiary gene pool or gene pool 3. *L. nigricans* is the most distantly related crop wild relative to the cultivated lentil, as their crosses are never successful beyond the F_1 generation, and both species form distinct groups in phylogenetic and STRUCTURE analyses of their genotyping-by-sequencing (GBS) data. *L. nigricans* has been suggested in some studies to be placed in the quaternary gene pool and considered the last option for use in lentil varietal improvement (Wong et al., 2015) (Figure 4.1).

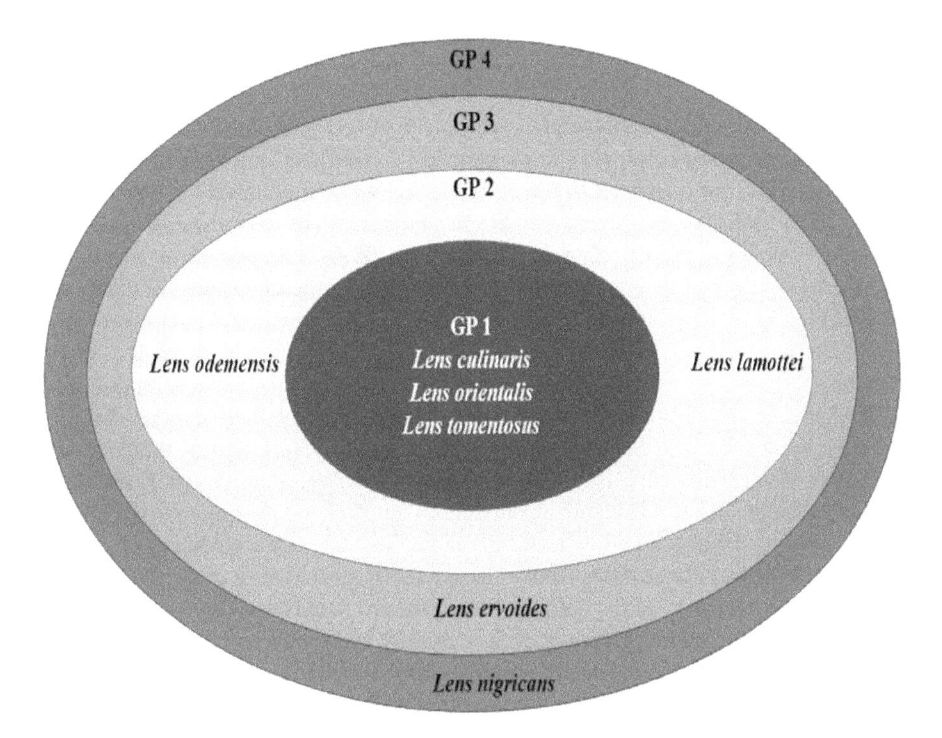

FIGURE 4.1 Gene pool of crop wild relatives of lentils.

Lentil cultivation started as early as 8500–600 BC in the region that today includes Syria, Turkey, and Iraq. Subsequently, it migrated to the Nile Valley, Greece, Central Europe, and eastwards to South Asia. Later, the crop dispersed to Afghanistan, China, Ethiopia, India, Pakistan, and Latin America. Old Indian scriptures like the *Brahadaranyaka* (c. 5500 BC), a commentary on the *Rigveda* (c. 8000 BC), and the *Yajurveda* (c. 7000 BC) all mention lentils, known in Sanskrit as *masura*, which means "cushion" (Nene, 2006). The distribution of wild relatives occurs in areas with similar as well as variable soil textures and climatic conditions. In addition to Jordan, Syria, Israel, Lebanon, and Cyprus, *Lens orientalis* is also found in Armenia, Azerbaijan, the Czech Republic, Iran, Russia, Turkmenistan, Tajikistan, and Uzbekistan in open or shaded habitats on rocky calcareous to basalt soils around 500–1,700 m. *L. tomentosus* is found in Turkey and Syria. *L. odemensis* is distributed in Israel, Syria, and Palestine in grassy habitats and in the calcareous soils of pine groves and basaltic gravel, at 700–1,400 m in Turkey. *L. ervoides* can be found in shaded and semi-shady habitats under trees and shrubs in Jordan, Syria, Palestine, Montenegro, Italy, and Croatia, as well as further north in Russia, Ukraine, Armenia, and Azerbaijan. *L. lamottei* grows in France and Spain. *L. nigricans* ranges up to 1,200 m in Bahrain, Crimea, Croatia, France, Italy, Montenegro, Spain, and Ukraine in terraced villages, plantations, and open or shaded stony habitats in granitic or limestone soil (Dikshit et al., 2022).

GENETIC RESOURCES AND CONSERVATION STATUS OF LENTIL CWRS

Ex-situ preservation is essential to preserve genetic variety and utilise the materials for improving agricultural crops. Numerous gene banks have been constructed worldwide for this purpose, but usually, CWRs are neglected without concern for their hidden genetic value. Most of the germplasm-storing institutes have less than 10% of their accessions belonging to wild relatives. There are 58,405 *Lens* accessions in gene banks worldwide, distributed among 103 different nations, with the percentage of wild relatives varying from 1% to 10% in different genebanks and/or institutes (Malhotra et al., 2019). The most diverse wild relative accessions are maintained by ICARDA, along with the most *Lens* accessions. ICARDA has 623 wild *Lens* accessions comprising all the wild species (http://www.genesys-pgr.org). The Vavilov Institute harbours the second most number of wild *Lens* accessions (285 accessions), followed by the Seed and Plant Improvement Institute, Iran (270 accessions) and Australian Grains Gene Bank (250 accessions). The ICAR-National Bureau of Plant Genetic Resources (NBPGR) of India maintains 108 accessions, whereas the Indian Institute of Pulses Research (IIPR), Kanpur, maintains 71 accessions from wild species (Guerra-García et al., 2021; Rajpal et al., 2023). A summary of the collection of wild *Lens* taxa is described in Table 4.1.

For utilising the stored germplasm, it is equally important to evaluate the materials for agro-morphological traits, quality traits, and biotic and abiotic stresses. A set of 405 diverse global wild annual *Lens* accessions comprising four cultivated *L. culinaris* ssp. *culinaris*, 171 of *L. culinaris* ssp. *orientalis*, 42 of *L. culinaris* ssp. *odemensis*, 20 of *L. culinaris* ssp. *tomentosus*, 35 of *L. nigricans*, 124 of *L. ervoides*, and 9 of *L. lamottei* were studied encompassing various aspects of genetic resource

TABLE 4.1
Genetic Resources of CWRs of *Lens* Worldwide

S. No	Genebank/Institute	Wild Taxa Accessions
1	International Centre for Research in Dry Areas	623
2	Vavilov Institute, Russia	285
3	Seed and Plant Improvement Institute, Iran	270
4	Australian Grains Gene bank (AGG), Agriculture Victoria	250
5	Plant Gene Resources of Canada	195
6	ICAR-National Bureau of Plant Genetic Resources, India	108
7	General Commission for Scientific Agricultural Research, Syria	75
8	Western Regional Plant Introduction Station, USDA-ARS, Washington State University	138
9	Research Centre for Agro-Botany, Hungary	42
10	Plant Genetic Resource Department Aegean Agricultural Research Institute, Turkey	10
11	Millenium Seed Bank, Royal Botanic Gardens Kew	94
12	Agrobiotechnology Scientific Center, Armenia	9
13	Institute of Plant Production n.a. V.Y. Yurja of UAAS, Ukraine	6

management. A core set, which constitutes about 10%–20% of the total amount of germplasm available, has a sizable amount of genetic variety in a manageable number of relatively few germplasms. Identification and construction of a core set are integral parts of evaluating and further utilising germplasm in crop improvement. A core set of wild *Lens* species consisting of 96 lines was created after the 405 lines were evaluated. In India, the core set was tested under field and controlled screening settings for earliness, pod number, seed yield, and resistance to rust, powdery mildew, and *Fusarium* wilt. The majority of wild germplasm was discovered to be only moderately resistant to rust, while some wild germplasm was reported to be resistant to powdery mildew and *Fusarium* wilt (Singh et al., 2014, 2020).

CWRS AS SOURCES OF NOVEL ALLELES FOR IMPORTANT TRAITS

The frequency and severity of various abiotic stresses, such as droughts, heat waves, cold snaps, and flooding, are alarmingly increasing due to global warming and climate change. This has an adverse effect on agricultural yields and contributes to food shortages. Additionally, the composition and behaviour of various insect and pathogen populations are changing due to climate change, which is increasing crop losses globally (Rivero et al., 2022). The limited genetic diversity of lentil varieties makes them susceptible to significant and increasing biotic and abiotic stresses. Their production on marginal soils, primarily in developing nations like India, restricts the expression of characteristics that increase output. Consolidated efforts are required to improve the nutrition and yield potential of lentils to satisfy the dietary needs of the expanding human population. Therefore, it is necessary to increase the genetic diversity of lentil cultivars by introducing a variety of genes from distantly related wild *Lens* species which are emerging as important sources of novel variation for stress resistance as well as agronomically important traits (Singh et al., 2018).

CWRS AS SOURCE FOR BIOTIC STRESS RESISTANCE

The constant change in climate has led to the evolution of new strains of pathogens including fungi, bacteria, and viruses. Additionally, it has allowed some pests and weeds to flourish where they formerly posed little threat. Among various diseases, fungal pathogens are the most threatening, reducing plant populations drastically at every growth period from seedling to the pod-bearing stage. For instance, *Ascochyta* blight infection resulted in yield reductions of 30%–70% in Canada, the United States, Australia, and northern India. *Colletotrichum truncatum* caused 60% yield reductions in Canada, while *Stemphylium* blight incurs nearly 95% yield loss in India (Roy et al., 2023). It has been discovered that *Lens orientalis* is a very potent source of resistance to *Ascochyta* blight. One of the accessions, ILWL180, was used to prepare an interspecific mapping population with a susceptible cultivar to map QTLs (quantitative trait loci). They identified three QTLs on linkage group 5, having 9.5%–11.5% of phenotypic variance for *Ascochyta* blight resistance (Ye et al., 2000; Dadu et al., 2018, 2021). Resistance sources are also available for *Stemphylium* blight in other gene pools of *Lens*, and the highest frequency of resistance to *Stemphylium* blight was found in *L. lamottei* followed by *L. ervoides* of the secondary gene pool.

Recently, a transcriptome sequencing approach has also been utilised to elucidate the DEGs and molecular pathways responsible for providing resistance in *L. ervoides* (Podder et al., 2013; Cao et al., 2019). *Colletotrichum lentis* is a major pathogen in lentils, causing anthracnose disease. For its virulent race 0, there are very limited sources of genetic resistance in the cultivated genotypes. Utilisation of wild relatives becomes necessary when dealing with this kind of virulent pathogenic race, and *L. ervoides* accessions have been shown to have high levels of resistance (Barilli et al., 2020). Sources for resistance against other diseases like Powdery mildew, Rust, *Fusarium* wilt, insect pests like Sitona weevil, bruchids, and weeds like broomrape have also been found in wild germplasm and have been summarised in Table 4.2.

CWRS AS SOURCE FOR ABIOTIC STRESS TOLERANCE

On marginal soil, lentil and its wild relatives evolved in regions with little precipitation. Drought is the most significant abiotic stress, which has been accelerating worldwide owing to continuous climate change. Compared to accessions in secondary and tertiary gene pools, those in the primary gene pool have worse water extraction and a slower growth rate. *Lens orientalis*, being native to low-rainfall areas like Syria, Azerbaijan, Jordan, Tajikistan, and Turkmenistan, has been proven to be a good source of water stress tolerance. Additionally, *L. odemensis* and *L. ervoides* have been shown to respond to drought by developing deep root systems. In response to the drought, *L. lamottei* showed delayed blossoming, but when it did flower, resources were directed toward seed formation, which can be seen as another type of avoidance. Additionally, *Lens tomentosus* tolerated moisture stress by decreasing its transpiration rates (Gorim and Vandenberg, 2017b). Other reports have also highlighted the potential of lentil CWRs with significant differences in the morphology of root traits for fine root distribution, variability in the number of nodules, and root biomass proportion in each soil layer (Gorim and Vandenberg, 2017a). Omar et al. (2019) examined the drought resistance of elite lentil varieties bred using CWRs. Cell membrane stability, an increase in the root-to-shoot ratio, pubescent leaves, relative leaf water content, and reduced transpiration and wilting were all associated with increased drought tolerance. This wide variation in lentil responses to drought suggests that wild species relatives will be crucial for future lentil improvement, subject to the creation of viable hybrids through successful cross-breeding between the wild and farmed species (Rajpal et al., 2023).

CWRS AS SOURCE FOR QUALITY AND IMPORTANT YIELD-ATTRIBUTING TRAITS

The primary objective in any breeding programme is to focus on improving the yield along with other assisting traits. Yield-attributing traits include the number of secondary branches, crop duration, biological yield, number of pods, seed size, etc., which are the primary focus of any plant breeder. Lentil is a nutritious food crop with high protein content (21%–30%), low-fat content (<1%), a good amount of vitamin B complex, prebiotic carbohydrates, and bioactive compounds like polyphenols. Lentil also has anti-nutritional factors like phytic acid, trypsin inhibitors, tannins, and deficiency

TABLE 4.2
List of CWRs of Lentils Having Sources for Resistance Against Biotic and Abiotic Stresses

Wild Species		Stress/Traits	Reference
Lens orientalis	Fungal diseases	*Ascochyta* blight, Powdery mildew, Rust, *Fusarium* wilt, *Stemphylium* blight	Bayaa et al. (1994); Ye et al. (2000); Gupta and Sharma (2006); Tull et al. (2010); Podder et al. (2013); Kumari et al. (2018); Dadu et al. (2018); Coyne et al. (2020); Singh et al. (2020); Barilli and Rubiales (2023)
	Insect pests	Sitona weevil, Seed weevil	El-Bouhssini et al. (2008); Laserna-Ruiz et al. (2012)
	Abiotic stress	Cold & Frost, Drought	Hamdi & Erskine (1996); Gorim and Vandenberg (2017a, 2017b); Omar et al. (2019)
L. tomentosus	Fungal diseases	Powdery mildew, Rust, *Fusarium* wilt, *Stemphylium* blight	Podder et al. (2013); Singh et al. (2020)
L. lamottei	Fungal diseases	*Ascochyta* blight, Powdery mildew, *Fusarium* wilt, *Stemphylium* blight, Anthracnose	Bayaa et al. (1994); Gupta and Sharma (2006); Tullu et al. (2006, 2010); Podder et al. (2013); Singh et al. (2020)
	Insect pests	Seed weevil/ Bruchids	Laserna-Ruiz et al. (2012)
	Abiotic	Drought	Gorim and Vandenberg (2017a, 2017b)
L. odemensis	Fungal diseases	*Ascochyta* blight, Powdery mildew, Rust, *Fusarium* wilt, *Stemphylium* blight	Bayaa et al. (1994); Gupta and Sharma (2006); Tullu et al. (2010); Singh et al. (2020)
	Insect pests	Sitona weevil, Bruchids	El-Bouhssini et al. (2008); Laserna-Ruiz et al. (2012)
	Weed	*Orobanche*/Broomrape	Fernández-Aparicio et al. (2009)
	Abiotic	Drought	Gorim and Vandenberg (2017a, 2017b)
L. ervoides	Fungal diseases	*Ascochyta* blight, Powdery mildew, *Fusarium* wilt, *Stemphylium* blight, Anthracnose, Rust	Gupta and Sharma (2006); Tullu et al. (2006, 2010); Vail et al. (2012); Podder et al. (2013); Bhaduria et al. (2017); Cao et al. (2019); Singh et al. (2020); Barilli and Rubiales (2023)
	Insect pests	Sitona weevil	El-Bouhssini et al. (2008)
	Weed	*Orobanche*/Broomrape	Fernández-Aparicio et al. (2009)
	Abiotic	Drought	Gorim and Vandenberg (2018)
L. nigricans	Fungal diseases	*Ascochyta* blight, Powdery mildew, *Fusarium* wilt, *Stemphylium* blight, Anthracnose, Rust	Bayaa et al. (1994); Gupta and Sharma (2006); Tullu et al. (2010); Podder et al. (2013); Singh et al. (2020); Barilli and Rubiales (2023)
	Insect pests	Sitona weevil, Bruchids	El-Bouhssini et al. (2008); Laserna-Ruiz et al. (2012); Gore et al. (2016)
	Abiotic	Drought	Gorim and Vandenberg (2017a, 2017b)

of minerals like Fe and Zn, which have compelled breeders and geneticists to improve the cultivars to provide an all-round nutrition to the population (Choukri et al., 2022). Promising donors for yield traits, such as 100-seed weight and pods per plant were observed in *L. lamottei* and *L. culinaris* ssp. *orientalis* and according to some reports, *L. ervoides* is an excellent source of alleles for plant architectural qualities like phenology, plant growth habit, and biomass in addition to seed traits (Gupta and Sharma, 2006; Tullu et al., 2011; Kumar et al., 2014; Pratap et al., 2021). According to research by Yuan et al. (2017), three *L. lamottei* genotypes (IG 110809, IG 110810, and IG 110813) and one *L. ervoides* genotype (IG 72646) maintained identical yield, biomass, and harvest index across variable light regimes, indicating better adaptability towards changes in the light environment and helping in the breeding of widely adaptable lentil cultivars. Quality traits like different micronutrient and protein content exhibited wide variability, either comparable levels to modern cultivars in *L. lamottei*, *L. nigricans*, L. *odemensis*, and *L. ervoides* (Fe, Zn, Ca, Cu, Mg, Mn, Mo) or higher in *L. orientalis* (protein, Fe, Zn) (Sen Gupta et al., 2016; Kumar et al., 2018; El Haddad et al., 2021). These sources can be used in the future for the bio-fortification of lentil genotypes and overall improvement of the nutritional quality of lentil.

GENOMIC RESOURCES

Recent advances in automation and high-throughput sequencing techniques have enabled researchers to generate a huge wealth of genome information. According to some reports, *L. ervoides* is an excellent source of alleles for plant architectural qualities like phenology, plant growth habit, and biomass in addition to seed traits. Breeders face difficulty in improving economical traits, most of which are quantitative and greatly influenced by the environment. To overcome the complex GxE interaction, it is necessary to understand the genetic nature of the trait. Genomic resource information can be utilised to better understand this complex genetic nature of economically important quantitative traits and thereby can be used to breed improved cultivars. The challenges posed by a changing global environment can be overcome by integrating traditional breeding techniques with a new era of molecular breeding, which will maintain agricultural productivity for future food and nutritional security. Crossing with CWRs is often riddled with the problem of negative alleles, whose downside sometimes overturns the benefits of the good alleles. In this era of smart and fast breeding, it becomes necessary to understand the genome of CWRs so as to be able to tactically integrate the positive alleles without being affected by the negative ones. Lentil, being a food legume, has been less focused on developing genomic resources and genomics-assisted breeding. However, a number of initiatives have been made over the past 10 years to pave the way for lentil breeding with full-fledged genomic assistance in the near future (Tiwari et al., 2022).

MOLECULAR MARKERS

During the start of the era of modern biotechnology, a number of molecular markers were used to study the relationship and taxonomic classification of wild lentil relatives with cultivated lentils. Molecular markers like RFLP (Restriction

Fragment Length Polymorphism) (Havey and Muehlbauer, 1989), RAPD (Random Amplified Polymorphic DNA) (Abo-elwafa et al., 1995; Sharma et al., 1995; Gupta et al., 2012b), AFLP (Amplified Fragment Length Polymorphism) (Sharma et al., 1996; Hawieh et al., 2005; Idrissi et al., 2015), SSR (Babayeva et al., 2009; Gupta et al., 2012b; Tewari et al., 2012; Kushwaha et al., 2015; Idrissi et al., 2015), ISSR (Inter Simple Sequence Repeat) (Fikiru et al., 2007; Gupta et al., 2012b), and SRAP (Bermajo et al., 2014; Mbasani-Mansi et al., 2019) have been used to study genetic diversity, phylogenetic relationships of wild lentil cultivars and exotic germplasms, construct genetic maps, localise disease resistance, and characterise germplasm (Toklu et al., 2009; Kumar et al., 2014; Singh et al., 2018). Advancement of transcriptomic technology has also enabled the mining of EST-SSRs (expressed sequence tag-derived simple sequence repeat markers) from RNA-Seq data to be used for genetic characterisation of wild lentil species (Gupta et al., 2012a; Singh et al., 2020).

GENOME AND TRANSCRIPTOME DATA

Advancements in sequencing technologies have enabled, in the previous decade, the sequencing of the genome of food-legumes like lentil. Lentil has a very large genome (~4 Gbp), which is larger than that of commonly cultivated pulses like chickpea, mungbean, urdbean, soybean, and the model legume *Medicago truncatula*. Researchers have thus faced difficulties in completely sequencing the whole genome. An initial draft assembly of the *L. culinaris* genome was developed using short-read technology (Ramsay et al., 2016; https://knowpulse.usask.ca/genome-assembly/Lc1.2). Different technologies (paired-end, mate-pair, and a few long-read libraries) were used to sequence the genome of a single plant generated from the cultivar CDC Redberry over a wide range of sizes. This method could only be used to assemble 2.7 Gbp due to the limits of short reads. However, with the advent of long-read sequencing technology, coupled with Hi-C proximity by ligation mapping, a new version of the CDC Redberry assembly was developed that covers 3.76 Gbp of an estimated 3.92-Gbp genome (Ramsay et al., 2019a, 2021; https://knowpulse.usask.ca/genome-assembly/Lcu.2RBY). A genome of *L. ervoides* accession IG 72815 was assembled from 52a× long reads, polished with the long reads and additional short reads. The finished assembly is 2.87 Gb, which is smaller than the cultivated lentil genome, and is arranged in seven pseudomolecules and 1,134 unplaced unitigs (Ramsay et al., 2019b, 2021; https://knowpulse.usask.ca/genome-assembly/Ler.1DRT). Kaur et al. (2016) also generated a whole-genome assembly of the Australian lentil cultivar PBA Blitz, assembling a total of 2.3 Gbp and generating 10 RNA-Seq libraries taken from different tissues of the cultivar. A GBS method based on restriction site-associated DNA (RAD) has also been used to re-sequence a lentil diversity panel (around 400 accessions from several lentil germplasm collections) (Bett, 2016). The vast majority of expressed genes can be identified via *de novo* assembly of RNA-Seq-derived data without the aid of a reference genome sequence, and a such reference unigene set for lentil has been developed, which had 58,986 contigs and scaffolds with an N50 length of 1,719 bp (Sudheesh et al., 2016). Recently, transcriptomes of each of the wild species were assembled, annotated, and evaluated along with the cultivated *Lens culinaris* cultivar 'Alpo' and the first pan-transcriptome of lentil was developed, indicating an acceleration in improving genomic resources of CWRs (Gutierrez-Gonzalez et al., 2022). The chloroplast (cp) genome, in addition to the nuclear genome, is a highly active genetic component of plants and can be used as a molecular marker for a variety of applications. Recently, de-novo assembly of the cp genome of *Lens ervoides* resulted in a single ~122 Kbp sequence as two separate coexisting structural haplotypes with similar lengths (Taysi et al., 2022). Increasing availability of reference genomes and transcriptomes will enable breeders to perform selection at earlier breeding cycles and integrate CWRs into molecular breeding programmes to easily access their novel positive alleles.

SUCCESSFUL EXAMPLES OF CWR UTILISATION IN CROP IMPROVEMENT

Other legumes, including chickpea, pigeon pea, scarlet bean, mungbean, and urdbean, have also benefited from the utilisation of wild relatives to improve resistance to nematodes and fungi, incorporate male sterility, and improve significant agro-morphological traits (Pratap et al., 2021). Screening of the global wild annual lentil collection has enabled the identification of mineral-rich accessions ILWL15 (*L. lamottei*), ILWL480 (*L. orientalis*), and ILWL401 (*L. ervoides*), which are being used in pre-breeding of biofortified lines. Resistance against *Colletotrichum truncatum* disease was transferred from an *L. ervoides* accession IG72815 by crossing it with *L. culinaris* 'Eston' and developing an interspecific recombinant inbred line. Additionally, it was found that the RIL population exhibited transgressive segregation on the positive side for other parameters such as podding ability, maturity, stand, and lodging (Fiala et al., 2009; Tullu et al., 2010; Tullu et al., 2013). Another successful example is the development of an advanced backcross population between *L. culinaris* and *L. ervoides*, which had individuals at the $BC_2F_{3:4}$ generation showing resistance against *Stemphylium botryosum* and *Colletotrichum lentis*. The population also segregated for other morphological traits, and it evidently introgressed resistance alleles into the CDC Redberry background (Gela et al., 2021). Early-maturing, high-yielding, and disease-resistant lentil cultivars (Jammu Lentil 144 and Jammu Lentil 71) have been developed for the Jammu region of India as a result of pre-breeding efforts by crossing cultivated lentils with *L. culinaris* ssp. *orientalis* and *L. ervoides* (Singh et al., 2022). ICARDA is currently running the DIIVA-PR (The ICARDA Dissemination of Interspecific ICARDA Varieties via Participatory Research) project, whose goals are to create novel cultivars through crosses with CWRs, evaluate how they react to climatic pressures in these places, and provide farmers with them through a participatory varietal selection technique for three crops, including lentil. Five hundred crop wild relative-derived elite lines were generated for durum wheat, barley, and lentil crops, and there is hope for future success in developing improved varieties (ICARDA Annual Report, 2020).

CONCLUSION

Advancements in the 21st century have led to an increase in the demand for food grains in climate-changing conditions. Lentil is one of the most popular pulses worldwide, and there is a need to safeguard its cultivation against emerging threats. Conventional breeding, taking into account the available genetic materials, has led to a narrowing of the genetic base of lentil and, in the near future, might exhaust the available sources of novel variations. It is imperative that CWRs, which are a reservoir of novel sources of positive alleles for important traits, be utilised in ongoing breeding programmes. CWRs of lentil have been observed in many studies to be superior in stress resistance compared to the present cultivars, although more screening studies need to fully understand the variability present in the germplasm. Efforts have been made to maintain genetic resources worldwide, and more effort should be made to use the existing germplasm samples. Recent advancements in sequencing technology have enabled the creation of a reservoir of genomic resources for cultivated species as well as their wild relatives. The creation of genomic resources will enable the integration of CWRs into advanced molecular studies and molecular breeding facilitated introgression programmes. Large-scale legume improvement projects, such as DIIVA-PR (ICARDA Annual Project, 2020), INCREASE – Intelligent Collections of Food Legumes Genetic Resources for European Agrofood Systems (https://www.pulsesincrease.eu/crops/lentil), and EVOLVES – Enhancing the Value of Lentil Variation for Ecosystem Survival (https://knowpulse.usask.ca/study/2691111), will provide a way forward for not just utilising CWRs but also for the overall development of the crop.

REFERENCES

Abo-elwafa, A., Murai, K., & Shimada, T. (1995). Intra-specific and interspecific variations in Lens, revealed by RAPD markers. *Theoretical and Applied Genetics*, 90, 335–340. https://doi.org/10.1007/BF0022974

Babayeva, S., Akparov, Z., Abbasov, M., Mammadov, A., Zaifizadeh, M., & Street, K. (2009). Diversity analysis of central Asia and Caucasian lentil (*L. culinaris* medik.) germplasm using SSR fingerprinting. *Genetic Resources and Crop Evolution*, 56, 293–298. https://doi.org/10.1007/s10722-009-9414-6

Barilli, E., & Rubiales, D. (2023). Identification and characterization of resistance to rust in lentil and its wild relatives. *Plants*, 12(3), 626.

Barilli, E., Moral, J., Aznar-Fernández, T., & Rubiales, D. (2020). Resistance to anthracnose (*Colletotrichum lentis*, race 0) in *Lens* spp. germplasm. *Agronomy*, 10(11), 1799.

Bayaa, B., Erskine, W., & Hamdi, A. (1994). Response of wild lentil to *Ascochyta fabae* f. sp. *lentis* from Syria. *Genetic Resources and Crop Evolution*, 41, 61–65.

Bermejo, C., Gatti, I., Caballero, N., Cravero, V., Martin, E., & Cointry, E. (2014). Study of diversity in a set of lentil RILs using morphological and molecular markers. *Australian Journal of Crop Science*, 8(5), 689–696.

Bett, K. (2016). *The Lentil Genome-from the Sequencer to the Field*. International conference on pulses, Marrakesh, Morocco.

Bhadauria, V., Ramsay, L., Bett, K. E., & Banniza, S. (2017). QTL mapping reveals genetic determinants of fungal disease resistance in the wild lentil species *Lens ervoides*. *Scientific Reports*, 7, 1–9.

Cao, Z., Li, L., Kapoor, K., & Banniza, S. (2019). Using a transcriptome sequencing approach to explore candidate resistance genes against stemphylium blight in the wild lentil species *Lens ervoides*. *BMC Plant Biology*, 19, 1–16.

Choudhary, M., Singh, V., Muthusamy, V., & Wani, S. (2017). Harnessing crop wild relatives for crop improvement. *LS International Journal of Life Sciences*, 6, 73.

Choukri, H., El Haddad, N., Aloui, K., Hejjaoui, K., El-Baouchi, A., Smouni, A., & Kumar, S. (2022). Effect of high temperature stress during the reproductive stage on grain yield and nutritional quality of lentil (*Lens culinaris* Medikus). *Frontiers in Nutrition*, 9, 857469.

Cokkizgin, A., & Shtaya, M. J. (2013). Lentil: origin, cultivation techniques, utilization and advances in transformation. *Agricultural Science*, 1(1), 55–62.

Coyne, C. J., Kumar, S., von Wettberg, E. J., Marques, E., Berger, J. D., Redden, R. J., ... & Smýkal, P. (2020). Potential and limits of exploitation of crop wild relatives for pea, lentil, and chickpea improvement. *Legume Science*, 2(2), e36.

Cubero, J. I., Perez De La Vega, M., & Frantini, R. (2009). Origin, phylogeny, domestication and spread. In: Erskine W, Muehlbauer FJ, Sarker A, Sharma B (eds) *Lentil – Botany, Production and Uses*. CABI, Wallingford, pp. 13–33.

Dadu, R. H., Ford, R., Sambasivam, P., & Gupta, D. (2017). A novel *Lens orientalis* resistance source to the recently evolved highly aggressive Australian *Ascochyta lentis* Isolates. *Frontiers in Plant Science*, 8, 1–7.

Dadu, R. H. R., Ford, R., Sambasivam, P., & Gupta, D. (2018). Evidence of early defence to *Ascochyta lentis* within the recently identified *Lens orientalis* resistance source ILWL180. *Plant Pathology*, 67(7), 1492–1501.

Dadu, R. H. R., Bar, I., Ford, R., Sambasivam, P., Croser, J., Ribalta, F., et al. (2021). *Lens orientalis* contributes quantitative trait loci and candidate genes associated with ascochyta blight resistance in lentil. *Frontiers in Plant Science*, 12, 703283. https://doi.org/10.3389/fpls.2021.703283

Dikshit, H. K., Mishra, G. P., Aski, M. S., Singh, A., Tripathi, T., Bansal, R., et al. (2022) Lentil breeding. In: Yadava DK et al. (eds) *Fundamentals of Field Crop Breeding*. Springer, Singapore, pp. 1181. https://doi.org/10.1007/978-981-16-9257-4_24

ElHaddad, N., Sanchez-Garcia, M., Visioni, A., Jilal, A., El Amil, R., Sall, A. T., & Bassi, F. M. (2021). Crop wild relatives crosses: multi-location assessment in durum wheat, barley, and lentil. *Agronomy*, 11(11), 2283.

El-Bouhssini, M., Sarker, A., Erskine, W., & Joubi, A. (2008). First sources of resistance to *Sitona* weevil (*Sitona crinitus* Herbst) in wild *Lens* species. *Genetic Resources and Crop Evolution*, 55, 1–4.

Fernández-Aparicio, M., Sillero, J. C., & Rubiales, D. (2009). Resistance to broomrape in wild lentils (*Lens* spp.). *Plant Breeding*, 128, 266–270.

Fiala, J. V., Tullu, A., Banniza, S., Séguin-Swartz, G., & Vandenberg, A. (2009). Interspecies transfer of resistance to anthracnose in lentil (*Lens culinaris* Medic.). *Crop Science*, 49(3), 825–830.

Foodand Agriculture Organization (FAO). (2023). *FAOSTAT*. Retrieved from https://www.fao.org/faostat/en/?#data (Accessed on 19.06.2023).

Fikiru, E., Tesfaye, K., & Bekele, E. (2007). Genetic diversity and population structure of Ethiopian lentil (*Lens culinaris* medikus) landraces as revealed by ISSR marker. *African Journal of Biotechnology*, 6, 1460–1468.

Garkoti, A., Kumar, S., Lal, M., & Singh, V. (2013). Major diseases of lentil: epidemiology and disease management – a review. *Agriways*, 1(1), 62–64.

Gela, T. S., Adobor, S., Khazaei, H., & Vandenberg, A. (2021). An advanced lentil backcross population developed from a cross between *Lens culinaris* × *L. ervoides* for future disease resistance and genomic studies. *Plant Genetic Resources*, 19(2), 167–173.

Genesys. (2023). *Genesys-PGR*. Retrieved from https://www.genesys-pgr.org (Accessed on 26.06.2023).

Gore, P. G., Tripathi, K., Chauhan, S. K., Singh, M., Bisht, I. S., & Bhalla, S. (2016). Searching for resistance in wild Lens species against pulse beetle, Callosobruchus chinensis (L.). *Legume Research-An International Journal*, 39(4), 630–636.

Gorim, L. Y., & Vandenberg, A. (2017a). Root traits, nodulation and root distribution in soil for five wild lentil species and *Lens culinaris* (medik.) grown under well-watered conditions. *Frontiers in Plant Science*, 8, 1–13.

Gorim, L. Y., & Vandenberg, A. (2017b). Evaluation of wild lentil species as genetic resources to improve drought tolerance in cultivated lentil. *Frontiers in Plant Science*, 8, 1–12.

Gorim, L. Y., & Vandenberg, A. (2018). Can wild lentil genotypes help improve water use and transpiration efficiency in cultivated lentil? *Plant Genetic Resources*, 16(5), 459–468.

Guerra-García, A., Gioia, T., von Wettberg, E., Logozzo, G., Papa, R., Bitocchi, E., & Bett, K. E. (2021). Intelligent characterization of lentil genetic resources: evolutionary history, genetic diversity of germplasm, and the need for well-represented collections. *Current Protocols*, 1(5), e134.

Gupta, D., & Sharma, S. K. (2006). Evaluation of wild Lens taxa for agro-morphological traits, fungal diseases and moisture stress in northwestern Indian hills. *Genetic Resources and Crop Evolution*, 53, 1233–1241.

Gupta, D., Taylor, P. W. J., Inder, P., Phan, H. T. T., Ellwood, S. R., Mathur, P. N., et al. (2012a). Integration of EST-SSR markers of medicago truncatula into intraspecific linkage map of lentil and identification of QTL conferring resistance to ascochyta blight at seedling and pod stages. *Molecular Breeding*, 30, 429–439. https://doi.org/10.1007/s11032-011-9634-2

Gupta, M., Verma, B., Kumar, N., Chahota, R. K., Rathour, R., Sharma, S. K., et al (2012b). Construction of intersubspecific molecular genetic map of lentil based on ISSR, RAPD and SSR markers. *Journal of Genetics*, 91, 279–287. https://doi.org/10.1007/s12041-012-0180-4

Gutierrez-Gonzalez, J. J., García, P., Polanco, C., González, A. I., Vaquero, F., Vences, F. J., & Sáenz de Miera, L. E. (2022). Multi-species transcriptome assemblies of cultivated and wild lentils (Lens sp.) provide a first glimpse at the lentil pangenome. *Agronomy*, 12(7), 1619.

Hamdi, A., & Erskine, W. (1996). Reaction of wild species of the genus *Lens* to drought. *Euphytica*, 91, 173–179.

Hamdi, A., Küsmenoglu, I., & Erskine, W. (1996). Sources of winter hardiness in wild lentil. *Genetic Resources and Crop Evolution*, 43, 63–67.

Hamwieh, A., Udupa, S. M., Choumane, W., Sarker, A., Dreyer, F., Jung, C., et al. (2005). A genetic linkage map of lens sp. based on microsatellite and AFLP markers and the localization of fusarium vascular wilt resistance. *Theoretical and Applied Genetics*, 110, 669–677. https://doi.org/10.1007/s00122-004-1892-5

Havey, M. J., & Muehlbauer, F. J. (1989). Variability for restriction fragment lengths and phylogenies in lentil. *Theoretical and Applied Genetics*, 77, 839–843. https://doi.org/10.1007/BF00268336

Idrissi, O., Udupa, S. M., Houasli, C., De Keyser, E., Van Damme, P., & De Riek, J. (2015). Genetic diversity analysis of Moroccan lentil (*Lens culinaris* medik.) landraces using simple sequence repeat and amplified fragment length polymorphisms reveals functional adaptation towards agro-environmental origins. *Plant Breeding*, 134, 322332. https://doi.org/10.1111/pbr.12261

InternationalCenter for Agricultural Research in the Dry Areas (ICARDA). (2020). *ICARDA Annual Report 2020: A Unified Approach*. ISSN 0254-8313. Retrieved from ICARDA Annual Report 2020

Laskar, R. A., Khan, S., Deb, C. R., Tomlekova, N., Wani, M. R., Raina, A., & Amin, R. (2019). Lentil (Lens culinaris Medik.) diversity, cytogenetics and breeding. *Advances in Plant Breeding Strategies: Legumes*, 7, 319–369.

Kaale, L. D., Siddiq, M., & Hooper, S. (2023). Lentil (*Lens culinaris* Medik) as nutrient-rich and versatile food legume: a review. *Legume Science*, 5(2), e169. https://doi.org/10.1002/leg3.169

Kapazoglou, A., Gerakari, M., Lazaridi, E., Kleftogianni, K., Sarri, E., Tani, E., & Bebeli, P. J. (2023). Crop wild relatives: a valuable source of tolerance to various abiotic stresses. *Plants*, 12(2), 328.

Kaur, S., Webster, T., Sudheesh, S., Pembleton, L., Sawbridge, T., Rodda, M., & Cogan, N. (2016). *Lentil Genome Sequencing Effort: A Comprehensive Platform for Genomics Assisted Breeding*. Australian Pulse Conference, Tamworth, Australia.

Koul, P. M., Sharma, V., Rana, M., Chahota, R. K., Kumar, S., & Sharma, T. R. (2017). Analysis of genetic structure and interrelationships in lentil species using morphological and SSR markers. *3 Biotech*, 7(1), 83. https://doi.org/10.1007/s13205-017-0683-z

Kumar, S., Barpete, S., Kumar, J., Gupta, P., & Sarker, A. (2013). Global lentil production: constraints and strategies. *SATSA Mukhapatra-Annual Technical*, 17, 1–13.

Kumar, J., Srivastava, E., Singh, M., Kumar, S., Nadarajan, N., & Sarker, A. (2014). Diversification of indigenous gene-pool by using exotic germplasm in lentil (*Lens culinaris* medikus ssp. *culinaris*). *Physiology and Molecular Biology of Plants*, 20, 125–132. https://doi.org/10.1007/s12298-013-0214-2

Kumar, S., Choudhary, A. K., Rana, K. S., Sarker, A., & Singh, M. (2018). Biofortification potential of global wild annual lentil core collection. *PLoS One*, 13, e0191122. https://doi.org/10.1371/journal.pone.0191122

Kumari, M., Mittal, R. K., Chahota, R. K., Thakur, K., Lata, S., & Gupta, D. (2018). Assessing genetic potential of elite interspecific and intraspecific advanced lentil lines for agronomic traits and their reaction to rust (*Uromyces viciae-fabae*). *Crop & Pasture Science*, 69, 999–1008.

Kushwaha, U. K. S., Ghimire, S. K., Yadav, N. K., Ojha, B. R., & Niroula, R. K. (2015). Genetic characterization of lentil (*Lens culinaris* L.) germplasm by using SSR markers. *Journal of Agricultural and Biological Science*, 1, 16–26.

Laserna-Ruiz, I., De-Los-Mozos-Pascual, M., Santana-Méridas, O., Sánchez-Vioque, R., & Rodríguez-Conde, M. F. (2012). Screening and selection of lentil (*Lens* Miller) germplasm resistant to seed bruchids (*Bruchus* spp.). *Euphytica*, 188(2), 153–162.

Liber, M., Duarte, I., Maia, A. T., & Oliveira, H. R. (2021). The history of lentil (*Lens culinaris* subsp. culinaris) domestication and spread as revealed by genotyping-by-sequencing of wild and landrace accessions. *Frontiers in Plant Science*, 12, 628439.

Malhotra, N., Panatu, S., Singh, B., Negi, N., Singh, D., Singh, M., & Chandora, R. (2019). Genetic resources: collection, conservation, characterization and maintenance. In: Singh M (eds) *Lentils*. Academic Press, London, pp. 21–41.

Mbasani-Mansi, J., Ennami, M., Briache, F. Z., Gaboun, F., Benbrahim, N., Triqui, Z. E. A., & Mentag, R. (2019). Characterization of genetic diversity and population structure of Moroccan lentil cultivars and landraces using molecular markers. *Physiology and Molecular Biology of Plants*, 25(4), 965–974.

Nene, Y. L. (2006). Indian pulses through the millennia. *Asian Agri-History*, 10(3), 179–202.

Omar, I., Ghoulam, S. B., Abdellah, E. A., & Sahri, A. (2019). Evaluation and utilization of lentil crop wild relatives for breeding in Morocco: Towards development of drought and herbicide tolerant varieties. In *First International Experts Workshop on Pre-breeding utilizing Crop Wild Relatives*. ICARDA, Rabat, Morocco.

Podder, R., Banniza, S., & Vandenberg, A. (2013). Screening of wild and cultivated lentil germplasm for resistance to stemphylium blight. *Plant Genetics Resources*, 11, 26–35.

Pratap, A., Das, A., Kumar, S., & Gupta, S. (2021). Current perspectives on introgression breeding in food legumes. *Frontiers in Plant Science*, 11, 589189.

Rajpal, V. R., Singh, A., Kathpalia, R., Thakur, R. K., Khan, M., Pandey, A., & Raina, S. N. (2023). The prospects of gene introgression from crop wild relatives into cultivated lentil for climate change mitigation. *Frontiers in Plant Science*, 14, 1127239.

Ramsay, L., Chan, C., Sharpe, A. G., Cook, D. R., Penmetsa, R. V., Chang, P., & Bett, K. E. (2016). *Lens culinaris* CDC Redberry Genome Assembly v1.2. Retrieved from https://knowpulse.usask.ca/genome-assembly/Lc1.2

Ramsay, L., Koh, C., Konkin, D., Cook, D., Penmetsa, V., Dongying, G., & Bett, K. E. (2019a). *Lens culinaris* CDC Redberry Genome Assembly v2.0. Retrieved from https://knowpulse.usask.ca/genome-assembly/Lcu.2RBY

Ramsay, L., Koh, K., Konkin, D., Stonehouse, R., Banniza, S., & Bett, K. E. (2019b). Lens ervoides IG 72815 Genome Assembly 1.0. Retrieved from https://knowpulse.usask.ca/genomeassembly/Ler.1DRT

Ramsay, L., Koh, C. S., Kagale, S., Gao, D., Kaur, S., Haile, T., & Bett, K. E. (2021). Genomic rearrangements have consequences for introgression breeding as revealed by genome assemblies of wild and cultivated lentil species. *Biorxiv*, 2021, 07.

Rivero, R. M., Mittler, R., Blumwald, E., & Zandalinas, S. I. (2022). Developing climate-resilient crops: improving plant tolerance to stress combination. *The Plant Journal: For Cell and Molecular Biology*, 109(2), 373–389. https://doi.org/10.1111/tpj.15483

Roy, A., Sahu, P. K., Das, C., Bhattacharyya, S., Raina, A., & Mondal, S. (2023). Conventional and new-breeding technologies for improving disease resistance in lentil (*Lens culinaris* Medik). *Frontiers in Plant Science*, 13, 1001682.

SenGupta, D., Thavarajah, D., McGee, R. J., Coyne, C. J., Kumar, S., & Thavarajah, P. (2016). Genetic diversity among cultivated and wild lentils for iron, zinc, copper, calcium and magnesium concentrations. *Australian Journal of Crop Science*, 10, 1381–1387. https://doi.org/10.21475/ajcs.2016.10.10.pne6

Sharma, S. K., Dawson, I. K., & Waugh, R. (1995). Relationships among cultivated and wild lentils revealed by RAPD analysis. *Theoretical and Applied Genetics*, 91, 647–654. https://doi.org/10.1007/BF00223292

Sharma, S. K., Knox, M. R., & Ellis, T. N. (1996). AFLP analysis of the diversity and phylogeny of Lens and its comparison with RAPD analysis. *Theoretical and Applied Genetics*, 93, 751–758. https://doi.org/10.1007/BF00224072

Singh, M., Bisht, I. S., Kumar, S., Dutta, M., Bansal, K. C., Karale, M., & Datta, S. K. (2014). Global wild annual Lens collection: a potential resource for lentil genetic base broadening and yield enhancement. *PLoS One*, 9(9), e107781.

Singh, D., Singh, C. K., Singh, Y. P., Singh, V., Singh, R., Tomar, R. S. S., & Sharma, P. C. (2018). Evaluation of cultivated and wild genotypes of Lens species under alkalinity stress and their molecular collocation using microsatellite markers. *PLoS One*, 13(8), e0199933.

Singh, M., Sharma, S. K., Singh, B., Malhotra, N., Chandora, R., Sarker, A., & Gupta, D. (2018). Widening the genetic base of cultivated gene pool following introgression from wild Lens taxa. *Plant Breeding*, 137(4), 470–485.

Singh, D., Singh, C. K., Tribuvan, K. U., Tyagi, P., Taunk, J., Tomar, R. S. S., & Pal, M. (2020). Development, characterization, and cross species/genera transferability of novel EST-SSR markers in lentil, with their molecular applications. *Plant Molecular Biology Reporter*, 38, 114–129.

Singh, M., Kumar, S., Basandrai, A. K., Basandrai, D., Malhotra, N., Saxena, D. R., & Singh, K. (2020). Evaluation and identification of wild lentil accessions for enhancing genetic gains of cultivated varieties. *PLoS One*, 15(3), e0229554.

Singh, M., Kumar, S., Mehra, R., Sood, S., Malhotra, N., Sinha, R., & Gupta, V. (2022). Evaluation and identification of advanced lentil interspecific derivatives resulted in the development of early maturing, high yielding, and disease-resistant cultivars under Indian agro-ecological conditions. *Frontiers in Plant Science*, 13, 936572.

Sudheesh, S., Verma, P., Forster, J. W., Cogan, N. O., & Kaur, S. (2016). Generation and characterisation of a reference transcriptome for lentil (*Lens culinaris* Medik.). *International Journal of Molecular Sciences*, 17(11), 1887.

Tayşi, N., Kaymaz, Y., Ateş, D., Sari, H., Toker, C., & Tanyolaç, M. B. (2022). Complete chloroplast genome sequence of lens ervoides and comparison to *Lens culinaris. Scientific Reports*, 12(1), 15068.

Tewari, K., Dikshit, H. K., Jain, N., Kumari, J., & Singh, D. (2012). Genetic differentiation of wild and cultivated Lens based on molecular markers. *Journal of Plant Biochemistry and Biotechnology*, 21, 198–204.

Tiwari, M., Singh, B., Min, D., & Jagadish, S. V. (2022). Omics path to increasing productivity in less-studied crops under changing climate-lentil a case study. *Frontiers in Plant Science*, 13, 813985.

Toklu, F. A., Karaköy, T., Hakl, E., Bicer, T., Brandolini, A., Kilian, B., et al. (2009). Genetic variation among lentil (*Lens culinaris* medik) landraces from southeast Turkey. *Plant Breeding*, 128, 178–186. https://doi.org/10.1111/j.1439-0523.2008.01548.x

Tullu, A., Buchwaldt, L., Lulsdorf, M., Banniza, S., Barlow, B., Slinkard, A. E., & Vandenberg, A. (2006). Sources of resistance to anthracnose (*Colletotrichum truncatum*) in wild *Lens* species. *Genetic Resources and Crop Evolution*, 53, 111–119.

Tullu, A., Banniza, S., Tar'an, B., Warkentin, T., & Vandenberg, A. (2010). Sources of resistance to ascochyta blight in wild species of lentil (*Lens culinaris* Medik.). *Genetic Resources and Crop Evolution*, 57, 1053–1063.

Tullu, A., Diederichsen, A., Suvorova, G., & Vandenberg, A. (2010). Genetic and genomic resources of lentil: status, use and prospects. *Plant Genetic Resources*, 9(01), 19–29. https://doi.org/10.1017/s1479262110000353

Tullu, A., Banniza, S., Bett, K., Vandenberg, A. (2011). A walk on the wild side: exploiting wild species for improving cultivated lentil. *Grain Legume*, 56, 13–14.

Tullu, A., Bett, K., Banniza, S., Vail, S., & Vandenberg, A. (2013). Widening the genetic base of cultivated lentil through hybridization of *Lens culinaris* 'Eston' and *L. ervoides* accession IG 72815. *Canadian Journal of Plant Science*, 93, 1037–1047. https://doi.org/10.4141/cjps2013-072

Vail, S., Strelioff, J. V., Tullu, A., & Vandenberg, A. (2012). Field evaluation of resistance to *Colletotrichum truncatum* in *Lens culinaris*, *Lens ervoides*, and *Lens ervoides*×*Lens culinaris* derivatives. *Field Crops Research*, 126, 145–151.

Wong, M. M., Gujaria-Verma, N., Ramsay, L., Yuan, H. Y., Caron, C., Diapari, M., & Bett, K. E. (2015). Classification and characterization of species within the genus Lens using genotyping-by-sequencing (GBS). *PLoS One*, 10(3), e0122025.

Ye, G., McNeil, D. L., & Hill, G. D. (2000). Two major genes confer Ascochyta blight resistance in *Lens orientalis. New Zealand Plant Protection*, 53, 109–113.

Yuan, H. Y., Saha, S., Vandenberg, A., & Bett, K. E. (2017). Flowering and growth responses of cultivated lentil and wild Lens germplasm toward the differences in red to far-red ratio and photosynthetically active radiation. *Frontiers in Plant Science*, 8, 386.

Zohary, D. (1972). The wild progenitor and the place of origin of the cultivated lentil: *Lens culinaris. Economic Botany*, 26, 326–332. https://doi.org/10.1007/BF02860702

5 Common Bean (*Phaseolus vulgaris*) Crop Wild Relatives
Their Role in Improving Climate-Resilient Common Bean

Sougata Bhattacharjee, Krishnayan Paul, Dipro Sinha, and Uday Chand Jha

INTRODUCTION

The common bean, scientifically known as *Phaseolus vulgaris*, is a widely cultivated legume and one of the most important staple foods worldwide. It belongs to the family *Fabaceae* and is native to the Americas, where it has been cultivated for thousands of years. Common beans are commonly referred to by various names, including kidney beans, pinto beans, navy beans, black beans, and many others, depending on their shape, colour, and regional preferences.

Some key points about common beans:

Varieties: Common beans come in a variety of shapes, sizes, and colours. They can be classified into three main types: bush beans, pole beans, and runner beans. Each type has its growth habit and preferred growing conditions.

Nutritional value: Common beans are highly nutritious and provide essential nutrients. They are a good source of dietary fibre, plant-based protein, complex carbohydrates, vitamins (such as folate), and minerals (including iron, potassium, and magnesium). Beans are also low in fat and cholesterol-free.

Culinary uses: Common beans have a versatile nature and are used in various cuisines around the world. They can be used in both savoury and sweet dishes. Common bean recipes include soups, stews, chilli, salads, dips, and even desserts. They are often cooked before consumption, although some varieties can be eaten raw when young.

Health benefits: Consuming common beans as part of a balanced diet can have several health benefits. Their high fibre content promotes digestive health, helps maintain healthy cholesterol levels, and contributes to weight management.

DOI: 10.1201/9781003434535-5

The protein content in beans makes them a valuable plant-based protein source, especially for vegetarians and vegans.

Cultivation: Common beans are relatively easy to grow and adapt to various climates. They require warm temperatures to germinate and thrive in well-drained, fertile soils. They are commonly cultivated as an annual crop, and the beans are harvested when the pods mature and dry. Common bean cultivation practices differ depending on the variety and regional conditions.

Importance: Common beans are a crucial food source, especially in developing countries. They play a vital role in providing nutrition and food security for millions of people worldwide. Additionally, common beans contribute to sustainable agriculture by fixing nitrogen in the soil and reducing the need for synthetic fertilizers.

CROP GENETICS AND DIFFERENT SPECIES OF COMMON BEANS ALONG WITH THEIR BOTANICAL FEATURES

Phaseolus vulgaris L. (common bean) is one of the most important legume crops in the world after soybean. The common bean taxonomic hierarchy is as follows:

Order: Fabales

Family: Fabaceae

Genus: *Phaseolus L.*

Species: *Phaseolus vulgaris L.*

Common synonyms: French bean, haricot bean, salad bean, snap bean, string bean

Low-income countries, however, are unable to produce more than 600 kg of beans per hectare due to environmental stresses. The current low yields lead to food insecurity, while increased yields are needed to keep up with population growth amidst climate change threats. It is imperative to continue exploring untapped genetic diversity in crop wild relatives (CWRs) and closely related species to achieve novel and significant crop genetic improvements. This chapter considers the current understanding of the genetic resources available for common bean improvement and the progress achieved through the introgression of genetic diversity from wild relatives of common bean and closely related species, *including P. acutifolius*, *P. coccineus*, *P. costaricensis*, and *P. dumosus*. Several new genomic tools are presented along with their potential applications. These genetic resources are being used for common bean improvement in a multidisciplinary effort.

DIFFERENT SPECIES OF COMMON BEANS AND THEIR BOTANICAL FEATURES

There are approximately 80 species in the genus Phaseolus, but *P. vulgaris* is the most widely cultivated (Porch et al., 2013). Several closely related species are found to *P. vulgaris*, *including P. albescens*, *P. coccineus*, *P. costaricensis*, *P. dumosus*, *P. parvifolius*, and *P. persistentus* (Table 5.1) (Chacón et al., 2007; Delgado-Salinas et al., 2006; Spataro et al., 2011). The *Phaseolus* genus includes four other cultivated species: *P. coccineus* (scarlet runner), *P. acutifolius* (tepary bean), and *P. lunatus* (lime bean) (Lioi and Piergiovanni, 2013).

TABLE 5.1
Different Species of *Phaseolus* Worldwide

Name	Description
P. vulgaris	Common bean
P. coccineus	Runner bean
P. lunatus	Lima, butter beans, sieva beans
P. dumosus	Acatalete, botil, or year beans
P. albescens	The native range of this species is SW Mexico. It is a liana and grows primarily in the seasonally dry tropical biome.
P. parvifolius	The native range of this species is Mexico
P. persistentus	From Guatemala
P. costaricensis	From Costa Rica and Panama, Central America
P. filiformis	Its common names include slim-jim bean and slender-stem bean. This plant resembles other beans in appearance, with leaves composed of lobed triangular leaflets and pink pea-like flowers.
P. acutifolius	Known as the tepary bean, it is a legume native to the southwestern United States and Mexico and has been grown there by native peoples since pre-Columbian times. It is more drought-resistant than the common bean (Phaseolus vulgaris) and is grown in desert and semi-desert conditions from Arizona through Mexico to Costa Rica.
P. talamancensis	From Montane Forests of Eastern Costa Rica
P. glabellus	The wild, red-flowered *Phaseolus glabellus* is generally considered either as a species of the *P. coccineus* complex or as a subspecies of *P. coccineus*.
P. macrolepis	The native range of this species is Guatemala. It grows primarily in the wet tropical biome.
P. microcarpus	It is a climbing perennial and grows primarily in the seasonally dry tropical biome.
P. oaxacanus	Information not defined.
P. amblyosepalus	It is found in the western part of the Pacific Ocean, namely in the Andaman Sea south of Japan, Taiwan, China, the Philippines, and Vietnam.
P. chiapasanus	Information not defined.
P. esperanzae	Naturally carry rhizobia on their testa
P. grayanus	Known as Gray's Bean; Perennial herb.
P. hintonii	The native range of this species is Mexico. It grows primarily in the seasonally dry tropical biome.
P. jaliscanus	Information not defined.
P. leptostachyus	The native range of this species is Mexico to Central America.
P. marechalii	The native range of this species is Central and S. Mexico.
P. micranthus	This genus has small flowers in a spike that are blue, white, or purple. Commonly known as comb flowers. Also found in India.
P. nelsonii	The native range of this species is Central and S. Mexico.
P. neglectus	The native range of this species is northeast Mexico.
P. oligospermus	Herbaceous to woody annual and perennial vines.
P. parvulus	It is a viney pea with solitary or nearly solitary pink-purple flowers and lanceolate, trifoliate dark green leaves. There is a small, globose bulb just under the soil surface. It is found in the understory of the upper elevation Ponderosa Pine Forest.

(*Continued*)

TABLE 5.1 (*Continued*)
Different Species of *Phaseolus* Worldwide

Name	Description
P. pedicellatus	Viney herb with pink flowers and trifoliate leaves with lobed leaflets.
P. pluriflorus	Information not defined.
P. polymorphus	The native range of this species is North and West Mexico.
P. scabrellus	Red flower, Viney nature.
P. tuerckheimii	The native range of this species is southeast Mexico to Venezuela. It is a climbing perennial and grows primarily in the wet tropical biome.
P. xanthotrichus	The native range of this species is Mexico to Central America. It grows primarily in the seasonally dry tropical biome.
P. acinacifolius	East Indian legume having hairy foliage and small yellow flowers followed by cylindrical pods
P. albiflorus	White-seeded runner bean
P. albinervus	Information not defined.
P. gladiolatus	The native range of this species is Northeast Mexico
P. juquilensis	The native range of this species is Mexico (Oaxaca); Yellowish-brown flower.
P. leptophyllus	The native range of this genus is the United States of America to northsouth America and northwest Argentina, Tanzania.
P. longiplacentifer	The native range of this species is Mexico.
P. macvaughii	The native range of this species is West Mexico; purple flower.
P. magnilobatus	The native range of this species is Mexico (Durango, Jalisco).
P. nudosus	The year bean or year-long bean is an annual to perennial herbaceous vine.
P. opacus	Information not defined.
P. palmeri	This name is a synonym of *Phaseolus pedicellatus*.
P. pauciflorus	It is a species of wild bean native to Mexico and Guatemala.
P. perplexus	The native range of this species is Mexico.
P. plagyosilix	Information not defined.
P. purpusii	Synonym of Phaseolus polymorphus
P. pyramidalis	The native range of this species is Mexico (Chihuahua). It grows primarily in the desert or dry shrubland biome.
P. rotundatus	A new bean species for western Mexico.
P. salicifolius	Molecular evidence for an Andean origin and a secondary gene pool for the Lima bean.
P. scrobiculatifolius	Information not defined.
P. sonorensis	Origin is Mexico.
P. tenellus	The native range of this species is northeast and Central Mexico.
P. teulensis	This heavy-producing genotype offers rich, fleshy, medium pods with beautiful white beans inside.
P. trifidus	Information not defined.
P. xolocotzii	It occurs on well-drained volcanic or limestone rocky hillsides.
P. zimapanensis	Information not defined.

A Brief History of the Domestication Process of Common Beans from Their CWRs

Historically, Phaseolus has been domesticated and evolved uniquely. Domesticated species include *P. vulgaris* (common bean), *P. coccineus* (runner bean), *P. lunatus* (Lima, butter beans, sieva beans), *P. acutifolius* (tepary beans), and *P. dumosus* (acatalete, botil, or year beans). Varieties differ in growth habits, size and shape of seeds, maturity days, and many other factors. In addition to pest and disease resistance, seeds are also nutritionally diverse. In both wild and cultivated populations, the common bean is mainly self-pollinated; however, many authors have documented outcrossing or natural hybridization. To maintain genetic integrity during regeneration, it is necessary to take some precautions during outcrossing.

GEOGRAPHIC DISTRIBUTION, ECOSYSTEMS, AND DIVERSITY

The first wild common bean populations were found in northern Mexico and northern Argentina. Central and South American countries are the origins of common beans. In northern Argentina and Central America, small-seeded and climbing ecotypes are found in the wild. Both Central America (Mexico and Guatemala) and the South American Andes (primarily Peru) independently domesticated the common bean. According to archaeology, 6000 BC is the earliest date of domestication in Peru and 5000 BC in Mexico. However, the species' distribution is affected by climatic differences, such as excessive rainfall or altitudes below 700 m or above 3,000 m (Chacón et al., 2007). Since the 16th century, common beans have been exported to other parts of the world. Common bean was probably introduced to Africa by Portuguese traders through Sofala (Mozambique), Zanzibar, and Mombasa, from where slave trading caravans and merchants carried it to higher altitude areas of the interior. In parts of Africa before colonization, the common bean had become an established pulse crop. As habitats have been destroyed throughout the species' range, the need to identify and preserve ancestral varieties has increased (Beebe et al., 1997; Chacón et al., 2005). Wild common beans are found across northern Mexico and northern Argentina, and distinct differences have been observed in both morphological characteristics and molecular markers between northernmost and southernmost populations (Graham and Ranalli, 1997; Singh et al., 1991). In the region where common beans originated, the climate is subtropical to temperate, with defined wet and dry seasons, and the bean prefers regions with moderate rainfall rather than areas with excessive rainfall (Beebe et al., 1997). Despite their inability to tolerate frost or elevations above 3,000 m, bean plants can grow as annuals in temperate climates and as annuals or short-lived perennials in tropical climates. High temperatures can cause flowers to abscise, while low temperatures can delay pod production and result in empty pods (Liebenberg et al., 2009). Beans prefer well-drained, sandy clay or sandy loam soils with a pH of 5.8–6.5 and moderate fertility. In regions outside the species' native range, cultivated varieties of beans rarely persist as feral populations. Genetic analyses of individual bean plants selected from feral populations and cultivated varieties indicate that the cultivated varieties are descended from feral populations rather than vice versa (Porch et al., 2013; Toro and Ocampo, 2004).

A total of 50 species of Phaseolus are known, most of which are found in North America. In addition to Phaseolus vulgaris, other Phaseolus species include *Phaseolus coccineus L.* (runner bean) and *Phaseolus acutifolius A. Gray* (tepary bean), enabling interspecific hybridization.

There are three reasons to study the wild ancestor of the common bean. First, since the wild common bean requires moderate temperatures and a long dry season, it thrives at altitudes where rainfall coincides with favourable temperatures (summer). Therefore, wild *P. vulgaris* is absent from Mediterranean climates in the Americas (e.g., California, Chile). Second, wild common beans' fragmented distribution is reflected in their genetic makeup (e.g., phaseolins). Since they are ancient, it is possible to explain these genetic differences at the molecular level by past trans-isthmic migrations during the tertiary period. In the central Andes (with the "C" haplotype), the first migration from Mesoamerica allowed future populations to become domesticated. As the Pacific clade migrated again, it would have reached Costa Rica, where the "H" haplotype is found today, not in Costa Rican cultivated beans. By exploitation of a wider longitude gradient, the original wild stock of *P. vulgaris* differentiated further, producing up to six haplotypes, of which four were found in cultivated varieties in Mesoamerica. The third trans-isthmic migration, made possible by past climatic oscillations, produced a third clade, whose expansion is visible in the Colombian eastern highlands today.

As a third factor, these genetic differences are also reflected in proteins, enzymes, secondary metabolites, and physiological differences. Among these seed proteins are arcelins, which are stored in cotyledons and possess insecticidal properties specific to western Mexico. Contrary to expectations, the presence of arcelin variants causing antibiosis in bruchids does not result in a natural selection advantage—since all wild populations along their entire range would carry the trait if it did—but rather reflects the species' natural variation. In contrast, Peruvian landraces have lower carbon dioxide exchange rates than wild counterparts, while photosynthetic nitrogen use efficiency is high in wild forms from Mexico. In addition to wild forms contributing QTLs (quantitative trait loci), this approach has also been successful in breeding tomato and rice varieties.

A living wild ancestor of the common bean was discovered relatively late, in the 1960s. It was also recently discovered that it was domesticated twice, independently, from different populations in Mesoamerica and the Andes. However, it was only recently discovered that the common bean has sister species in the same genus. According to Marshal et al., *P. coccineus* L and some of its wild relatives, as well as *P. dumosus*, are closely related to the common bean genetically. As recommended by Schmit et al., *P. albescens* should be added to that list as well. It would be helpful to study additional material on *P. persistentus* Freytag and Debouck to determine if it is part of the same lineage as *P. persistentus* Freytag and Debouck. To date, the section Phaseoli has six wild species: *P. albescens*, *P. coccineus*, *P. costaricensis*, *P. dumosus*, *P. persistentus*, and *P. vulgaris*. The only Phaseoli species found in South America is *P. vulgaris*, based on the evidence available today. Nevertheless, *P. dumosus*, a wild plant found in western Guatemala, can be a weed in the northern Andes, as can some escapes of *P. coccineus*. There are wild populations of

P. coccineus in Mexico and Guatemala, ranging from Chihuahua to Jalapa. *The natural distribution of P. costaricensis is in eastern Costa Rica and western Panamá. In western Mexico, P. albescens is found in montane forests.* Therefore, it is not impossible that the section Phaseoli has undergone most or all of its speciation process in Central America. *P. vulgaris* itself appears to have a Mesoamerican origin as well.

It is interesting to compare the duration of the common bean as cultivated and as wild, even though we are still missing paleobotanical and archaeological evidence. According to the latest evidence, the common bean was domesticated up to 2,000 years ago in Mesoamerica and 4,000–5,000 years ago in the central Andes. While the breeder is always striving for a perfect cultivated mutant in the gene pool, she/he is breeding, from the standpoint of mere accumulation of potentially interesting traits, the wild stock is likely to be the largest! Additionally, domestication events were few (one to three in Mesoamerica and one in the Andes), which means that most of the diversity was left untouched. It may have been because the clever early domesticators were not interested in repeating the process, but rather exchanging their early successes, that most of the wild diversity was not domesticated. In addition, there were limited technologies to deal with antinutritional factors in wild ancestors' seeds.

Why CWRs of the Common Bean are Important

There is tremendous potential for the introgression of novel variation into the elite gene pool through CWRs, exotic germplasm, and landraces. It allows for continuous incremental gains over breeding cycles and the discovery of cryptic genetic variations previously undiscovered. CWRs can also improve agronomic traits through QTLs. To transfer polygenic traits into elite backgrounds, the "specialized population concept" has been advocated.

Introgression breeding has successfully developed improved cultivars of legumes including common beans. These cultivars were developed by unleashing new gene recombination and hidden variability even in late filial generations. Vertical gene transfer has resulted in the most useful gene introgressions in food legumes, but horizontal gene transfer through transgenic technology, somatic hybridization, and intragenesis also offers promise.

To develop commercial food vegetable cultivars, it is imperative to capitalize on the resurgence of interest in introgression breeding. Over 27.5 million tonnes of dried beans were produced worldwide in 2020, almost 20% of which came from India.

The common bean is an annual herb up to 3m long that has three rounded leaflets, white, pale purple, or red-purple flowers, and a long pod that can be green, yellow, red, or purple and contains up to 12 seeds. The mature seed is mostly eaten as a pulse, and the immature pods and seeds as vegetables. The young leaves can be cooked and eaten as a vegetable. *P. vulgaris* belongs to the Fabaceae family and has five-petaled flowers with a distinctive papilionaceous or butterfly-like shape. They have ten stamens and a tube surrounding the ovary. Phaseolus vulgaris shares many features with the rest of the family, but it also has two distinctive features: a coil-shaped flower keel and uncinate hairs. The common bean is the most consumed legume worldwide and is the most important legume produced for direct human consumption. It is an important source of dietary protein for millions of people throughout the tropics. Dry beans are processed before consumption, usually by cooking in water, but some beans are consumed after roasting or milling into flour.

Due to extensive plant-breeding efforts, *P. vulgaris* has numerous cultivars with a wide range of morphological and agronomic characteristics, including different seed sizes and colours. Determinate growth habits are more commonly selected and are better suited to shorter growing seasons. Determinate varieties of common bean have terminal inflorescences, while indeterminate varieties have axillary racemes. There are indeterminate cultivars of *P. vulgaris* with a twining or climbing growth habit, as well as many cultivars with a partially erect and partially trailing intermediate growth habit, although they are less frequently grown than the determinate cultivars.

Cultivated *P. vulgaris* has a taproot-based root system with lateral roots located within the top 15 cm of soil. The stems are hairy and play a role in both disease and insect resistance. The leaves are trifoliolate and alternate on the stems. The flowers are borne on axillary or terminal racemes, and the seed pods are narrow, 8–20 cm × 1–2 cm, with up to 12 seeds per pod, but most varieties have 4–6 seeds. Wild *P. vulgaris* differs from the cultivated types in several characteristics, including being indeterminate climbers with shorter main stems and more numerous main stem branches with fewer nodes. The wild species of the plant have more flowers, seed pods, and seeds and a longer flowering period than cultivated varieties.

Its appearance varies depending on the variety. Generally, it is a perennial herb growing up to 3 m long that spreads across the ground, climbs up supports, or grows as a bush or shrub. In addition to the three rounded leaflets, the leaves are arranged alternately along the stem. It has white, pale purple, or red-purple flowers that resemble butterflies (known as papilionaceous). There are up to 12 seeds in the long pod, which can be green, yellow, red, or purple. There are several varieties of seeds, ranging in shape from rounded to kidney-shaped to oblong. The seeds come in solid black, brown, yellow, and red colours, as well as speckled and flecked variations. Among the most popular beans are kidney beans, pinto beans, and haricot beans, typically used to make baked beans.

The mature seeds are mostly eaten as pulses, and the immature pods and seeds are eaten as vegetables. As well as being boiled with seasoning and oil, the seeds are often mashed or made into soup in tropical Africa. The young leaves are edible as a vegetable but are not widely used due to their toughness. As a salad, the young leaves of this plant are eaten in Java. The common bean's immature seed pods, known as French or green beans, are harvested in temperate regions. Raw, boiled, steam-fried, or pickled, they can be eaten in a variety of ways. Numerous varieties of common beans are moderately rich in vitamins, with good levels of protein and iron. Unless cooked before eating, common beans may cause nausea, diarrhoea, and vomiting if eaten in excessive quantities due to lectins and antinutrients.

Although *P. vulgaris* shares many characteristics with its family, it is distinguished by two features: the flower's keel is a coil containing one to two turns, and uncinate hairs are present on both the vegetative and reproductive structures. *P. vulgaris var. vulgaris* (Gepts et al., 1986), *P. vulgaris var. mexicanus* (Debouck, 1999; Delgado-Salinas et al., 1999), and *P. vulgaris subsp. arborigineus* (Brücher, 1988) are considered the wild ancestor of *P. vulgaris* (Table 5.2).

TABLE 5.2
Species Closely Related to *Phaseolus vulgaris*

Species	Geographic Distribution
P. acutifolius	Mexico, southwestern United States
P. albescens	Western Mexico
P. coccineus	Guatemala, Honduras, Mexico
P. costaricensis	Eastern Costa Rica, western Panama
P. dumosus	Western Guatemala, Mexico
P. parvifolius	Southwestern United States, Guatemala, the Pacific coast of Mexico, and Central America
P. persistentus	Guatemala

Sources: Porch et al. (2013); Spataro et al. (2011).

P. vulgaris dried seeds, or "pulses", are a source of dietary protein for millions of people throughout the tropics despite their low levels of methionine and cysteine, supplementing the amino acids lacking in diets based on maize, rice, or other cereals (Miklas et al., 2006). The amino acids lysine and tryptophan, as well as the minerals iron, copper, and zinc, and phytochemicals, antioxidants, and flavonoids, are particularly abundant in beans. Most dry beans are processed before consumption, typically by cooking in water, but some beans are consumed after roasting or after being ground into flour.

Some regions consume snap beans as vegetables, and straw from the plants is used as forage. In developing countries in Latin America and Africa, most beans are produced by smallholder farmers. As a result of intercropping bean production, the total area planted is overestimated and global yield is underestimated (Myers and Kmiecik, 2017).

Even though the FAO reported that dry bean production was over 26 million tonnes in 2014 (https://www.fao.org/4/i3591e/i3591e00.htm), this number may also include other bean species and minor food legumes. Plant breeding has resulted in numerous cultivars of *P. vulgaris* with a variety of morphological and agronomic characteristics, including seed size, colour, and growth habit. As a result of determinate growth (Blair et al., 2012; Singh and Schwartz, 2010, 2011), branching is reduced, internodes are shorter, twining is reduced, day length is insensitive, and biomass is allocated more appropriately to reproductive growth. Specific agronomic circumstances also favour the use of varieties with a determinate growth habit: they are better adapted to shorter growing seasons because they mature earlier; they produce pods over a shorter, more consistent period, which simplifies the harvest of green beans; and determinate varieties are more amenable to mechanized cultivation and harvest (Bitocchi et al., 2013). In the determinate variety, the main stem is replaced by a terminal inflorescence, whereas in the indeterminate variety, the main stem continues to produce axillary racemes.

Cool, highland areas with short day lengths are more suited to prostrate or semi-climbing indeterminate varieties (Singh and Schwartz, 2011). Plants produce more seed pods when their main stem is longer, which is positively correlated with the number of nodes on each stem (García et al., 1997). Several other traits are selected due to domestication of *P. vulgaris*, including increased pod size and fleshiness, reduced pod dehiscence, larger seeds, and increased seed permeability to water. Typically, the lateral roots of cultivated *P. vulgaris* are located within 15 cm of soil above the taproot. As a result of the colonization of the roots by Rhizobium bacteria, irregular root nodules form. Cultivars vary in both the density and length of hair on stems. Younger stems, however, are always covered with short, hooked hairs (uncinate hairs). As well as assisting in disease resistance, hairs also play an important role in insect resistance. Fungal spores are interrupted by hairs, which reduce secondary inoculum (e.g., bean rust). Insects can also be physically wounded (such as leafhoppers), reducing predation.

Wild *P. vulgaris* stems can grow to a diameter of 1.5 cm and may develop a corky outer layer when the climate permits semi-perennial growth. Unlike cultivated varieties, wild species produce flowers throughout their lifespan. Physiologically, cultivated and wild species differ as well. Among wild populations, Porch et al. found greater efficiency in nitrogen uptake and carbon dioxide exchange (Porch et al., 2013).

Various Novel Traits Biotic Stress Tolerance, Abiotic Stress Tolerance, Quality Traits, etc. Exist across Common Bean and Success Stories

Agricultural activities are first and foremost a disturbance to the environment. Crops have lost a significant amount of genetic variability as a result of domestication, especially for bean cultivars in the Andean gene pool. The common bean has less genetic variability than wild beans. Compared to the Andean gene pool, the Middle American gene pool had a threefold greater reduction in genetic variability than the wild members. In the wild Andean gene pool, there is less genetic variation because of limited migration from Middle America. There is an almost untapped source of traits for improving common beans in wild common beans (primary gene pool). In addition to endemic pests and pathogens, wild beans are often intensely virulent against domesticated varieties. According to Acevedo et al., monitoring rust populations (caused by *Uromyces appendiculatus*) on wild beans in native habitats can provide opportunities to track pathogen virulence patterns and develop resistance strategies.

Droughts and high temperatures during flowering and pod development are associated with climate change. *P. filiformis* is capable of tolerating salinity as well as extreme temperatures. *P. acutifolius*, another species, has also proven to be resistant to high temperatures, droughts, and salinity. Considering future evaluations, it is necessary to establish a baseline for the number of Phaseoli populations that exist and how many of them are represented in gene banks.

There is a sharp contrast between the numbers of *P. coccineus* and *P. vulgaris* populations in the wild and the other wild Phaseoli. This may reflect the former taxa's capacity to colonize habitats modified by humans. Although wild and domesticated beans can be easily crossed, Blair et al. noted that differences in flowering patterns and growth cycles may complicate the task of synchronizing flowering. According to Keneni et al., wild species generally have poor agronomic performance, and breeding for traits such as disease and pest resistance may be challenging if the mechanism of resistance affects productivity or crop quality. It has been possible to recover acceptable seed size and agronomic traits for wild beans using the inbred backcross breeding method at CIAT, despite their indeterminacy and small seeds.

It has already been proven that wild beans are resistant to economically important diseases and pests. Arcelin, previously mentioned as a seed protein that confers moderate resistance against bruchids (*Acanthoscelides obtectus* and *Zabotes subfasciatus*), was first discovered in Mexican accessions of wild beans. A variety of arcelin alleles and levels of resistance to bruchids were developed by crossing common bean germplasm with wild bean accessions. As per the literature, a single dominant gene controls the different forms of arcelin.

The PR0650-31 black bean germplasm line was created from the cross BAT 93/PI 417662/VAX 6 between a web blight-resistant and a common bacterial blight-resistant (caused by *Thanatephorus cucumeris*) black bean. Wild-type bean germplasm PI 417662 was collected in Mexico's Jalisco region in 2009 by United States Department of Agriculture-Agricultural Research Service (USDA-ARS). In their study, Acevedo et al. identified two wild bean accessions from Honduras that had high levels of resistance to rust. An inbred backcross line population derived from a cross between the black bean cultivar "Tacana" as the recurrent parent and the Mexican wild bean accession PI 318695 was reported to have QTLs for white mould resistance (caused by *Sclerotinia sclerotiorum*).

We have made less progress in using wild beans to evaluate quantitatively inherited traits like abiotic stress tolerance or seed yield. Based on a cross between G 24423, a wild bean accession from Colombia, and the Mexican black bean cultivar "Negro Tacana", Acosta-Gallegos et al. developed an inbred backcross population. Based on linkage group Pv10, Wright and Kelly identified a major QTL for seed yield in a Phaseolus population. A cross between the ICA Cerinza cultivar of Andean origin and a wild bean accession, G24404, was used for the analysis of QTLs associated with seed yield by Blair et al. From field trials conducted in Colombia, several QTLs were identified for seed yield and seed yield components, including a QTL for larger seed sizes. Breeding methods must be developed to exploit wild bean genetic variability, according to the authors.

It is also possible to cross the common bean with the scarlet runner bean (*Phaseolus coccineus* L.), which represents a secondary gene pool. Blair et al. identified differences in photoperiod sensitivity, flowering pattern, and other factors as barriers to crossing wild beans between common beans and scarlet runner beans. A crinkle leaf dwarf, a dwarf trait, and a blocked cotyledon are lethal and have been identified in crosses between common beans and scarlet runner beans. These traits are controlled by complementary dominant genes. According to their report, the black bean line 5–593 and the snap bean cultivar "Regalfin" could be useful bridging lines between scarlet runner beans and common beans.

A variety of fungal diseases can affect scarlet runner beans, including rust, anthracnose (caused by Colletotrichum lindemuthianum), and web blight. In addition, wild populations of common bean and scarlet runner bean are often found growing together. Due to this, scarlet runner beans are exposed to endemic populations of highly virulent common bean pathogens in their native habitat. Interspecific crosses between the common bean and scarlet runner bean have produced disease-resistant common bean germplasm and cultivars.

Singh et al. found that scarlet runner bean accessions G35172 and G35005 were resistant to multiple diseases, including bean golden yellow mosaic virus (BGYMV) and white mould. Using congruity backcrossing between ICA Pijao and G35172, Singh et al. developed white mould-resistant VCW 54 and VCW 55 bean germplasm lines. The genes bgm-2 and bgp-2 in G35172 confer resistance to leaf chlorosis and pod deformation caused by BGYMV, respectively, according to Osorno et al.

In a study by Schwartz et al., two populations derived from interspecific crosses between the common bean and the scarlet runner bean were tested for white mould resistance. PI 433246 and PI 439534 of scarlet runner bean were both found to have a single dominant gene controlling white mould resistance. A straw test conducted in seven different university laboratories in the United States found that two interspecific bean lines, first screened for web blight and then screened for white mould resistance, were among the most resistant to white mould.

A root rot-resistant line, Cornell 2114-12, was developed by crossing a common bean with a scarlet runner bean. An interspecific cross between scarlet runner bean and common bacterial blight-resistant bean germplasm lines ICB-3, IBC-6, ICB-8, and ICB-10 was developed and released by Miklas et al. Colombian accessions of *P. coccineus* and *P. dumosus* (syn. *P. polyanthus*) were found to be resistant to Ascochyta blight. Mahuku et al. reported that *P. dumosus* has a higher resistance to anthracnose than *P. coccineus* in Colombia. Lines derived from crosses between scarlet runner bean and common bean also exhibited a higher resistance level.

From an interspecific cross between the common bean and scarlet runner bean, Freytag et al. developed the common bacterial blight-resistant germplasm line XR-235-1-1. From interspecific crosses with scarlet runner beans, W-BB-11, W-BB-20-1, W-BB-35, W-BB-52, and W-BB-56 germplasm lines were developed. To develop the common blight-resistant cultivar "Verano", WBB-20-1 was used as a parent.

Aside from matching flowering times and choosing the species used as the female parent in crosses with *P. coccineus*, obtaining traits from Phaseolus' primary and secondary gene pools has been relatively routine. Hybrids of *P. vulgaris* and *P. coccineus* occur naturally and can be easily made by controlled pollination, whereas reciprocal crosses have met with limited success due to unidirectional compatibility, postzygotic barriers, and F1 hybrid sterility. Even across gene pools, crosses with wild accessions of *P. vulgaris* have been routine. The Dl incompatibilities that arise in inter-gene pool crosses of cultivated *P. vulgaris* also exist in wild members. It appears that Dosage-dependent lethal (DL) genes developed before domestication due to geographic isolation rather than selection pressure to eliminate unadopted hybrids among cultivars.

Plant breeders have used scarlet runner beans less to improve tolerance to abiotic stress. Common bean's aluminium tolerance is currently being improved with interspecific crosses with scarlet runner beans. Scarlet runner beans may be useful for improving climbing beans due to their adaptation to higher altitudes and viney growth habits.

It is difficult to hybridize the common bean with the tepary bean (*Phaseolus acutifolius* A. Gray), which is a more distant relative of the common bean. Compared to common beans, tepary beans are native to warmer, more arid climates. Heat and drought tolerance are superior in some tepary landraces. It is possible to produce similar seed yields to common beans when adapted tepary bean lines are grown under favourable conditions. A special breeding method, embryo rescue, was used to secure viable embryos in crosses between the tepary bean in the tertiary gene pool and the common bean. Heat and drought tolerance, insect resistance, and disease resistance are some of the valuable traits in tepary beans, however early attempts to transfer these traits have had limited success because of their genetic complexity. It was first reported that congruity backcrossing produced fertile intermediate hybrids between *P. acutifolius* and *P. vulgaris.* As opposed to traditional recurrent backcrossing to a single recurrent parent, this method involves an alternate generation of backcrossing to each parent. Through congruity-backcross (CBC), substantial recombination can occur between distant species, leading to the emergence of new phenotypes. In bean breeding programmes, the method has been used to force introgression between distant species and to eliminate barriers such as embryo abortion and hybrid sterility. A transfer of *P. parvifolius* into common beans was also accomplished using CBC. Several authors have reported successful transfers of resistance to common bacterial blight from tepary to the common bean.

Tepary beans provide bacterial blight resistance to the bean breeding line XAN 159. A bacterial blight-resistant Great Northern bean cultivar, ABC Weihing was also developed from XAN 159 and a pinto bean germplasm line, ABCP-8, was developed from XAN 159. Speckled sugar and small white beans were introgressed with CBB resistance from XAN 159 using marker-assisted selection (SCAR marker). A high level of resistance to common bacterial blight was introduced into common bean breeding lines VAX 1 to VAX 6 through intensive screening for resistance. As a parent, VAX 6 was used to develop the bacterial blight-resistant white bean cultivar Verano, and as a parent, VAX 3 was used to develop the small red cultivar, Rio Rojo. Several other traits have been identified in tepary bean accessions that may be useful to common bean breeders, such as resistance to BGYMV, ashy stem blight, and Fusarium wilt (*F. oxysporum*). Interestingly, common beans nodulate with Rhizobium species, while tepary beans nodulate with Bradyrhizobium species. The selection of interspecific lines with more promiscuous nodulation and/or improved biological nitrogen fixation may thus be possible. *P. vulgaris* has hybridized successfully with a small number of wild Phaseolus species. Interspecific hybrids between *P. costaricensis* and other species have been reported, and VRW 32 is the first interspecific breeding line to exhibit white mould resistance. Future improvements to common beans will require the transfer of complex traits from other species, such as heat and drought tolerance. In the quaternary gene pool, no successful crosses have been reported between *P. vulgaris* and *P. lunatus.* A Caribbean germplasm collection of Lima beans shows broad agronomic, genetic, and cyanogen diversity that can be a source of traits for common bean production in lowland tropical areas.

Status of Genomic Resources of Common Bean CWRs

Common bean relatives in the wild represent an underexploited source of diversity exhibiting beneficial traits for abiotic and biotic stress tolerance, such as drought, heat, and disease resistance. It has been determined that there are three wild gene pools for the common bean: the Mesoamerican gene pool, the Andean gene pool, and the northern Peru-Ecuador gene pool. Nearly 117,000 Phaseolus accessions are listed in Genesys: 86% are *P. vulgaris*, 4% are *P. coccineus*, 6% are *P. lunatus*, 1% are *P. acutifolius*, and 0.5% are *P. dumosus*. Traditional cultivars or landraces account for about 65% of accessions, while improved cultivars make up 15%, and wild relatives make up 3%. A total of almost 40,000 Phaseolus accessions are held by the International Center for Tropical Agriculture (CIAT), followed by nearly 18,000 samples by the USDA-ARS gene banks, and more than 10,000 samples by the Leibniz Institute for Plant Genetics and Crop Plant Research (IPK).

Among developing countries in Central and South America, East and Southern Africa, and Central and South America, the common bean, Phaseolus vulgaris L. is considered a valuable source of nutrients. Approximately 20 million tonnes of common beans are produced annually on nearly 28 million ha of land worldwide. Bean production is influenced by both biotic and abiotic factors. Despite efforts made by breeders to develop better varieties that can withstand multiple biotic and abiotic constraints, maintaining bean production requires continuous research on finding new genes. Underexploited wild relatives of common beans offer valuable traits for abiotic and biotic stress tolerance and resistance, including heat tolerance, drought tolerance, waterlogging, and disease resistance.

GENETICS AND HYBRIDIZATION

1. Common beans and wild forms are both diploid ($2n=22$), meaning they have the same number of chromosomes.
2. Hybridization between cultivated and wild beans occurs readily, indicating that they can successfully reproduce and produce offspring.
3. Crossing Middle American and Andean gene pools (different populations of common beans) is relatively easy, despite differences in flowering time (Porch et al., 2013).
4. Hybridization between the two gene pools can result in dwarfism or lethality, possibly due to genetic differences.
5. Two "dosage-dependent lethal" (DI) genes are suspected to be responsible for the hybrid weakness, where the homozygosity of these genes determines the lethality of the hybrids (Singh and Schwartz, 2010).
6. Natural crosses between common beans and other Phaseolus species are inhibited by various mechanisms such as incomplete chromosome pairing, sterility of F1 hybrids, and embryo abortion (Broughton et al., 2003).
7. However, there are instances of wild-collected plants that represent hybrids between *P. vulgaris* and *P. coccineus*, indicating that some natural hybridization can occur (Escalante et al., 1994).
8. Experimental crosses have been attempted between *P. vulgaris* and several closely related species to harness disease resistance traits and abiotic stress tolerance (Beebe et al., 2014).
9. Hybrids resulting from these crosses often exhibit dwarfism and partial or complete sterility due to partial incompatibility between the species.

10. *P. vulgaris* plants can pollinate nearby wild plants, leading to fertile hybrids and the introduction of domestication traits into wild populations (Zizumbo-Villarreal et al., 2005).
11. Gene flow to cultivated varieties can also occur when wild plants act as the male parent, albeit at a lower frequency than when the cultivated variety acts as the male parent (Papa and Gept, 2003).
12. Hybridization between cultivated bean varieties and *P. vulgaris* results in viable, fertile offspring.
13. Natural outcrossing with wild species can occur at low to moderate levels under specific conditions, facilitated by insect pollinators like bumblebees (Delgado-Salinas et al., 1999).
14. Wild populations generally have high homozygosity (Kwak et al., 2009; Kwak and Gepts, 2009), making outcrossing rare (Pedrosa-Harand et al., 2009).

Overall, the occurrence and challenges of hybridization between cultivated and wild bean species, as well as the potential benefits and limitations associated with such hybridization, are significant areas of study.

HOW NOVEL BREEDING APPROACHES, INCLUDING GENOME EDITING, SPEED BREEDING, AND GENOMIC SELECTION, CAN BE USED FOR IMPROVING CULTIVATED COMMON BEAN

In the last century, 75% reduction in crop diversity in farmers' fields has been recorded, and climate change is making this problem more severe (Massawe et al., 2016).

ADVANCED BREEDING STRATEGY FOR COMMON BEAN

Uncertainty regarding the behaviour of the trait in the genetic background of domesticated bean germplasm, and a lack of information on the genetic foundation of the trait in the CWR were the major causes for under-utilization of the CWRs in common bean breeding (Porch et al., 2013). In this post-NGS era, we are blessed with the immense power of genomic resources (Table 5.3). Genome-wide high-throughput SNPs have been generated using DArTseq technology in common bean. This SNP collection may be utilized in genotyping platforms for a variety of purposes, including molecular breeding techniques for common beans (Valdisser et al., 2017a). NGS platforms have generated a huge wealth of genomic information for the common bean, which can be directly used for its improvement. Using genotyping-by-sequencing (GBS) techniques, unique genes/alleles for any given characteristic may be found (Schmutz et al., 2014; Valdisser et al., 2017). Genome-wide association studies (GWAS), which statistically correlate DNA polymorphisms with trait variations in a variety of germplasms that have been genotyped and phenotyped for traits of interest, can also be used to pinpoint the genomic regions governing particular traits. NGS and GWAS together improve mapping resolution for pinpointing the location of genes, alleles, and QTL (Ma et al., 2012; Varshney et al., 2014). Genomic resources of common bean and its related species are available at LegumeInfo (*https://legumeinfo.org*) and Phytozome (*https://phytozome-next.jgi.doe.gov*).

TABLE 5.3
Genome assemble information of different Phaseolus species.

Common name	Organism	Assembly
Tepary bean	*Phaseolus acutifolius*	HA v1.0
Tepary bean	*Phaseolus acutifolius WLD*	U Saskatchewan
Lima bean	*Phaseolus lunatus*	Uniandes v1
Common bean	*Phaseolus vulgaris*	JGI v2.0
Common bean	*Phaseolus vulgaris 5-593*	HA v1.0
Common bean	*Phaseolus vulgaris Labor Ovalle*	HA v1.0
Common bean	*Phaseolus vulgaris UI111*	JGI v1.0

Two major strategies have been envisioned to improve common beans. Firstly, employing a collection of 30 wild common bean genotypes or sister species to introduce abiotic stress tolerance into common beans. The second strategy will involve working concurrently with existing common bean populations and germplasm derived from earlier produced hybrids with wild cousins and sister species. This dual strategy offers both higher- and lower-risk options, which might lead to comparable improvements over a very brief period of 10 years (Porch et al., 2013). A Comprehensive genomic-assisted breeding strategy has been suggested by (Varshney et al., 2021), for futuristic crops. This strategy utilizes superior germplasm identified by the multi-omics approach and uses it for haplotype-based breeding. It also considers genomic prediction and optimum contributions selection (OCS), which maintain a balance between the rate of genetic gain and genetic diversity/inbreeding, along with speed breeding to hasten the breeding time.

GENOMIC SELECTION

Genomic selection (GS) uses two separate but related populations, the so-called training and breeding populations, to quantify marker effects over the whole genome of the target population. Depending on the results of genomic estimated breeding values (GEBVs), the breeding population will be the subject of the selection choice (Desta and Ortiz, 2014). GS employs all molecular markers for GP of the performance of the candidates for selection, in contrast to QTL and association mapping. In order to anticipate breeding and/or genetic qualities, GS is used. To determine the GEBVs of individuals in a breeding population (BP) that have been genotyped but not phenotyped, GS integrates molecular and phenotypic data from a training population (TP). In a GS programme, there are two fundamental populations: the TP data, whose genotype and phenotype are known, and the BP data, whose genetic values must be hypothesized. For a few rounds of selection, GS is utilized in place of phenotyping. The cost per cycle and the amount of time needed for variety generation are significantly reduced for GS over phenotype-based selection in breeding (Crossa et al., 2017). Genomic prediction has been employed in a panel of elite Andean breeding lines of common bean for diverse agronomic traits in different locations under irrigated, drought, and low phosphorus conditions to optimize GS model parameters, G × E effect has been considered and stimulation was done for breeding application (Keller et al., 2020). In another study genetic diversity, population structure, and linkage disequilibrium (LD) of Brazilian common bean accessions have been analyzed, which could further lay the path for GWAS and GS (Delfini et al., 2021). The predictive accuracy of genetic prediction models for cooking time is assessed using genomic prediction methods, with good prediction accuracy for the Mesoamerican MAGIC population (Diaz et al., 2021). In an African bean panel of 358 genotypes, the GS was utilized to forecast GEBVs for grain yield, cooking time, Fe, and Zn content (Saradadevi et al., 2021).

HIGH-THROUGHPUT PHENOTYPING/PHENOMICS

The development of high-throughput phenotyping (HTP) in CB has frequently lagged behind genetic advancement. It is still difficult and expensive to do exact phenotyping of basic and complex characteristics like plant height, biomass, blooming, and yield for a large BP with repeatable experiments across many settings. Many of the plant measurements needed for this are time-sensitive and depend on the development stage (Assefa et al., 2019). Some high throughput and nondestructive methods of phenotyping have been developed, creating great opportunities in crop improvement. Genetic gains in plants could be enhanced severalfold by combining targeted genomics approaches and high throughput phenotyping (Varshney et al., 2018). To conduct phenotyping at a ground level, it may be simple to employ ground-based HTP, such as tractor-based systems or specialized vehicles like pheno-mobiles and pheno-carts equipped with GPS and sensors. Some of the key developments in common bean phenotyping in recent times are listed below (Table 5.4).

TABLE 5.4
The key developments in the common bean phenotyping in recent time

Phenotyping Platform/ Technique Used	Features	References
Legume shovelomics	• High-throughput phenotyping of common bean and cowpea root architecture in the field • Integrated protocol made up of visual scoring, manual measurements, and image analysis	Burridge et al. (2016)
Chlorophyll fluorescence imaging	A method of image analysis for measuring the amount of leaf surface affected by the disease in common bean	Rousseau et al. (2013)
Multispectral imaging, 3D scanning, and chlorophyll fluorescence measurements	Nutrients (nitrogen, phosphorus, potassium, magnesium, and iron) deficiency in common bean at early and prolonged deficiency stage	Lazarević et al. (2022)
Scanalyzer PL semi-automated platform imaging system with a visible or RGB platform	Nondestructive phenotyping in water stress condition	Padilla-Chacón et al. (2019)

(*Continued*)

TABLE 5.4 (*Continued*)
The key developments in the common bean phenotyping in recent time

Phenotyping Platform/ Technique Used	Features	References
TSWIFT (Tower Spectrometer on Wheels for Investigating Frequent Timeseries)	For high-throughput phenotyping, continuous and automated hyperspectral reflectance monitoring is used to detect variations in plant structure and function at high spatial and temporal resolutions.	Wong et al. (2023)
Unmanned aerial system and satellite-based high-resolution imagery	Remote sensing methods for high-throughput plant phenotyping in dry bean	Sankaran et al. (2019)
Magnetic resonance imaging (MRI), automated fluorescence imaging (PAM fluorometry) in combination with automated shape detection, imaging spectrometers	The efficiency of resource usage in plants, mechanisms involved in resource allocation, transport, and below-ground roots were analyzed non-invasively.	Rascher et al. (2011)

SPEED BREEDING

Through quick generation progress, "speed breeding" (SB) shortens the breeding cycle and speeds up agricultural research. There are several ways to implement SB, one of which is to lengthen the amount of time that plants are exposed to light each day together with early seed harvesting to promote rapid seed-to-seed cycling and shorten generation lengths for some crops (Ghosh et al., 2018). SB protocols have been optimized in various grain legume, such as chickpea and lentil, and faba bean by extending photoperiod with light-emitting diode (LED) illumination (Mitache et al., 2023). Seed-to-seed cycle has also been reduced for temperate annual pasture legumes by pod drying at higher temperatures, along with scarification and hormonal treatment to break physiological dormancy. These kinds of manipulations of photoperiod and reduction in dormancy could greatly assist common bean improvement. An *in vitro–in vivo* SB technique in pea (Pisum sativum L.) has been reported, consisting of hydroponics, temperature control, hormonal control, 22-hour photoperiod, and early harvest, resulting in up to five generations per year (Cazzola et al., 2020). Rapid generation advancement, up to seven generations, has also been reported in chickpea (Samineni et al., 2020). In corporation of speed, breed is very much needed for common bean improvement programmes. Seed breeding programmes are still not well optimized for common bean and its CWR.

CRISPR-CAS GENOME EDITING IN *P. VALGARIS*

CRISPR (clustered regularly interspaced short palindromic repeats)-associated protein (Cas), a bacterial or archaeal protein, has been optimized to target and edit specific genomic sequences with a small guide RNA (sgRNA) (Jinek et al., 2012). This system virtually can target any sequence having only NGG, PAM (Protospacer adjacent motif) for target recognition, making genome editing the simplest, most versatile, and precise method of genetic manipulation in plants (Rasheed et al., 2022). A solid basis has been created for using innovative breeding methods such as GS and genome editing for crop improvement by recent improvements in genomic resources for numerous legume crops (Bhowmik et al., 2021). However, the significant application of CRISPR-Cas and transgenic technologies in common beans is seriously lagging. One of the reasons behind this lag is the lack of dependable regeneration and transformation protocol for both common bean and most of its CWR. Transformation protocol by particle bombardment of apical meristems, *in vitro* culture, and selection of transgenic plants has been reported in common bean (Rech et al., 2008). Recently, an indirect regeneration method has also been reported for common beans (Xiong et al., 2023). CRISPR/Cas9 and its associated technology can be utilized for nutritionally enriched, abiotic, and biotic stress-resistant varieties of common beans to ensure nutritional and food security. *Agrobacterium rhizogenes*-mediated nodule transformation for functional validation of genes related to the ureide biosynthesis pathway has been performed with success (Voß et al., 2022). Bean CWR can be utilized to diversify our food basket by altering domestication-related genes by CRISPR/Cas technologies to attain de novo domestication of CWR (Curtin et al., 2022; Yu and Li, 2022). Various domestication-related genes could be manipulated in this regard. For example, pectin acetylesterase-8 that induces seed dormancy in wild beans (Soltani et al., 2021), *COL2* related to flowering repression (González et al., 2021), and *PHYA3* in photosensitivity (Weller et al., 2019). Hard-to-cook (HTC) defect is characterized by prolonged cooking of common beans controlled by pectin methyl-esterase (PME) and PME inhibitor (PMEI) along with other factors. PMEs promote fast-cooking phenotype in *P. vulgaris*, and PMEI inhibits PMEs resulting in slow-cooking. The inhibition of PMEI activity by CRISPR-Cas9 constructs has been attempted to overcome HTC defects (Toili, 2022). CRISPR/Cas9 also opens up new possibilities for improving the sulfur-containing amino acids methionine (Met) and cysteine (Cys) in pulses (Warsame et al., 2018). Increasing nutrient bioavailability and removal of antinutritional factors from common beans could be interesting areas for future common bean improvement by CRISPR/Cas technology (Losa et al., 2022). Further, CRISPR-related genome editing toolbox based on transgene-free approaches could successfully eliminate regulatory hurdles related to genome-edited plants, thereby smoothing common bean improvement (Bhattacharjee et al., 2023).

CONCLUSION

Several factors limit the yield of common beans, including insect predation, diseases, and abiotic stressors. In the course of numerous ongoing research efforts, biotechnological approaches are being explored to address these factors. Despite the challenge of transforming and regenerating common beans (Veltcheva et al., 2005; Bonfim et al., 2007), beans have been successfully transformed using *Agrobacterium tumefaciens* and biolistic methods (Faria et al., 2014; Faria, Carneiro and Aragão, 2010; Kwapata, Nguyen and Sticklen, 2012). There have also been attempts to transform beans to be resistant to the bean golden mosaic virus (Faria et al., 2014; Aragão et al., 2013). RNA interference was used to mediate resistance, and the target of interference was the AC1 viral gene, which encodes a protein involved in virus replication (Bonfim et al., 2007). A transgenic bean event resistant to bean golden mosaic virus was approved for commercial cultivation in Brazil in 2011 (Calvalho et al., 2015). A significant amount of work has been performed on sequencing and assembling the bean genome, with approximately 80% of the genome already sequenced (Schmutz et al., 2014).

To expedite the introduction of CWR traits for common bean improvement, sufficient background information has been developed. To facilitate the breeding of wild relatives, gaps in knowledge must be filled. Wild relatives need to be phenotyped in an improved way to identify traits associated with stress tolerance. It is, therefore, necessary to develop both low and high-throughput methods of evaluating drought and heat, as well as associated traits, such as root, photosynthetic, and seed and pod formation characteristics. Defining appropriate environments for testing abiotic stress-related traits in the field and greenhouses/growth chambers is essential. It is essential to develop reproducible, effective, and efficient transformation methods for common beans. To identify the wild genome during breeding programmes, genomic methods need to be developed. The application of genomic tools to the breeding process needs to be determined through the development of decision-making protocols. It will be possible to identify and characterize the molecular traits from sister species or wild relatives that can be transmitted to the common bean, as opposed to the molecular traits that are difficult to transfer. Hybridization will be facilitated by assembling a set of effective parents for wild crosses and interspecific crosses, and their corresponding wild relative genotypes. There are several potential limitations to these approaches, including the lack of passport information for most of these accessions, the long generation time for seed production, and the limited supply of seeds. The progress in introgression from wild relatives can be facilitated by research efforts to fill these gaps in knowledge or techniques. A wide range of expertise is required for this research, including systematics, botany, breeding, genetics, and molecular biology. It is expected that several valuable resources and tools will be generated as a result of this research direction. The research involves identifying characterized germplasm derived from wild relatives, developing germplasm lines that have novel traits for release and use in breeding programmes, genotyping wild accessions, structured populations (e.g., NILs, biparental populations), advanced populations with wild relative introgressions, and identifying drought, heat, and disease-related markers.

REFERENCES

Assefa, T., Assibi Mahama, A., Brown, A. V., Cannon, E. K. S., Rubyogo, J. C., Rao, I. M., Blair, M. W., & Cannon, S. B. (2019). A review of breeding objectives, genomic resources, and marker-assisted methods in common bean (*Phaseolus vulgaris* L.). *Molecular Breeding*, *39*(2), 1–23. https://doi.org/10.1007/S11032-018-0920-0/FIGURES/3

Aragão, F. J., Nogueira, E. O., Tinoco, M.L.P. & Faria, J. C. (2013). Molecular characterization of the first commercial transgenic common bean immune to the Bean golden mosaic virus. *Journal of Biotechnology*, *166*(1-2), 42–50. https://doi.org/10.1016/j.jbiotec.2013.04.009

Beebe, S., Toro Ch., O., González, A. V., Chacón, M. I., & Debouck, D. G. (1997). Wild-weed-crop complexes of common bean (*Phaseolus vulgaris* L., Fabaceae) in the Andes of Peru and Colombia, and their implications for conservation and breeding. *Genetic Resources and Crop Evolution*, *44*(1), 73–91. https://doi.org/10.1023/A:1008621632680/METRICS

Beebe, S. E., Rao, I. M., Devi, M. J., & Polania, J. (2014). Common beans, biodiversity, and multiple stresses: challenges of drought resistance in tropical soils. *Crop and Pasture Science*, *65*(7), 667–675. https://doi.org/10.1071/CP13303

Bhattacharjee, S., Bhowmick, R., Kant, L., & Paul, K. (2023). Strategic transgene-free approaches of CRISPR-based genome editing in plants. *Molecular Genetics and Genomics*, *298*(3), 507–520. https://doi.org/10.1007/S00438-023-01998-3/TABLES/1

Bhowmik, P., Konkin, D., Polowick, P., Hodgins, C. L., Subedi, M., Xiang, D., Yu, B., Patterson, N., Rajagopalan, N., Babic, V., Ro, D. K., Tar'an, B., Bandara, M., Smyth, S. J., Cui, Y., & Kagale, S. (2021). CRISPR/Cas9 gene editing in legume crops: opportunities and challenges. *Legume Science*, *3*(3), e96. https://doi.org/10.1002/LEG3.96

Bitocchi, E., Bellucci, E., Giardini, A., Rau, D., Rodriguez, M., Biagetti, E., Santilocchi, R., Spagnoletti Zeuli, P., Gioia, T., Logozzo, G., Attene, G., Nanni, L., & Papa, R. (2013). Molecular analysis of the parallel domestication of the common bean (Phaseolus vulgaris) in Mesoamerica and the Andes. *New Phytologist*, *197*(1), 300–313. https://doi.org/10.1111/J.1469-8137.2012.04377.X

Blair, M. W., Soler, A., & Cortés, A. J. (2012). Diversification and population structure in common beans (*Phaseolus vulgaris* L.). *PLoS One*, *7*(11), e49488. https://doi.org/10.1371/JOURNAL.PONE.0049488

Bonfim, K., Faria, J.C., Nogueira, E. O., Mendes, É. A. & Aragão, F. J., (2007). RNAi-mediated resistance to Bean golden mosaic virus in genetically engineered common bean (Phaseolus vulgaris). *Molecular Plant-Microbe Interactions*, *20*(6), 717–726. https://doi.org/10.1094/MPMI-20-6-0717

Broughton, W. J., Hernández, G., Blair, M., Beebe, S., Gepts, P., & Vanderleyden, J. (2003). Beans (Phaseolus spp.) - model food legumes. *Plant and Soil*, *252*(1), 55–128. https://doi.org/10.1023/A:1024146710611/METRICS

Brücher, H. (1988). The Wild Ancestor of *Phaseolus Vulgaris* in South America. In: Gepts, P. (eds) *Genetic Resources of Phaseolus Beans. Current Plant Science and Biotechnology in Agriculture*. Springer, Dordrecht, pp. 185–214. https://doi.org/10.1007/978-94-009-2786-5_10

Burridge, J., Jochua, C. N., Bucksch, A., & Lynch, J. P. (2016). Legume shovelomics: high-Throughput phenotyping of common bean (Phaseolus vulgaris L.) and cowpea (Vigna unguiculata subsp, unguiculata) root architecture in the field. *Field Crops Research*, *192*, 21–32. https://doi.org/10.1016/J.FCR.2016.04.008

Cazzola, F., Bermejo, C. J., Guindon, M. F., & Cointry, E. (2020). Speed breeding in pea (Pisum sativum L.), an efficient and simple system to accelerate breeding programs. *Euphytica*, *216*(11), 1–11. https://doi.org/10.1007/S10681-020-02715-6/TABLES/8

Carvalho, J. L., de Oliveira Santos, J., Conte, C., Pacheco, S., Nogueira, E. O., Souza, T. L., Faria, J.C. and Aragão, F.J. (2015). Comparative analysis of nutritional compositions of transgenic RNAi-mediated virus-resistant bean (event EMB-PV051-1) with its non-transgenic counterpart. *Transgenic Research*, *24*(5), 813–819. https://doi.org/10.1007/s11248-015-9877-5

Chacón, S. M. I., Pickersgill, B., & Debouck, D. G. (2005). Domestication patterns in common bean (Phaseolus vulgaris L.) and the origin of the Mesoamerican and Andean cultivated races. *Theoretical and Applied Genetics*, *110*(3), 432–444. https://doi.org/10.1007/S00122-004-1842-2/FIGURES/4

Chacón, S. M. I., Pickersgill, B., Debouck, D. G., & Arias, J. S. (2007). Phylogeographic analysis of the chloroplast DNA variation in wild common bean (Phaseolus vulgaris L.) in the Americas. *Plant Systematics and Evolution*, *266*(3), 175–195. https://doi.org/10.1007/S00606-007-0536-Z

Crossa, J., Pérez-Rodríguez, P., Cuevas, J., Montesinos-López, O., Jarquín, D., de los Campos, G., Burgueño, J., González-Camacho, J. M., Pérez-Elizalde, S., Beyene, Y., Dreisigacker, S., Singh, R., Zhang, X., Gowda, M., Roorkiwal, M., Rutkoski, J., & Varshney, R. K. (2017). Genomic selection in plant breeding: methods, models, and perspectives. *Trends in Plant Science*, *22*(11), 961–975. https://doi.org/10.1016/j.tplants.2017.08.011

Curtin, S., Qi, Y., Peres, L. E. P., Fernie, A. R., & Zsögön, A. (2022). Pathways to de novo domestication of crop wild relatives. *Plant Physiology*, *188*(4), 1746–1756. https://doi.org/10.1093/PLPHYS/KIAB554

Debouck, D. G. (1999). Diversity in phaseolus species in relation to the common bean. In: Singh, S. P. (ed) *Common Bean Improvement in the Twenty-First Century*. Kluwer Academic Publishers, Dordrecht, The Netherlands, pp. 25–52. https://doi.org/10.1007/978-94-015-9211-6_2

Delfini, J., Moda-Cirino, V., dos Santos Neto, J., Ruas, P. M., Sant'Ana, G. C., Gepts, P., & Gonçalves, L. S. A. (2021). Population structure, genetic diversity and genomic selection signatures among a Brazilian common bean germplasm. *Scientific Reports*, *11*(1), 1–12. https://doi.org/10.1038/s41598-021-82437-4

Delgado-Salinas, A., Turley, T., Richman, A., & Lavin, M. (1999). Phylogenetic analysis of the cultivated and wild species of Phaseolus (Fabaceae). *Systematic Botany*, *24*(3), 438–460. https://doi.org/10.2307/2419699

Delgado-Salinas, A., Bibler, R., & Lavin, M. (2006). Phylogeny of the genus Phaseolus (Leguminosae): a recent diversification in an ancient landscape. *Systematic Botany*, *31*(4), 779–791. https://doi.org/10.1600/036364406779695960

Desta, Z. A., & Ortiz, R. (2014). Genomic selection: genome-wide prediction in plant improvement. *Trends in Plant Science*, *19*(9), 592–601. https://doi.org/10.1016/j.tplants.2014.05.006

Diaz, S., Ariza-Suarez, D., Ramdeen, R., Aparicio, J., Arunachalam, N., Hernandez, C., Diaz, H., Ruiz, H., Piepho, H. P., & Raatz, B. (2021). Genetic architecture and genomic prediction of cooking time in common bean (Phaseolus vulgaris L.). *Frontiers in Plant Science*, *11*, 622213. https://doi.org/10.3389/FPLS.2020.622213/BIBTEX

Escalante, A. M., Coello, G., Eguiarte, L. E., & Piñero, D. (1994). Genetic structure and mating systems in wild and cultivated populations of Phaseolus coccineus and P. vulgaris (Fabaceae). *American Journal of Botany*, *81*(9), 1096–1103. https://doi.org/10.1002/J.1537-2197.1994.TB15603.X

de Faria, L. C., Melo, P. G. S., Pereira, H. S., Wendland, A., Borges, S. F., Filho, I.A.P., Diaz, J.L.C., Calgaro, M. & Melo, L.C. (2014). Genetic progress during 22 years of black bean improvement. *Euphytica*, *199*(3), 261–272. https://doi.org/10.1007/s10681-014-1135-z

Faria, J. C., Carneiro, G. E. & Aragão, F. J. (2010). Gene flow from transgenic common beans expressing the bar gene. *GM Crops*, *1*(2), 94–98. https://doi.org/10.4161/gmcr.1.2.11609

García, E. H., Peña-Valdivia, C. B., Aguirre, J. R. R., & Muruaga, J. S. M. (1997). Morphological and agronomic traits of a wild population and an improved cultivar of common bean (Phaseolus vulgaris L.). *Annals of Botany*, *79*(2), 207–213. https://doi.org/10.1006/ANBO.1996.0329

Gepts, P., Osborn, T. C., Rashka, K., & Bliss, F. A. (1986). Phaseolin-protein variability in wild forms and landraces of the common bean (Phaseolus vulgaris): evidence for multiple centers of domestication. *Economic Botany*, *40*(4), 451–468. https://doi.org/10.1007/BF02859659/METRICS

Ghosh, S., Watson, A., Gonzalez-Navarro, O. E., Ramirez-Gonzalez, R. H., Yanes, L., Mendoza-Suárez, M., Simmonds, J., Wells, R., Rayner, T., Green, P., Hafeez, A., Hayta, S., Melton, R. E., Steed, A., Sarkar, A., Carter, J., Perkins, L., Lord, J., Tester, M., & Hickey, L. T. (2018). Speed breeding in growth chambers and glasshouses for crop breeding and model plant research. *Nature Protocols*, *13*(12), 2944–2963. https://doi.org/10.1038/s41596-018-0072-z

González, A. M., Vander Schoor, J. K., Fang, C., Kong, F., Wu, J., Weller, J. L., & Santalla, M. (2021). Ancient relaxation of an obligate short-day requirement in common bean through loss of CONSTANS-like gene function. *Current Biology*, *31*(8), 1643–1652.e2. https://doi.org/10.1016/j.cub.2021.01.075

Graham, P. H., & Ranalli, P. (1997). Common bean (Phaseolus vulgaris L.). *Field Crops Research*, *53*(1–3), 131–146. https://doi.org/10.1016/S0378-4290(97)00112-3

Jinek, M., Chylinski, K., Fonfara, I., Hauer, M., Doudna, J. A., & Charpentier, E. (2012). A programmable dual-RNA-guided DNA endonuclease in adaptive bacterial immunity. *Science*, *337*(6096), 816–821. https://doi.org/10.1126/SCIENCE.1225829/SUPPL_FILE/JINEK.SM.PDF

Keller, B., Ariza-Suarez, D., de la Hoz, J., Aparicio, J. S., Portilla-Benavides, A. E., Buendia, H. F., Mayor, V. M., Studer, B., & Raatz, B. (2020). Genomic prediction of agronomic traits in common bean (Phaseolus vulgaris L.) under environmental stress. *Frontiers in Plant Science*, *11*, 543352. https://doi.org/10.3389/FPLS.2020.01001/BIBTEX

Kwak, M., & Gepts, P. (2009). Structure of genetic diversity in the two major gene pools of common bean (Phaseolus vulgaris L., Fabaceae). *Theoretical and Applied Genetics*, *118*(5), 979–992. https://doi.org/10.1007/S00122-008-0955-4/TABLES/4

Kwak, M., Kami, J. A., & Gepts, P. (2009). The putative mesoamerican domestication center of phaseolus vulgaris is located in the lerma-santiago basin of Mexico. *Crop Science*, *49*(2), 554–563. https://doi.org/10.2135/CROPSCI2008.07.0421

Kwapata, K., Nguyen, T. & Sticklen, M. (2012). Genetic transformation of common bean (Phaseolus vulgaris L.) with the gus color marker, the bar herbicide resistance, and the barley (Hordeum vulgare) HVA1 drought tolerance genes. *International Journal of Agronomy*, *2012*(1), 198960. https://doi.org/10.1155/2012/198960

Lazarević, B., Carović-Stanko, K., Živčak, M., Vodnik, D., Javornik, T., & Safner, T. (2022). Classification of high-throughput phenotyping data for differentiation among nutrient deficiency in common bean. *Frontiers in Plant Science*, *13*, 931877. https://doi.org/10.3389/FPLS.2022.931877/BIBTEX

Liebenberg, M. M., Madubanya, L. A., Mienie, C. M. S., & Kelly, J. D. (2009.). A closer look at the resistance gene cluster on common bean chromosome 11. Naldc.Nal.Usda.Gov. Retrieved July 7, 2023, from https://naldc.nal.usda.gov/download/IND44207135/PDF

Lioi, L., & Piergiovanni, A. R. (2013). European common bean. *Genetic and Genomic Resources of Grain Legume Improvement*, *2013*, 11–40. https://doi.org/10.1016/B978-0-12-397935-3.00002-5

Losa, A., Vorster, J., Cominelli, E., Sparvoli, F., Paolo, D., Sala, T., Ferrari, M., Carbonaro, M., Marconi, S., Camilli, E., Reboul, E., Waswa, B., Ekesa, B., Aragão, F., & Kunert, K. (2022). Drought and heat affect common bean minerals and human diet-what we know and where to go. *Food and Energy Security*, *11*(1), e351. https://doi.org/10.1002/FES3.351

Ma, Y., Qin, F., & Tran, L. S. P. (2012). Contribution of genomics to gene discovery in plant abiotic stress responses. *Molecular Plant*, *5*(6), 1176–1178. https://doi.org/10.1093/mp/sss085

Massawe, F., Mayes, S., & Cheng, A. (2016). Crop diversity: an unexploited treasure trove for food security. *Trends in Plant Science*, *21*(5), 365–368. https://doi.org/10.1016/j.tplants.2016.02.006

Miklas, P. N., Kelly, J. D., Beebe, S. E., & Blair, M. W. (2006). Common bean breeding for resistance against biotic and abiotic stresses: from classical to MAS breeding. *Euphytica*, *147*(1), 105–131. https://doi.org/10.1007/S10681-006-4600-5

Mitache, M., Baidani, A., Houasli, C., Khouakhi, K., Bencharki, B., & Idrissi, O. (2023). Optimization of light/dark cycle in an extended photoperiod-based speed breeding protocol for grain legumes. *Plant Breeding*, *142*(1), 1–14. https://doi.org/10.1111/PBR.13112

Myers, J. R., & Kmiecik, K. (2017). Common bean: economic importance and relevance to biological science research. In: Pérez de la Vega, M., Santalla, M., Marsolais, F. (eds) *The Common Bean Genome. Compendium of Plant Genomes*. Springer, Cham, pp. 1–20. https://doi.org/10.1007/978-3-319-63526-2_1

Padilla-Chacón, D., Peña Valdivia, C. B., García-Esteva, A., Cayetano-Marcial, M. I., & Kohashi Shibata, J. (2019). Phenotypic variation and biomass partitioning during post-flowering in two common bean cultivars (Phaseolus vulgaris L.) under water restriction. *South African Journal of Botany*, *121*, 98–104. https://doi.org/10.1016/J.SAJB.2018.10.031

Papa, R., & Gepts, P. (2003). Asymmetry of gene flow and differential geographical structure of molecular diversity in wild and domesticated common bean (Phaseolus vulgaris L.) from Mesoamerica. *Theoretical and Applied Genetics*, *106*(2), 239–250. https://doi.org/10.1007/S00122-002-1085-Z/METRICS

Pedrosa-Harand, A., Kami, J., Gepts, P., Geffroy, V., & Schweizer, D. (2009). Cytogenetic mapping of common bean chromosomes reveals a less compartmentalized small-genome plant species. *Chromosome Research*, *17*(3), 405–417. https://doi.org/10.1007/S10577-009-9031-4/FIGURES/4

Porch, T. G., Beaver, J. S., Debouck, D. G., Jackson, S. A., Kelly, J. D., & Dempewolf, H. (2013). Use of wild relatives and closely related species to adapt common bean to climate change. *Agronomy*, *3*(2), 433–461. https://doi.org/10.3390/AGRONOMY3020433

Rascher, U., Blossfeld, S., Fiorani, F., Jahnke, S., Jansen, M., Kuhn, A. J., Matsubara, S., Mrtin, L. L. A., Merchant, A., Metzner, R., Mller-Linow, M., Nagel, K. A., Pieruschka, R., Pinto, F., Schreiber, C. M., Temperton, V. M., Thorpe, M. R., Van Dusschoten, D., Van Volkenburgh, E., & Schurr, U. (2011). Non-invasive approaches for phenotyping of enhanced performance traits in bean. *Functional Plant Biology*, *38*(12), 968–983. https://doi.org/10.1071/FP11164

Rasheed, A., Barqawi, A. A., Mahmood, A., Nawaz, M., Shah, A. N., Bay, D. H., Alahdal, M. A., Hassan, M. U., & Qari, S. H. (2022). CRISPR/Cas9 is a powerful tool for precise genome editing of legume crops: a review. *Molecular Biology Reports*, *49*(6), 5595–5609. https://doi.org/10.1007/S11033-022-07529-4

Rech, E. L., Vianna, G. R., & Aragão, F. J. L. (2008). High-efficiency transformation by biolistics of soybean, common bean and cotton transgenic plants. *Nature Protocols*, *3*(3), 410–418. https://doi.org/10.1038/nprot.2008.9

Rousseau, C., Belin, E., Bove, E., Rousseau, D., Fabre, F., Berruyer, R., Guillaumès, J., Manceau, C., Jacques, M. A., & Boureau, T. (2013). High throughput quantitative phenotyping of plant resistance using chlorophyll fluorescence image analysis. *Plant Methods*, *9*(1), 1–13. https://doi.org/10.1186/1746-4811-9-17/FIGURES/5

Samineni, S., Sen, M., Sajja, S. B., & Gaur, P. M. (2020). Rapid generation advance (RGA) in chickpea to produce up to seven generations per year and enable speed breeding. *The Crop Journal*, *8*(1), 164–169. https://doi.org/10.1016/J.CJ.2019.08.003

Sankaran, S., Quirós, J. J., & Miklas, P. N. (2019). Unmanned aerial system and satellite-based high resolution imagery for high-throughput phenotyping in dry bean. *Computers and Electronics in Agriculture*, *165*, 104965. https://doi.org/10.1016/J.COMPAG.2019.104965

Saradadevi, R., Mukankusi, C., Li, L., Amongi, W., Mbiu, J. P., Raatz, B., Ariza, D., Beebe, S., Varshney, R. K., Huttner, E., Kinghorn, B., Banks, R., Rubyogo, J. C., Siddique, K. H. M., & Cowling, W. A. (2021). Multivariate genomic analysis and optimal contributions selection predicts high genetic gains in cooking time, iron, zinc, and grain yield in common beans in East Africa. *The Plant Genome*, *14*(3), e20156. https://doi.org/10.1002/TPG2.20156

Schmutz, J., McClean, P. E., Mamidi, S., Wu, G. A., Cannon, S. B., Grimwood, J., Jenkins, J., Shu, S., Song, Q., Chavarro, C., Torres-Torres, M., Geffroy, V., Moghaddam, S. M., Gao, D., Abernathy, B., Barry, K., Blair, M., Brick, M. A., Chovatia, M., & Jackson, S. A. (2014). A reference genome for common bean and genome-wide analysis of dual domestications. *Nature Genetics*, *46*(7), 707–713. https://doi.org/10.1038/NG.3008

Singh, S. P., & Schwartz, H. F. (2010). Breeding common bean for resistance to diseases: a review. *Crop Science*, *50*(6), 2199–2223. https://doi.org/10.2135/CROPSCI2009.03.0163

Singh, S. P., & Schwartz, H. F. (2011). Review: breeding common bean for resistance to insect pests and nematodes. *Canadian Journal of Plant Science*, *91*(2), 239–250. https://doi.org/10.4141/CJPS10002/ASSET/IMAGES/CJPS10002TAB1.GIF

Singh, S. P., Nodari, R., & Gepts, P. (1991). Genetic diversity in cultivated common bean: I. Allozymes. *Crop Science*, *31*(1), 19–23. https://doi.org/10.2135/CROPSCI1991.0011183X003100010004X

Soltani, A., Walter, K. A., Wiersma, A. T., Santiago, J. P., Quiqley, M., Chitwood, D., Porch, T. G., Miklas, P., McClean, P. E., Osorno, J. M., & Lowry, D. B. (2021). The genetics and physiology of seed dormancy, a crucial trait in common bean domestication. *BMC Plant Biology*, *21*(1), 1–17. https://doi.org/10.1186/S12870-021-02837-6/FIGURES/8

Spataro, G., Tiranti, B., Arcaleni, P., Bellucci, E., Attene, G., Papa, R., Zeuli, P. S., & Negri, V. (2011). Genetic diversity and structure of a worldwide collection of phaseolus coccineus L. *Theoretical and Applied Genetics*, *122*(7), 1281–1291. https://doi.org/10.1007/S00122-011-1530-Y/FIGURES/3

Toili, M. E. M. (2022). *Insights into the molecular mechanism of the hard-to-cook defect towards genetic improvement of common bean (phaseolus vulgaris L.) through crispr-cas9 gene editing optimization.* https://researchportal.vub.be/en/publications/insights-into-the-molecular-mechanism-of-the-hard-to-cook-defect-

Toro C. O., & Ocampo, C. H. (2004). *Additional evidence about wild-weed-crop complexes of common bean in different parts of Colombia.* Bean Improvement Cooperative. Annual Report (USA).

Valdisser, P. A. M. R., Pereira, W. J., Almeida Filho, J. E., Müller, B. S. F., Coelho, G. R. C., de Menezes, I. P. P., Vianna, J. P. G., Zucchi, M. I., Lanna, A. C., Coelho, A. S. G., de Oliveira, J. P., Moraes, A. da C., Brondani, C., & Vianello, R. P. (2017a). In-depth genome characterization of a Brazilian common bean core collection using DArTseq high-density SNP genotyping. *BMC Genomics*, *18*(1), 423. https://doi.org/10.1186/S12864-017-3805-4

Varshney, R. K., Terauchi, R., & McCouch, S. R. (2014). Harvesting the promising fruits of genomics: applying genome sequencing technologies to crop breeding. *PLoS Biology*, *12*(6), e1001883. https://doi.org/10.1371/JOURNAL.PBIO.1001883

Varshney, R. K., Thudi, M., Pandey, M. K., Tardieu, F., Ojiewo, C., Vadez, V., Whitbread, A. M., Siddique, K. H. M., Nguyen, H. T., Carberry, P. S., & Bergvinson, D. (2018). Accelerating genetic gains in legumes for the development of prosperous smallholder agriculture: integrating genomics, phenotyping, systems modelling and agronomy. *Journal of Experimental Botany*, *69*(13), 3293–3312. https://doi.org/10.1093/JXB/ERY088

Varshney, R. K., Bohra, A., Yu, J., Graner, A., Zhang, Q., & Sorrells, M. E. (2021). Designing future crops: genomics-assisted breeding comes of age. *Trends in Plant Science*, *26*(6), 631–649.https://doi.org/10.1016/J.TPLANTS.2021.03.010

Veltcheva, M., Svetleva, D., Petkova, S. P. & Perl, A. (2005). In vitro regeneration and genetic transformation of common bean (Phaseolus vulgaris L.)—Problems and progress. *Scientia Horticulturae*, *107*(1), 2–10. https://doi.org/10.1016/j.scienta.2005.07.005

Voß, L., Heinemann, K. J., Herde, M., Medina-Escobar, N., & Witte, C. P. (2022). Enzymes and cellular interplay required for flux of fixed nitrogen to ureides in bean nodules. *Nature Communications*, *13*(1), 1–13. https://doi.org/10.1038/s41467-022-33005-5

Warsame, A. O., O'Sullivan, D. M., & Tosi, P. (2018). Seed storage proteins of faba bean (Vicia faba L): current status and prospects for genetic improvement. *Journal of Agricultural and Food Chemistry*, *66*(48), 12617–12626. https://doi.org/10.1021/ACS.JAFC.8B04992/SUPPL_FILE/JF8B04992_SI_001.PDF

Weller, J. L., Vander Schoor, J. K., Perez-Wright, E. C., Hecht, V., González, A. M., Capel, C., Yuste-Lisbona, F. J., Lozano, R., & Santalla, M. (2019). Parallel origins of photoperiod adaptation following dual domestications of common bean. *Journal of Experimental Botany*, *70*(4), 1209–1219. https://doi.org/10.1093/JXB/ERY455

Wong, C. Y. S., Jones, T., McHugh, D. P., Gilbert, M. E., Gepts, P., Palkovic, A., Buckley, T. N., & Magney, T. S. (2023). TSWIFT: tower spectrometer on wheels for investigating frequent timeseries for high-throughput phenotyping of vegetation physiology. *Plant Methods*, *19*(1), 1–15. https://doi.org/10.1186/S13007-023-01001-5/FIGURES/7

Xiong, L., Liu, C., Liu, D., Yan, Z., Yang, X., & Feng, G. (2023). Optimization of an indirect regeneration system for common bean (Phaseolus vulgaris L.). *Plant Biotechnology Reports*, *1*, 1–13. https://doi.org/10.1007/S11816-023-00830-Z/FIGURES/11

Yu, H., & Li, J. (2022). Breeding future crops to feed the world through de novo domestication. *Nature Communications*, *13*(1), 1–4. https://doi.org/10.1038/s41467-022-28732-8

Zizumbo-Villarreal, D., Colunga-GarcíaMarín, P., Cruz, E. P. de la, Delgado-Valerio, P., & Gepts, P. (2005). Population structure and evolutionary dynamics of wild-weedy-domesticated complexes of common bean in a mesoamerican region. *Crop Science*, *45*(3), 1073–1083. https://doi.org/10.2135/CROPSCI2004.0340

6 Widening the Genetic Base Potential of Soybean Harnessing Wild Relatives

A Multidimensional Approach

Ajaz A. Lone, Asmat Ara, Munezeh Rashid, Z. A. Dar, Shamshir ul Hussan, M. H. Khan, M. Habib, Sanjay Gupta, V. Rajesh, L. Ahmad, Anshuman Singh, and S. A. Dar

INTRODUCTION

Changing dietary preferences, climatic change, and scarce water and land resources along with an expanding population are placing increased pressure on the food supply (Considine et al., 2017). Despite their modest output, grain legumes have superior nutritional content to cereals, making them a distinctive and crucial part of a balanced diet (Foyer et al., 2016). Plant breeding techniques desired for the selective breeding of preferred features have significantly reduced the genetic diversity of grain legumes. In the farmer's field, the beneficial traits of the improved kinds and landraces have genetically distinguished themselves from those of their ancestors or wild progenitors (Zhou et al., 2015). Agricultural development is constantly required to cope with the climatic changes brought on by global warming and limited water resources; the agricultural potential has been shrunk because of domestication, genetic bottlenecks, and selective breeding. Crop wild relatives are a sensible option for examining additional genes for different systems of production partly because there are few or no breeding barriers (Munoz et al., 2017). Sometimes, it is simple to find the wild ancestors of crops, but not always because some wild species have become extinct or because the genome of domesticated plants, like wheat, was influenced by several progenitors. Some species may be significantly extending the domesticated crops' genomes because they are related to the wild species that served as their ancestors or wild relatives (Brozynska et al., 2015). Wild crop relatives have evolved to endure biotic and abiotic pressures and are primarily adapted to more extreme climatic changes (Zhang et al., 2017).

DOI: 10.1201/9781003434535-6

The cultivated soybean (*Glycine max* L.) is a crop with significant economic worth that is grown worldwide for diverse purposes in addition to the creation of biofuels and the consumption of oil and protein for different life forms. Due to *G. max's* restricted genetic variation, it is currently not possible to breed soybean cultivars with significant levels of environmental stress tolerance and resilience. These issues with the existing cultivars are becoming more problematic due to the ecological changes brought on by climatic changes, as well as other causes like drought or salty conditions. Meanwhile, the production of soybeans faces challenging obstacles due to the quickly evolving pests and pathogens. Owing to the preceding reasons, soybean production must be raised promptly using the pioneer genetic potential from the wild ancestors of the domesticated soybean, *Glycine max*, descended from the wild soybean, *Glycine soja* (Siebold & Zucc.) *Glycine soja* can thrive in a variety of settings and is a native of East Asia with a wide geographic range that stretches from East Russia to South China. Despite having differing behavioural traits, *G. soja* and *G. max* have an identical number of chromosomes ($2n = 40$), display typical meiotic chromosomal pairing, and are cross-compatible (Figure 6.1) (Carter et al., 2004). Additionally, it was shown that *G. soja* has prominent genetic worth, but G. max may

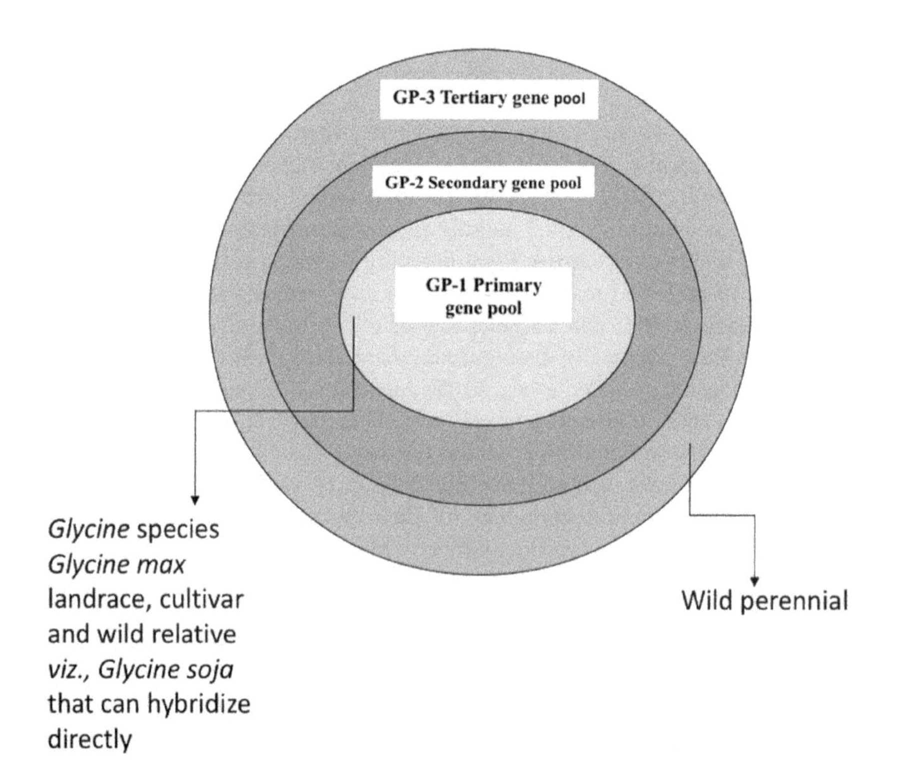

FIGURE 6.1 Using Harlan and de Wet's (1971) classification system, the gene pools for the cultivated soybean.

have lost this trait during domestication and refinement (Hyten et al., 2006). Given the larger amount of genetic potential still present in *G. soja* and its environmental adaptations, *G. soja* has significant potential to outperform its domesticated relative, which is crucial for agriculture (Stupar, 2010; Qiu et al., 2013; Qi et al., 2014; Zhang et al., 2017d). Utilizing crop wild relatives has helped tomato, barley, and wheat gain traits, especially for improved pest and disease resistance (Zhang and Batley, 2020).

Similar to this, scientists have worked to increase the gene pool (GP) of soybeans using wild relatives to circumvent the bottleneck effect of cultivation. The three main categories of GPs were identified by Harland and de Wet (1971) by considering the success rate of species hybridization (Figure 6.1). All landraces and cultivars of soybean, including *G. soja* Sieb. & Zucc., which is their wild parent, are included in GP-1, which has easily hybridizable germplasm. According to Dwivedi et al. (2008), the species in GP-2 can cross with GP-1 without producing sterile F1 offspring, but this category does not yet include any soybean species. The 26 perennial wild species that make up GP-3 are thought to be exclusive to Australia. These serve as the outer limits for potential GPs, as the GP-3 species and soybean F1 hybridization seeds created by crossing GP-3 species and soybeans need *in vitro* techniques to rescue lethality (Singh, 2017). Williams 82, a cultivar of soybean, was the subject of a reference genome released in 2010 by Schmutz et al. Reference genomes for its annual relatives, *G. soja* accessions IT182932 and W05, were published in 2011 by Kim et al. and in 2019 by Xie et al. As a result of the recent, rapid development of genome sequencing technology and the reduction in sequencing prices, more pan-genome analyses have been carried out for both farmed soybean plants and their wild counterparts, providing a plethora of genomic data for soybean improvement (Figure 6.2). In this section, we highlight recent genomic advances and helpful genes discovered in both annual and perennial wild soybean species (Table 6.1) that may be employed as important resources in soybean breeding efforts.

WILD SOYBEAN

Two subgenera, *Glycine Willd* and *Soja* (Moench) F. J. Hermann, make up the genus *Glycine.* Only two annual species, *G. soja* Sieb & Zucc (wild) and *G. max* (cultivated), out of the 28 species included in the two subgenera, are directly or indirectly used as food or feed (Guo et al., 2010). The wild species *G. soja* is thought to be the ancestor of the cultivated species G. max based on cytological, proteomic, and genomic data. Both annual herbaceous plants are of Asian ancestry, with the majority of their range being in Southeast and Far East Asia, which includes China, Korea, Japan, Taiwan, and Russia. Wild soybean contains a priceless genetic resource and a crucial GP, including genes and gene families responsible for the plant's increased levels of protein and oil as well as its tolerance to disease and extreme heat (Lam et al., 2015).

Domestication History

According to various studies (Carter et al., 2004; Li et al., 2010; Han et al., 2016), *G. soja* was domesticated between 6,000 and 9,000 years ago in areas of Central

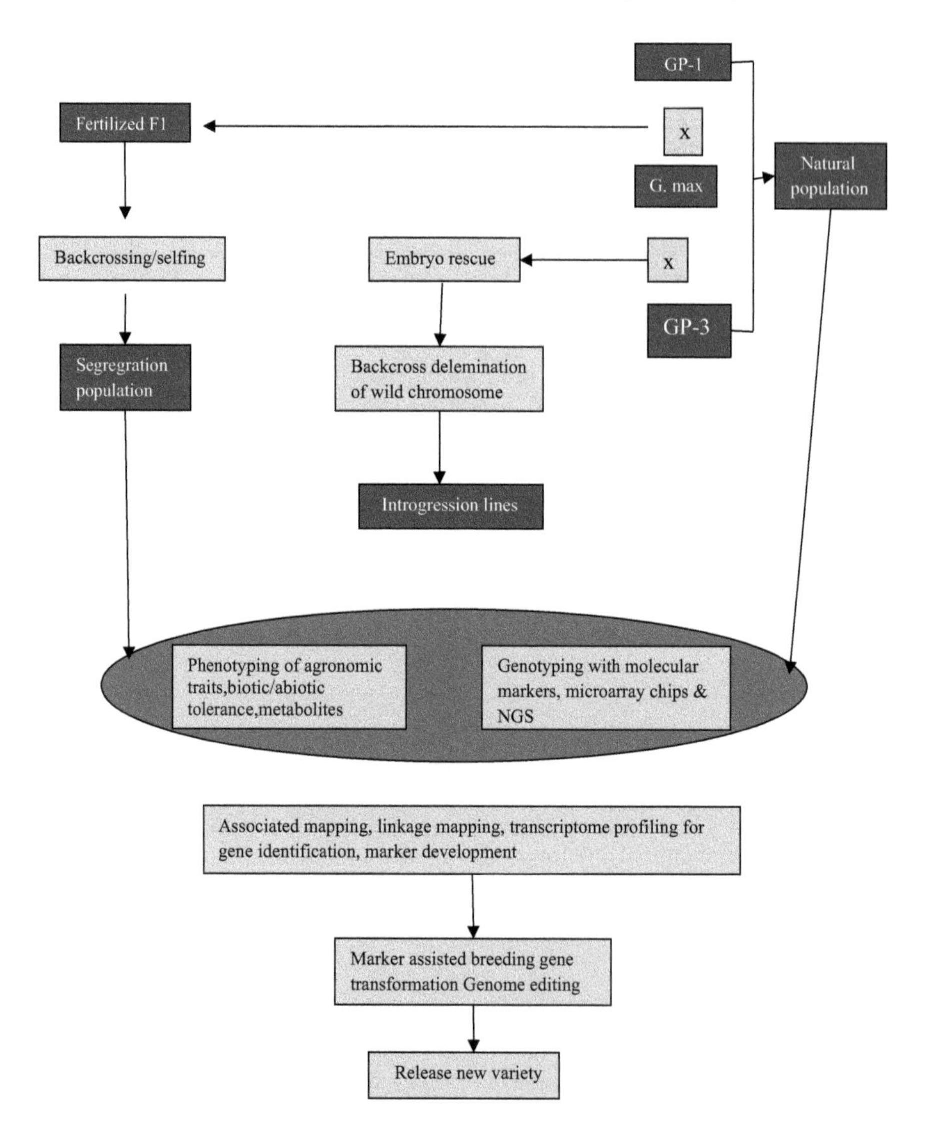

FIGURE 6.2 The use of wild soybean in *G. max* genetic improvement.

China near the Yellow River or Huang-Huai Valley. As a result, *G. max* landraces evolved, and subsequent selection produced the current (elite) farmed material. Three main hypotheses—the single origin theory, the multiple origin hypothesis, and the complex hypothesis—were recently summarized in a study of the history of domestication by Sedivy et al. The single origin hypothesis, which holds that *G. max* and *G. soja* diverged from one another from a single domestication event in Central China, no earlier than 9,000 years ago, is supported by the discovery that all the domesticated soybeans were clustered together by analyzing whole-genome single nucleotide polymorphism (SNP) of 302 wild, landrace, and cultivated soybeans (Zhou et al., 2015b).

TABLE 6.1
Genes and QTL's for Different Traits

Gene/Locus	Role	References
Rhg1	Resistance to Soybean cyst nematode	Melito et al. (2010)
cqSCN-006	Resistance to Soybean cyst nematode	Kim and Diers (2013)
Rag3b/c	Resistance to Soybean aphid	Zhang et al. (2017a)
Rag6	Resistance to Soybean aphid	Zhang et al. (2017b)
GsWRKY20	Drought stress	Ning et al. (2017)
GmSAIT3	Salinity stress	Guan et al. (2014)
Ncl2	Salinity stress	Lee et al. (2009)
GmCHX1	Salinity stress	Qi et al. (2014)
GmTIP2	Salinity stress	Zhang et al. (2017c)
Tof5	Regulation of flowering	Dong et al. (2022)
Tof11/12	Regulation of flowering	Lu et al. (2020)
Tof18	Regulation of flowering	Kou et al. (2022)
Tof1111	Regulation of flowering	Lu et al. (2017)
Tof112	Regulation of flowering	Lu et al. (2017)

Genetic Architecture of Soybean's Wild Relatives

The genomes of Glycine-grouped species were split into seven groups based on their ability to produce fertile hybrids and the degree to which meiotic chromosomes pair (Sherman-Broyles et al., 2014). *Glycine soja* Sieb. & Zucc., an annual wild relative of the cultivated soybean *G. max* (L.) Merr., belongs to the G genome group and has the same number of chromosomes ($2n=40$). The soybean species *G. soja* "IT182932" was revealed to have the first sequenced genome by Kim et al. (2010). Illumina short reads were used to build a genome that has 97.65% coverage of the published *G. max* "Williams 82" genome sequence. The two genomes showed 0.31% sequence variance, including single nucleotide changes and minor insertions/deletions, based on the comparison.

For the accession *G. soja* "W05," Xie et al. (2019) created the first *G. soja* reference-grade genome. This facilitated the discovery of significant structural differences and variations in gene copy number by combining PacBio long reads and Illumina short reads. They were able to successfully identify an inversion linked to seed coat colour during domestication through a genome-wide comparison of the wild species "W05" and the cultivated soybean accession "Williams 82," highlighting the significance of high-quality genome assembly. A significant step towards understanding the genetic diversity of annual soybeans as a species was made post the release of full genome sequences for 26 representative accessions. Also deeply re-sequenced were 2,898 accessions inclusive of 103 annual wild soybeans (Liu et al., 2020). Annual wild relatives of soybeans were compared to their domesticated counterparts, and this revealed various structural changes and potential genes related to domestication, such as seed coat colouration, offering useful information for future soybean breeding.

Wild Soybean Genome: Rediscovering the Lost Diversity

With the help of whole-genome sequencing of the wild soybean genome, we have acquired an improved knowledge of the history of soybean domestication, bottlenecks, lost diversity, and a future route towards its potential application in expanding the GP of soybean. The considerable and practical genetic differences between wild and domesticated soybeans reveal phenotypic variations in addition to domestication-related features. Kim et al. (2010) showed that the wild soybean genome encompassed 97.65% of the soybean genome with a difference of 35.2 Mb (3.76% of 937.5 Mb) after aligning the 915.4 Mb wild soybean genomic sequence with the soybean reference genome while removing the gaps. The different region contained 3.45% of big deleted sequences, 0.043% of insertion/deletions (indels), and 0.267% of substitution bases. Along with SNPs and indels, complex genome rearrangement is mostly produced by indels, inversions, and translocations (up to thousands of base pairs) (Kim et al., 2012). The allelic diversity of the wild soybean genome is higher than that of soybean. Higher allelic diversity was found after resequencing 17 wild and 14 farmed soybean genomes to an average depth of 5 and >90% coverage (Lam et al., 2015).

The genetic diversity (π) decreased from 2.94 103 in G. soja to 1.40 103 in landraces and to 1.05 103 in improved cultivars, indicating that nearly half of the annotated resistance-related sequences were lost during domestication, according to findings from the whole-genome sequencing of 302 wild and cultivated soybeans to an average depth of > x11. It is noteworthy that the overall number of SNPs in wild soybeans (5,924,662) and cultivated soybeans (4,127,942) was equivalent (35% and 5%, respectively), and the ratio of non-synonymous SNP to synonymous SNP was higher in cultivated soybeans (1.38) than in wild soybeans (1.36). These results imply that due to the domestication bottleneck, low-frequency alleles should exist in domesticated soybean. On the other hand, wild soybean has a high abundance of low-frequency alleles, indicating that the wild soybean habitat has shrunk and the cultivated soybean population has increased (Lam et al., 2015).

SOYBEAN IMPROVEMENT: GENE POOL ENHANCEMENT

The study of alternative production methods is considerably aided using wild soybeans as a possible genetic resource for the development of farmed soybeans. Similar to another wild ancestor of a crop species, wild soybeans have had more time to evolve and endure in a variety of environmental conditions without human interference, thus they have more genetic diversity (Munoz et al., 2017; Lam et al., 2015). Easily accessible or the main GP for soybean improvement, wild soybeans are cross-fertile with cultivated soybeans (Chung and Sigh, 2008). But as the population of the world and the environment change, it has become possible to secure, conserve, identify, and use wild soybeans as a resource for soybean improvement. Because of the fall of genetic potential experienced amid the domestication of soybeans and the existence of the domestication bottleneck known as domestication syndrome, the growth habits and mechanisms for dispersing seeds have changed (Munoz et al., 2017). This domestication has also made it possible for crop plants to endure and comply with modern farming and agriculture practices being extremely hopeful. However, the loss of genetic variety in farmed soybeans necessitates a review of natural diversity reservoirs, such as wild soybeans, to look for potential genes or alleles for increased production. There have been several sequences found that are specific to wild soybeans; however, a paper from Korea by Chung et al. (2014) also noted instances of gene loss in wild soybeans.

Resistance to Biotic Stress

The Asian-native soybean aphid (*Aphis glycines*) was brought to the United States in 2000. *A. glycines* infestations can directly damage soybean biomass and yield while also having an indirect impact on production due to the spread of the soybean mosaic virus (SMV). Three *G. soja* genotypes were found to be resistant to the soybean aphid after being screened for this trait (Hesler, 2013). Subsequent investigations used linkage mapping to find two more quantitative trait loci (QTLs), Rag3c and Rag6, linked to aphid resistance (Zhang et al., 2017e, f). Rag3c explains 12.5%–22.9% of the phenotypic variation across trials, whereas Rag6 explains 19.5%–46.4%. The creation of aphid-resistant soybean cultivars has been made possible by this unique aphid-resistance gene(s) obtained from *G. soja* (Zhang et al., 2017e). To determine the underlying mechanism, the two QTLs are required to be cloned using a map, and candidate genes are tested for functionality. Candidate genes inside the QTL regions are most likely to encode canonical nucleotide-binding site leucine-rich repeat (NBS-LRR)-containing proteins, which are a subset of the resistance (R) protein that has been discovered to be resistant to potato aphid (Rossi et al., 1998). Five additional aphid resistance genes had been discovered up to this point, all of which were mapped from *G. max* cultivars, but Raso2 differs from those obtained from *G. max* in that it exhibits potent antixenosis and antibiosis responses to the foxglove aphid (*Aulacorthum solani*) in the first report of its genetic identification in *G. soja* (Lee et al., 2015). This finding shows that *G. soja* may have a novel mechanism of *A. solani* resistance and calls for additional functional analysis.

It has been demonstrated that *G. soja* displays varied resistance to several soybean cyst nematode (SCN) populations presently harming crops in the United States (Kim et al., 2011; Zhang et al., 2016). The inquiry into the genetic processes giving resistance is currently ongoing, and a sizable portion of the *G. soja* accessions have not yet been evaluated for resistance. As of now, several studies using linkage mapping (Kim et al., 2011) and genome-wide association studies (Zhang et al., 2016, 2017d) have uncovered loci and candidate genes that may function distinctly from those so far identified in *G. max* conferring SCN resistance, suggesting species-specific resistance mechanisms in *G. soja*. A recent study found genotypes of *G. soja* resistant to soybean cyst nematode heterodera glycines (SCN HG) type 2.5.7 and identified novel QTLs in *G. soja* related to HG type 2.5.7 resistance using a genome-wide association assessment (Zhang et al., 2016).

An additional examination of the RNA-seq data using one of the resistant genotypes indicated a naturally vibrant defence regulation network involved in SCN resistance (Zhang and Song, 2017; Zhang et al., 2017b). It is essential to create soybean cultivars with broad-spectrum resistance to various SCN types due to the variety in virulence and quick progress of SCN populations. This can be done by finding additional resistance-causing agents and using gene stacking or genome editing to create soybean cultivars broadly resistant to SCN.

Abiotic Stress Tolerance

For present-day crops, soil salinity is becoming a bigger problem. The most significant crop for producing protein is the cultivated soybean, which has been grown in well-ploughed soil and is salt-sensitive. The finding of salt tolerance in *G. soja* was particularly important, and the first examination of this characteristic was conducted in 1997. However, only a shaky association between the genetic markers selected and the salt tolerance qualities has been established (Hu and Wang, 1997). Further analysis employing six randomly amplified polymorphic DNA markers (OPF05-213, OPF19-4361, OPF19-1727, OPF19-14000-, OPF19-700, and OPH02-1350) indicated relationships between salt tolerance in *G. soja* and these markers; however, the chromosomal locations of these markers are uncertain (Zhang et al., 1999). Ncl2, the first salt tolerance gene identified in *G. soja*, has not been found in *G. max* (Lee et al., 2009). A separate salt-tolerant gene, GmCHX1, was found after screening and sequencing of another wild soybean ecotype, W05 (Qi et al., 2014), revealing genotype-specific salt tolerance mechanisms. A wild soybean genotype named Tongyu06311 was shown to have a mechanism for salt tolerance controlled by amino acid and organic acid metabolism, where compatible solutes were stored rather than relying on the use of ATP (Yang et al., 2017). The aquaporin gene GmTIP2;1 was most recently discovered to be related to salt tolerance in *G. soja* (Zhang et al., 2017a). More aquaporin alleles may be associated with salt stress tolerance in *G. soja* given the significance of aquaporins in water transport.

Recently, the soybean overexpression method was used to discover and functionally confirm the drought tolerance gene GsWRKY20 from *G. soja*. In the same dry conditions, the transgenic soybean displays more yield, plant height, and root length than the non-transgenic plant. The stomatal density and closure rate are associated with the mechanisms controlling drought tolerance (Ning et al., 2017). This study offers viable options for farming in semi-arid and dry settings. Candidate SNPs connected to various environmental parameters, such as monthly precipitation, substrate sand percentage, and substrate silt %, were found in a landscape genomics study of *G. soja* (Anderson et al., 2016). These results infer that these loci may be involved in some types of abiotic stress tolerance; nevertheless, each of these genes must first undergo functional testing before being considered for soybean crop development.

Nutrition

Considering they are associated with cardiovascular disease, saturated fatty acids in soybeans should be minimized. In a GWAS study on the seed composition of *G. soja*, three novel markers related to the amounts of palmitic acid, a saturated fatty acid, were discovered; however, additional precision is needed to identify candidate genes. This result led to the identification of the possible candidate gene Glyma. 07G112100 for linoleic acid synthesis in *G. soja* (Leamy et al., 2017). Linoleic acid is a polyunsaturated fatty acid that is usually partially hydrogenated and associated with cardiovascular health. The nutritional profile of farmed soybeans may be greatly improved thanks to the recent discovery of a gene that controls both undesired fatty acids. Similar studies on *G. soja* found two candidate genes, Glyma.14G121400 and Glyma.16G068500, associated with the saturated fatty acid stearic acid as well as a putative gene linked to the "healthy" fatty acid unsaturated oleic acid (Leamy et al., 2017). These candidate genes were submitted with *G. max* gene IDs despite being discovered in *G. soja*. Additionally, it has been demonstrated that *G. soja* has higher seed protein content than *G. max* generally, probably as a result of selection for higher yield and oil content (Chen and Nelson, 2004; Leamy et al., 2017).

Traits Related to Yield

The primary agronomic trait selected during the domestication of soybeans was yield, and G. max is superior to its wild cousins in this regard. Additionally, by employing wild relatives of crops, the yields of a number of species have increased. Crop yield and photosynthetic efficiency are connected, particularly under salinity and cadmium stress conditions. Kao et al. (2003) investigated photosynthetic gas exchange and chlorophyll a fluorescence in three wild soybean species (*G. soja*) and two perennial *Glycine* species (*G. tomentella and G. tabacina*) under salt stress. These lines displayed variable levels of sensitivity; photosynthesis in perennial Glycine spp. was less vulnerable to salt stress than in *G. soja* and was least sensitive in *G. tomentella*. According to Chen et al. (2013) and Xue et al. (2014), the halophytic wild soybean "Dongying" (*G. soja* Sieb. & Zucc. "ZYD 03262") maintained better photosynthetic activity in a salt stress environment than farmed soybean. Furthermore, *G. soja* accumulated more Na in roots than various *G. max* accessions under salt stress, but noticeably less in leaves, according to Xue et al. (2014). In terms of photosynthetic activity, stomatal conductance, and carboxylation efficiency, salt stress therefore had a lesser effect on *G. soja* accessions than it did on *G. max* accessions.

Another significant agronomic characteristic is the control of photoperiodic flowering. Photoperiodic blooming is crucial for regional adaptation and production because early maturity is often linked to low yields (Lin et al., 2021). When the length of the day drops below a set threshold, short-day plants, including the cultivated soybean and its annual wild relative *G. soja*, flower (Sedivy et al., 2017). Narrow areas of latitude are where soybean cultivars are most suited for cultivation (Kenworthy, 1989; Ohigashi et al., 2019). When planted at low latitudes, varieties acclimated to high latitudes may blossom early; these plants are often short and have few pods. While cultivars acclimated to low latitude regions typically exhibit delayed blooming, which hastens vegetative growth for optimum production potential, such plants cannot complete their life cycles before winter when planted at high latitudes (Lin et al., 2021). At least 18 main loci, labelled E1 through E11, J, Tof5, Tof11, Tof12, Tof18, Tof1111, and Tof112, that are involved in the photoperiod response in soybean have been found using a combination of forward and reverse genetic methods (Xia et al., 2012; Watanabe et al., 2009, 2011; Cober, 2011; Kong et al., 2010; Zhai et al., 2014; Wang et al., 2019; Lu et al., 2020; Lin et al., 2021; Dong et al., 2022; Kou et al., 2022).

CONCLUSION AND FUTURE PROSPECTS

The wild soybean has made a significant contribution to crop improvement and supported studies on the evolution of the soybean. The domestication history of soybeans is now extensively known. Other beneficial genetic sources for the genetic improvement of legumes will become apparent with more research and the screening of new ecotypes for potential helpful features in *G. soja*. Collaborations and efforts should be made to complete what has already started, linking candidate gene identification with research on pertinent gene regulatory networks and finally, the practical and molecular techniques for crop development.

Soybean breeding faces a significant difficulty due to low genetic diversity. Only after developing mapping populations and identifying the alleles linked to advantageous agronomic features by QTL mapping, association analysis, combining multiple omics techniques, and comparative genomic analysis can wild relative resources be used to improve soybeans. Further in-depth investigation of the molecular processes underpinning phenotypic features is necessary for the use of wild soybeans, particularly the perennial *Glycine* spp., to improve soybean breeding. The development of reference sequences and resequencing of their genomes will enable us to more fully describe the genetic traits of wild soybean relatives. To create, evaluate, and distribute data, a concerted effort is needed. Only with the development of robust computing resources, bioinformatics tools, and the integration of databases for annual and perennial soybean will soy breeders be able to utilize the genetic resources that are already available. The preservation of wild soybean populations must also be encouraged to guarantee the survival of adaptable, wild features that could eventually be employed to enhance crop yields. To fully utilize this untapped genetic resource, international cooperation is required to conduct extensive, goal-directed research on *G. soja*.

REFERENCES

Anderson JE, Kono TJ, Stupar RM, Kantar MB, Morrell PL. 2016. Environmental association analyses identify candidates for abiotic stress tolerance in *Glycine soja*, the wild progenitor of cultivated soybeans. *G3 (Bethesda)*. 6, 835–843. https://doi.org/10.1534/g3.116.026914

Brozynska M, Furtado A, Henry RJ. 2015. Genomics of crop wild relatives: expanding the gene pool for crop improvement. *Plant Biotechnol. J.* 14, 1070–1085.

Carter TE, Hymowitz T, Nelson RL. 2004. Biogeography, local adaptation vavilov, and genetic diversity in soybean. In: Werner D. (ed) *Biological Resources and Migration*. Berlin, Heidelberg: Springer, pp. 47–59.

Chen YW, Nelson RL. 2004. Genetic variation and relationships among cultivated, wild, and semi-wild soybean. *Crop Sci.* 44, 316–325. https://doi.org/10.2135/cropsci2004.3160

Chen P, Yun K, Shao HB, Zhao SJ. 2013. Physiological mechanisms for high salt tolerance in wild soybean (*Glycine soja*) from Yellow River Delta, China: photosynthesis, osmotic regulation, ion flux and antioxidant capacity. *PLoS One*. 8, e83227.

Chung G, Singh, RJ. 2008. Broadening the genetic base of soybean: a multidisciplinary approach. *Crit. Rev. Plant Sci.* 5, 295–341.

Chung WH, Jeong N, Kim H, Lee WK, Lee YG, Lee SH, Yoon W, Kim JH, Choi IY, Choi H, Moon JK, Kim N, Jeong SC. 2014. Population structure and domestication revealed by high-depth resequencing of Korean cultivated and wild soybean genomes. *DNA Res.* 21, 153–167.

Cober ER. 2011. Long juvenile soybean flowering responses under very short photoperiods. *Crop Sci.* 51, 140–145.

Considine MJ, Siddique KHM, Foyer CH. 2017. Nature's pulse power: legumes, food security and climate change. *J. Exper. Bot.* 68, 1815–1818. https://doi.org/0.1093/jxb/erx099

Dong LD, Cheng Q, Fang C, Kong LP, Yang H, Hou ZH, Li YL, Nan HY, Zhang YH, Chen QS, Zhang CB, Kou K, Su T, Wang LS, Li SC, Li HY, Lin XY, Tang Y, Zhao XH, Lu SJ, Liu BH, Kong, FJ. 2022. Parallel selection of distinct Tof5 alleles drove the adaptation of cultivated and wild soybean to high latitudes. *Mol. Plant.* 15, 308–321.

Dwivedi S, Perotti E, Ortiz, R. 2008. Towards molecular breeding of reproductive traits in cereal crops. *Plant Biotechnol.* 6, 529–559.

Foyer CH, Lam HM, Nguyen HT, Siddique KHM, Varshney RK, Colmer TD, Cowling W, Bramley H, Mori TA, Hodgson JM, Cooper JW, Miller AJ, Kunert K, Vorster J, Cullis C, Ozga JA, Qahlqvist ML, Liang Y, Shou H, Shi K, Yu J, Fodor N, Kaiser BN, Wong FL, Valliyodan B, Considine MJ. 2016. Neglecting legumes has compromised human health and sustainable food production. *Nat. Plants.* 2, 16112. https://doi.org/10.1038/nplants.112

Guan RX, Chen JG, Jiang JH, Liu GY, Liu Y, Tian L, Yu LL, Chang RZ, Qiu LJ. 2014. Mapping and validation of a dominant salt tolerance gene in the cultivated soybean (*Glycine max*) variety Tiefeng 8. *Crop J.* 2, 358–365.

Guo J, Wang Y, Song C, Zhou J, Qiu L, Huang H, Wang Y. 2010. A single origin and moderate bottleneck during domestication of soybean (*Glycine max*): implications from microsatellites and nucleotide sequences. *Ann. Bot.* 106, 505–514.

Harlan JR, de Wet JMJ. 1971. Toward a rational classification of cultivated plants. *Taxon.* 20, 509–517.

Han Y, Zhao X, Liu D, Li Y, Lightfoot DA, Yang Z. 2016. Domestication footprints anchor genomic regions of agronomic importance in soybeans. *New Phytol.* 209, 871–884. https://doi.org/10.1111/nph.13626

Hesler LS. 2013. Resistance to soybean aphid among wild soybean lines under controlled conditions. *Crop Prot.* 53, 139–146. https://doi.org/10.1016/j.cropro.2013.06.016

Hu ZA, Wang HX. 1997. Salt tolerance of wild soybean (*Glycine soja*) in populations evaluated by a new method. *Soybean Genet. Newsl.* 24, 79–80.

Hyten DL, Song Q, Zhu Y, Choi IY, Nelson RL, Costa JM. 2006. Impacts of genetic bottlenecks on soybean genome diversity. *Proc. Natl. Acad. Sci. U.S.A.* 103, 16666–16671. https://doi.org/10.1073/pnas.0604379103

Kao WY, Tsai TT, Shih CN. 2003. Photosynthetic gas exchange and chlorophyll a fluorescence of three wild soybean species in response to NaCl treatments. *Photosynthetica.* 41, 415–419.

Kenworthy WJ. 1989. Potential genetic contributions of wild relatives to soybean improvements. In: Pascale AJ (ed) World Research Conference. Argentina: Assoc. Soja, Buenos Aires, pp. 883–888.

Kim M, Diers BW. 2013. Fine mapping of the SCN resistance QTL cqSCN-006 and cqSCN-007 from *Glycine soja* PI 468916. *Crop Sci.* 53, 775–785.

Kim MY, Lee S, Van K, Kim TH, Jeong SC, Choi IY, Kim DS, Lee YS, Park D, Ma JX, Kim WY, Kim BC, Park S, Lee KA, Kim DH, Kim KH, Shin JH, Jang YE, Kim KD, Liu WX, Chaisan T, Kang YJ, Lee YH, Kim KH, Moon JK, Schmutz J, Jackson SA, Bhak J, Lee SH. 2010. Whole-genome sequencing and intensive analysis of the undomesticated soybean (*Glycine soja* Sieb. and Zucc.) genome. *Proc. Natl. Acad. Sci. USA.* 107, 22032–22037

Kim M, Hyten DL, Niblack TL, Diers, BW. 2011. Stacking resistance alleles from wild and domestic soybean sources improves soybean cyst nematode resistance. *Crop Sci.* 51, 934–943.

Kim MY, Van K, Kang YJ, Kim KH, Lee SH. 2012. Tracing soybean domestication history: From nucleotide to genome. *Breed. Sci.* 61, 445–452.

Kong FJ, Liu BH, Xia ZJ, Sato S, Kim BM, Watanabe S, Yamada T, Tabata S, Kanazawa A, Harada K., Abe J. 2010. Two coordinately regulated homologs of flowering locus T are involved in the control of photoperiodic flowering in soybean. *Plant Physiol.* 154, 1220–1231.

Kou K, Yang H, Li HY, Fang C, Chen LY, Yue L, Nan HY, Kong LP, Li XM, Wang F, Wang JH, Du HP, Yang ZY, Bi YD, Lai YC, Dong LD, Cheng Q, Su T, Wang LS, Li SC, Hou ZH, Lu SJ, Zhang YH, Che ZJ, Yu DY, Zhao XH, Liu BH, Kong FJ. 2022. A functionally divergent SOC1 homolog improves soybean yield and latitudinal adaptation. *Curr. Biol.* 32, 1–15.

Lam HM, Xu X, Liu X, Chen W, Yang G, Wong FL, Li MW, He W, Qin N, Wang B, Li J, Jian M, Wang J, Shao G, Wang J, SSM S, Zhang G. 2015. Resequencing of 31 wild and cultivated soybean genomes identifies patters of genetic diversity and selection. *Nat. Genet.* 42, 1053–1059.

Leamy LJ, Zhang H, Li C, Chen CY, Song BH. 2017. A genome wide association study of seed composition traits in wild soybean (*Glycine soja*). *BMC Genomics.* 18, 18. https://doi.org/10.1186/s12864-016-3397-4

Lee JD, Shannon JG, Vuong TD, Nguyen HT. 2009. Inheritance of salt tolerance in wild soybean (*Glycine soja* Sieb. and Zucc.) accession PI483463. *J. Hered.* 100, 798–801.

Lee JS, Yoo MH, Jung JK, Bilyeu KD, Lee JD, Kang S. 2015. Detection of novel QTLs for foxglove aphid resistance in soybean. *Theor. Appl. Genet.* 128, 1481–1488. https://doi.org/10.1007/s00122-015-2519-8

Li YH, Li W, Zhang C, Yang L, Chang RZ, Gaut BS. 2010. Genetic diversity in domesticated soybean (*Glycine max*) and its wild progenitor (*Glycine soja*) for simple sequence repeat and single-nucleotide polymorphism loci. *New Phytol.* 188, 242–253.

Lin XY, Liu BH, Weller JL, Abe J, Kong FJ. 2021. Molecular mechanisms for the photoperiodic regulation of flowering in soybean. *J. Integr. Plant Biol.* 63, 981–994

Liu YC, Du HL, Li PC, Shen YT, Peng H, Liu SL, Zhou GA, Zhang HK, Liu Z, Shi M, Huang XH, Li Y, Zhang M, Wang Z, Zhu BG, Han B, Liang CZ, Tian ZX. 2020. Pan-genome of wild and cultivated soybeans. *Cell.* 182, 162–176.

Lu SJ, Zhao XH, Hu YL, Liu SL, Nan HY, Li XM, Fang C, Cao D, Shi XY, Kong LP, Su T, Zhang FG, Li SC, Wang Z, Yuan XH, Cober ER, Weller JL, Liu BH, Hou XL, Tian ZX, Kong FJ. 2017. Natural variation at the soybean J locus improves adaptation to the tropics and enhances yield. *Nat. Genet.* 49, 773–779.

Lu SJ, Dong LD, Fang C, Liu SL, Kong LP, Cheng Q, Chen LY, Su T, Nan HY, Zhang D, Zhang L, Wang ZJ, Yang YQ, Yu DY, Liu XL, Yang QY, Lin XY, Tang Y, Zhao XH, Yang XQ, Tian CE, Xie QG, Li X, Yuan XH, Tian ZX, Liu BH, Weller JL, Kong FJ. 2020. Stepwise selection on homeologous PRR genes controlling flowering and maturity during soybean domestication. *Nat. Genet.* 52, 428–436.

Melito S, Heuberger AL, Cook D, Diers BW, MacGuidwin AE, Bent AF. 2010. A nematode demographics assay in transgenic roots reveals no significant impacts of the Rhg1locus LRRKinase on soybean cyst nematode resistance. *BMC Plant Biol.* 10, 104

Munoz N, Liu A, Kan L, Li MW, Lam HM. 2017. Potential uses of wild germplasms of grain legumes for crop improvement. *Inter. J. Mol. Sci.* 18, 328.

Ning WF, Zhai H, Liang Yu JQS, Yang X, Xing XY, Huo JL, Pang T, Yang YL, Bai X. 2017. Overexpression of *Glycine soja* WRKY20 enhances drought tolerance and improves plant yields under drought stress in transgenic soybean. *Mol. Breed.* 37, 19.

Ohigashi K, Mizuguti A, Nakatani K, Matsuo K. 2019. Modeling the flowering sensitivity of five accessions of wild soybean (*Glycine soja*) to temperature and photoperiod, and its latitudinal cline. *Breed. Sci.* 69, 84–93.

Qi XP, Li MW, Xie M, Liu X, Ni M, Shao GH, Song C, Yim AKY, Tao Y, Wong FL, Isobe S, Wong CF, Wong KS, Xu CY, Li CQ, Wang Y, Guan R, Sun FM, Fan GY, Xiao ZX, Zhou F, Phang TH, Liu X, Tong SW, Chan TF, Yiu SM, Tabata S, Wang J, Xu X, Lam HM. 2014. Identification of a novel salt tolerance gene in wild soybean by whole-genome sequencing. *Nat. Commun.* 5, 4340.

Qiu LJ, Xing LL, Guo Y,Wang J, Jackson SA, Chang RZ. 2013. A platform for soybean molecular breeding: the utilization of core collections for food security. *Plant Mol. Biol.* 83, 41–50. https://doi.org/10.1007/s11103-013-0076-6

Rossi M, Goggin FL, Milligan SB, Kaloshian I, Ullman DE, Williamson VM. 1998. The nematode resistance gene Mi of tomato confers resistance against the potato aphid. *Proc. Natl. Acad. Sci. U.S.A.* 95, 9750–9754. https://doi.org/10.1073/pnas.95.17.9750

Schmutz J, Cannon SB, Schlueter J, Ma JX, Mitros T, Nelson W, Hyten DL, Song QJ, Thelen JJ, Cheng JL, Xu D, Hellsten U, May GD, Yu Y, Sakurai T, Umezawa T, Bhattacharyya MK, Sandhu D, Valliyodan B, Lindquist E, Peto M, Grant D, Shu SQ, Goodstein D, Barry K, Futrell-Griggs M, Abernathy B, Du JC, Tian ZX, Zhu LC, Gill N, Joshi T, Libault M, Sethuraman A, Zhang XC, Shinozaki K, Nguyen HT, Wing RA, Cregan P, Specht J, Grimwood J, Rokhsar D, Stacey G, Shoemaker RC, Jackson SA. 2010. Genome sequence of the palaeopolyploid soybean. *Nature.* 463, 178–183.

Sedivy EJ, Wu FQ, Hanzawa, Y. 2017. Soybean domestication: the origin, genetic architecture and molecular bases. *New Phytol.* 214, 539–553.

Sherman-Broyles S, Bombarely A, Powell AF, Doyle JL, Egan AN, Coate JE, Doyle, JJ. 2014 The wild side of a major crop: soybean's perennial cousins from Down Under. *Am. J. Bot.* 101, 1651–1665.

Singh, RJ. 2017. Botany and cytogenetics of soybean. In: Nguyen MK, Bhattacharyya HT (eds) *The Soybean Genome*. Cham: Springer, pp. 11–14.

Stupar RM. 2010. Into the wild: the soybean genome meets its undomesticated relative. *Proc. Natl. Acad. Sci.* U.S.A. 107, 21947–21948.

Wang FF, Nan HY, Chen LY, Fang C, Zhang HY, Su T, Li SC, Cheng Q, Dong LD, Liu BH, Kong FJ, Lu, SJ. 2019. A new dominant locus, E11, controls early flowering time and maturity in soybean. *Mol. Breed.* 39, 70.

Watanabe S, Hideshima R, Xia ZJ, Tsubokura Y, Sato S, Nakamoto Y, Yamanaka N, Takahashi R, Ishimoto M, Anai T, Tabata S, Harada K. 2009. Map-based cloning of the gene associated with the soybean maturity Locus E3. *Genetics.* 182, 1251–1262.

Watanabe S, Xia ZJ, Hideshima R, Tsubokura Y, Sato S, Yamanaka N, Takahashi R, Anai T, Tabata S, Kitamura K, Harada, K. 2011. A map-based cloning strategy employing a residual heterozygous line reveals that the GIGANTEA gene is involved in soybean maturity and flowering. *Genetics.* 188, 395–407

Xia ZJ, Watanabe S, Yamada T, Tsubokura Y, Nakashima H, Zhai H, Anai T, Sato S, Yamazaki T, Lu¨ SX, Wu HY, Tabata S, Harada, K. 2012. Positional cloning and characterization reveal the molecular basis for soybean maturity locus E1 that regulates photoperiodic flowering. *Proc. Natl. Acad. Sci. USA.* 109, E2155–2164.

Xie M, Chung CYL, Li MW, Wong FL, Wang X, Liu AL, Wang ZL, Leung AKY, Wong TH, Tong SW, Xiao ZX, Fan KJ, Ng MS, Qi XP, Yang LF, Deng TQ, He LJ, Chen L, Fu AS, Ding Q, He JX, Chung G, Isobe S, Tanabata T, Valliyodan B, Nguyen HT, Cannon SB, Foyer CH, Chan TF, Lam, HM. 2019. A reference-grade wild soybean genome. *Nat. Commun.* 10, 1216.

Xue ZC, Zhao SJ, Gao HY, Sun, S. 2014. The salt resistance of wild soybean (*Glycine soja Sieb.* et Zucc. ZYD 03262) under NaCl stress is mainly determined by Na? Distribution in the plant. *Acta Physiol. Plant.* 36, 61–70.

Yang DS, Zhang J, Li MX, Shi LX. 2017. Metabolomics analysis reveals the salt-tolerant mechanism in *Glycine soja*. *J. Plant Growth Regul*. 36, 460–471. https://doi.org/10.1007/s00344-016-9654-6

Zhai H, Lu¨ SX, Liang S, Wu HY, Zhang XZ, Liu BH, Kong FJ, Yuan XH, Li J, Xia ZJ. 2014. GmFT4, a homolog of flowering locus T, is positively regulated by E1 and functions as a flowering repressor in soybean. *PLoS One*. 9, e89030.

Zhang FN, Batley, J. 2020. Exploring the application of wild species for crop improvement in a changing climate. *Curr.Opin. Plant Biol.* 56, 218–222.

Zhang H, Mittal N, Leamy LJ, Barazani O, Song BH. 2017. Back into the wild-apply untapped genetic diversity of wild relatives for crop improvement. *Evol. Appl.* 10, 5–24.

Zhang SC, Zhang ZN, Bales C, Gu CH, DiFonzo C, Li M, Song QJ,Cregan P, Yang ZY, Wang DC. 2017a. Mapping novel aphid resistance QTL from wild soybean. *Theor. Appl. Genet.* 130, 1941–1952.

Zhang SC, Zhang ZN, Wen ZX, Gu CH, An YC, Bales C, DiFonzo C, Song QJ, Wang DC. 2017b. Fine mapping of the soybean aphid-resistance genes Rag6 and Rag3c from Glycine soja 85-32. *Theor. Appl. Genet.* 130, 2601–2615.

Zhang DY, Kumar M, Xu L, Wan Q, Huang YH, Xu ZL, He XL, Ma JB, Pandey GK, Shao HB. 2017c. Genome-wide identification of major intrinsic proteins in Glycine soja and characterization of GmTIP2;1 function under salt and water stress. *Sci. Rep.* 7, 4106.

Zhang HY, Song QJ, Griffin JD, Song BH. 2017d. Genetic architecture of wild soybean (*Glycine soja*) response to soybean cyst nematode (*Heterodera glycines*). *Mol. Genet. Genom*. 292, 1257–1265. https://doi.org/10.1007/s00438-017-1345-x

Zhang S, Zhang Z, Bales C, Gu C, Difonzo C, Li M. 2017e. Mapping novel aphid resistance QTL from wild soybean, *Theor. Appl. Genet.* 130, 1941–1952. https://doi.org/10.1007/s00122-017-2935-z

Zhang S, Zhang Z, Wen Z, Gu C, An YC, Bales C. 2017f. Fine mapping of the soybean aphid-resistance genes Rag6 and Rag3c from *Glycine soja*. *Theor. Appl. Genet.* 130, 2601–2615. https://doi.org/10.1007/s00122-017-2979-0

Zhou Z, Jiang Y, Wang Z, Gou Z, Lyu J, Li W. 2015b. Resequencing 302 wild and cultivated accessions identifies genes related to domestication and improvement in soybean. *Nat. Biotechnol.* 33, 408–414. https://doi.org/10.1038/nbt.3096

7 Cowpea Wild Relatives for Cowpea Sustainability through Introgression Breeding

Latif Ahmad Peer, Mohd. Yaqub Bhat, Ajaz A. Lone, Zahoor A. Dar, Muneeb Ahmad Rather, and Saima Fayaz

INTRODUCTION

ORIGIN AND DOMESTICATION

Vigna unguiculata L. Walp., cowpea, is a diploid with a chromosome number ($2n=2x=22$) belonging to the tribe Phaseoleae of the Leguminosae family (Mahalakshmi et al., 2007). Phaseoleae includes several commercially significant arid-season oilseed and grain legumes, like soybean, snap bean, and mungbean. *Vigna unguiculata* is designated cowpea due to the plant's prominence as a significant fodder source for livestock in different regions across the globe. Cowpea has different regional names across the world, including "pink peas," "southern peas," "crowder," "blackeyed peas," and "field peas" in the United States "wake," "ewa," and "niebe" in most parts of West Africa, and "caupi" in Brazil (Timko et al., 2007). The origin of these names may be traced back to the traditional seed and market classes that have evolved in these regions gradually. Wild relatives of cowpeas have only been found in Madagascar and Africa, suggesting that cowpeas evolved in Africa (Smartt, 1985). West Africa is the center of domestication and diversification of cultivated cowpea (Ba et al., 2004), whereas Southeastern Africa is the epicenter of wild *Vigna* species diversity (Padulosi, 1997). The wild species, *Vigna unguiculata* ssp. unguiculata var. spontanea, is possibly the progenitor of the domesticated cowpea (Herniter et al., 2020). As the cowpea crop arrived in India, most likely during the Neolithic period, India also seems to represent the center of the genetic diversity of cowpea. Research implies that "yardlong beans," a unique cultivar group (Sesquipedialis) that produces considerably longer pods typically used as a fresh green or "snap" bean in Asia, originated in Asia and are rare in African germplasm. From at least the 8th century BC, and perhaps much earlier, cowpeas have been a staple crop across the Mediterranean and the southern parts of Europe (Gogile et al., 2013). Spanish traders introduced cowpea to the West Indies in the 16th century AD, and by the early 1700s, it had

DOI: 10.1201/9781003434535-7

made its way to the United States. It was most likely introduced into South America at about the same period (Timko et al., 2007).

There are presently ten distinct subspecies of cowpea within the gene pool. These subspecies can be divided into two distinct groupings based on Verdcourt's (1970) infraspecific treatment of *Vigna unguiculata.* The Mensensis group comprises four forest-dwelling subspecies: subsp. *letouzeyi* Pasquet of Central Africa, subsp. *aduensis* Pasquet from northern Ethiopia, subsp. *baoulensis* (A. Chev.) Pasquet of West Africa, and *subsp. pawekiae* Pasquet of the highlands extending from Ethiopia to South Africa. *Subsp. aduensis* Pasquet is related to subsp. *pawekiae*, but it is unclear whether it should be treated as a distinct subspecies or a subsp. *pawekiae* variety. The dekindtiana group has six savannah or grassland subspecies: southern Angola highlands's subsp. *dekindtiana* (Harms) Verdc. sensu stricto, Congo and Angola's subsp. *alba* (G. Don) Pasquet, *subsp. tenuis* (E. Mey.) Mare'chal et al. in Zimbabwe, Malawi highlands, and coastal South Africa and Mozambique, subsp. *pubescens* (R. Wilczek) Pasquet found in Mozambique and Kenya coastal plains, subsp. *stenophylla* (Harv.) Mare'chal et al. sensu lato found in eastern South Africa, the Kalahari area, and highveld around Pretoria (Pasquet, 1996). Taxonomists based their suggestions on primitive characteristics such as hairiness of plant parts, perenniality, pod shattering, unique exine on pollen, small seed size, and outcrossing, among other characteristics linked with wild relatives. Several taxonomists mark the Transvaal region of South Africa as the center of cowpea diversity, owing to the presence of the most primitive wild relatives, particularly varieties such as protracta, rhomboidea, stenophylla, and tenuis of the Catiang section. The rhomboidea variety shows a narrow range of geographical distribution and is found primarily in the area from 20°S to 27°S and 26°E to 32°E across the Cape Town region of South Africa. Primitive cowpea wild relatives obtained from different regions across Zimbabwe include *V. unguiculata subsp. stenophylla*, *V. gazensis*, *V. unguiculata* subsp. *tenuis*, *V. nuda*, *V. luteola*, *V. oblongifolia*, *V. nervosa*, *Vigna unguiculata subsp. dekindtiana*, *V. vexillata*, *V. frutescens*, *V. pygmaea*, and *V. reticulata* (Padulosi and Ng, 1990). Additionally, samples of *V. wittei*, *V. kirkii*, and *V. platyloba* collected from Tanzania suggest that the southern parts of Africa are the center of diversification for wild cowpeas. The center of domestication and diversification of cultivated cowpeas is still debated among taxonomists. Some suggest Senegal (West Africa) to Eritrea (East Africa) as the center of the diversity of cultivated cowpeas. The results of a single nucleotide polymorphisms (SNP) marker study conducted on 1,200 cowpea lines have led researchers to conclude that cowpea domestication occurred in two separate centers, one in East Africa and another in West Africa (Huynh et al., 2013). Further evidence from a molecular marker diversity analysis also points to West and Central Africa as the epicenter of cowpea domestication (Xiong et al., 2016). India and Southeast Asia are hypothesized as the center of the evolution of *V. unguiculata* ssp. *biflora* and *V. unguiculata* ssp. *sesquipedalis* (yard-long-bean). The domestication of cowpea experienced a double bottleneck one from its wild progenitor, resulting in cv.-gr. Biflora and cv.-gr (primitive cultivar group) and second from these primitive cultivar groups to the evolved cultivar groups (cv.-gr. Melanophthalmus and cv.-gr. Sesquipedalis) in West Africa and Asia, respectively (Pasquet, 1996). Nevertheless,

Vigna unguiculata ssp. *dekindtiana*, found throughout Africa, is largely considered the plant's direct progenitor (Padulosi, 1997). Archaeological findings in Ghana (West African sub-region) provide the earliest evidence that cowpea was domesticated before 1500 BC (D'Andrea et al., 2007).

Cowpea has a small genetic base despite the availability of numerous germplasm accessions in gene banks worldwide. Genetic diversity assessment across improved cultivars and cowpea breeding lines using SSR (Simple sequence repeats) markers found that modified cowpea varieties generally had a narrow genetic basis. The small genetic base of such an important crop can be ascribed to a highly self-pollinated crop derived from a wild ancestor and, in part, to breeders who continually utilize improved elite lines as parents in establishing segregated populations in their programs. To increase cowpea's genetic base, employing foreign germplasm, particularly among cross-compatible wild relatives, is necessary.

Genetic Modification

Researchers around different parts of the world are working on creating genetically modified (GM) cowpea lines. The genetically modified cowpea currently on the market has a gene called Cry1Ab that codes for a toxin produced by *Bacillus thuringiensis* (Bt). Research on transgenic Bt cowpea lines has been conducted in confined field settings in Nigeria, Ghana, Burkina Faso, and Malawi (Togola et al., 2017). Several attempts have been made to create cowpea lines resistant against the most destructive and commercially significant post-reproductive cowpea pest (*Maruca vitrata*), a legume pod borer. However, success was not achieved because it was impossible to make a compelling cross between cowpea and its wild relatives (Togola et al., 2017). The Cry1Ab Bt gene was the first transgene successfully introduced into cowpeas through transgenesis (Popelka et al., 2006). The field evaluation of the transgenic Bt cowpea lines exhibited resistance to *Maruca*. Bt genes have two significant drawbacks: their inability to be expressed in higher eukaryotes and their specificity, which is limited primarily to Lepidopteran species. This means that even after switching to transgenic cowpeas, farmers still need to take measures to safeguard their crops from insects resistant to the Bt gene (Mohammed et al., 2014).

In Nigeria, the cowpea variety resistant against legume pod borer (SAMPEA 20-T) was the first insect-resistant cowpea line to be generated and licensed for marketing (Mohammed et al., 2014). This marketed cultivar is resistant to *Maruca vitrata*, which is responsible for causing production losses of grains up to 55%–60% (Kedisso et al., 2022). This achievement was made possible through the collaboration of the African Agricultural Technology Foundation (AATF) in Zaria, the Institute for Agricultural Research (IAR) in Nigeria, as well as several other partners (Mohammed et al., 2014). Although the cowpea line PBR imparts resistance to pod borer (*M. vitrata*), it exhibits susceptiblity to other pest species that can cause significant damage to cowpea agriculture. Several insects attack the crop at various stages of its life cycle, each capable of inflicting significant grain production losses. Other insect pests that can cause significant yield losses in cowpea include flower bud thrips (*Megalurothrips sjostedti* Trybom), cowpea aphids (*Aphis craccivora* Koch), pod-sucking bugs (*Clavigralla tomentosicollis* Stål, *Riptortus dentipes* Fabricius), and bruchids (*Callosobruchus maculatus* Fabricius and *Bruchidius atrolineatus*) (Singh and Jackai, 1988). These pests are detrimental to cowpea output in the field and seed storage. The productivity of improved cowpea is continuously reduced owing to pest damage. This necessitates the generation of transgenic cowpea resistant to various pests using sophisticated transgenic and gene editing techniques.

COWPEA WILD RELATIVES (CWRS) FOR INTROGRESSION BREEDING

The wild cowpea relatives constitute a significant gene pool yet to be utilized in producing novel cowpea varieties. In the context of climate change, wild relative genotypes of cowpeas, which have independently evolved within varying environmental conditions, might be significant. A substantial germplasm collection that represents the genetic variability within the species and an understanding of agronomically important traits of the germplasm accessions are both essential for the success of a plant breeding effort. An extensive and diverse collection of germplasm accessions is a valuable source of parental lines for hybridization, leading to improved variety production as part of an ongoing crop improvement effort (Chheda and Fatokun, 1982). Since 1967, the International Institute of Tropical Agriculture (IITA) has collected over 15,000 domesticated cowpea cultivars from more than 100 countries. Over 1,500 wild *Vigna* varieties collected in different gene banks across the globe display variation in characteristics such as plant height, pigmentation, leaf morphology, growth habit, grain and fodder quality, photo-sensitivity or insensitivity, root architecture, drought tolerance, pod and seed characteristics, heat tolerance and resistance against parasitic weeds (*Striga* and Alectra), insect pests (aphids, bruchids, thrips), and root-knot nematodes. Thus, wild forms and closely related cowpea species promise new beneficial gene sources for enhancing cowpea (Baudoin and Marechal, 1989).

Characterizing germplasm by describing the composition and phenotypic traits of the accessions entails observing and documenting those traits that are heritable and readily observable in varying environmental conditions. Such characterization describes the plant morphology, either throughout the plant's life cycle or solely at maturity. The use of markers that emerged along with molecular genetics for characterizing crops' germplasms has become prevalent. These markers are numerous, unaffected by the environment, and consequently more effective in discriminating different germplasm lines. Markers like SSRs (Li et al., 2001), RFLP (Restriction Fragment Length Polymorphism) (Menancio-Hautea et al., 1993), RAPD (Randomly amplified polymorphic DNA) (Ba et al., 2004), inter-SSR analysis (Xiong et al., 2016), and AFLP (Amplified fragment length polymorphism) (Coulibaly and Lowenberg-DeBoer, 2002) have been used for cowpea characterization.

Though breeders have been reluctant to employ wild relatives in generating novel cowpea cultivars due to their undesirable traits, there are studies on the use of wild cowpea genotypes in producing better cultivars. Cross-compatible wild cowpea genotypes often exhibit traits such as pod shattering, unattractive seed coat color and texture, a weedy growth behavior, small seed size, susceptibility to viral diseases, and an uncertain plant life form (Rawal et al., 1976). Tiny seed size, a trait shared by wild genotypes of cowpea, is dominant over large seed size; the larger grain size has greater market demand. Therefore, efforts are being made to consider seed size while making selections (Rawal et al., 1976). Breeders may leverage the potential advantages of wild cowpea genotypes via pre-breeding programs owing to advances in genomic technology. The possible linkage drag of some undesired traits typical of wild genotypes may be readily reduced using molecular technology. On chromosomes Vu09 and Vu05, two QTLs (quantitative trait locus) impacting the number of seeds in cowpea per pod were found. (Lo et al., 2018). The wild parent contributed the CSp09 allele on chromosome Vu09, which caused a 21.09% difference in seed yield. Higher grain output should come from more seeds per pod, as found in peanuts (Songsri et al., 2009), soybean (Van Roekel and Purcell, 2016), and rapeseed (Yang et al., 2016). As a result, despite their tiny seed size, wild cowpea genotypes may contribute to increased grain output in cultivated cowpea by raising the seed number per pod, i.e., increasing the ovary number per pod. The QTLs linked with this trait indicate that the number of seeds in a pod is heritable and may therefore be selected for cowpea breeding to increase grain output. Prominent peduncles allow plants to hold pods above the canopy, minimizing *Maruca vitrata* damage and making pod harvesting easier (Boukar et al., 2020). This characteristic is present in the wild relative line of cowpea used to generate the linkage map. QTL mapping revealed a single locus on chromosome VU05 responsible for 71.83% of the observed phenotypic variance (Lo et al., 2018). By delving into the genetics of perenniality, we may increase our odds of success in perennializing farmed cowpeas, which might lead to the generation of new, perhaps more productive varieties.

BENEFICIAL TRAITS PRESENT IN SOME WILD COWPEA RELATIVES

Many studies have been conducted to study the feasibility of transferring beneficial traits from wild cousins of cowpeas to domesticated cowpeas. Some of the characteristics identified as a result of these attempts are discussed below:

Insect Resistance

Aphids (*Aphis craccivora* Koch) are the most devastating insect pests to cowpea plants during their initial stages of development. They harm seedlings by extracting sap from plants, particularly during drought. Seedlings might be destroyed by severe infection coupled with delayed rains (Abdou et al., 2013). Transgenic cowpea (IITA (The International Institute of Tropical Agriculture) breeding lines) generated by incorporating the dominant gene (Rac) from the cowpea line (TVu-3000) provided tolerance against this pest (Bata et al., 1987){Bata, 1987 #19}. Unfortunately, this gene is no longer effective. Plants possessing this gene get easily infested by pests, necessitating the search for additional sources of genes resistant to aphids. TVNu-1158, a wild relative of cowpea, demonstrated aphid resistance at the seedling stage (Souleymane et al., 2013). This wild cowpea genotype was efficiently crossed with cultivated cowpea, yielding a collection of RILs (Recombinant inbred lines) to generate a cowpea linkage map.

Moreover, utilizing this RIL population, QTLs impacting domestication-related characteristics were discovered (Lo et al., 2018). The group of RILs was tested for aphid tolerance, and a few were shown to be tolerant; thus they are being utilized as parental sources in a breeding program to impart resistance to the crop. Three domesticated cowpea germplasms, TVu15445, TVu-6464, and TVu-1583, with remarkable tolerance levels against *A. craccivora* similar to TVu-801, were described (Togola et al., 2020). These novel aphid tolerance sources in wild and cultivated cowpea varieties must be examined for allelism. Pyramiding non-allelic sources into superior genetic backgrounds would develop novel aphid-tolerant cowpea types. Additionally, the resistance will persist in such kinds for a longer period of time. Low sucrose and high quercetin and kaempferol (phenolic compounds) are associated with the tolerance mechanism in the aforementioned cowpea germplasms (Wani et al., 2022). Aphid resistance and high flavonoid glycoside levels exhibited a positive correlation during flavonoid HPLC fingerprinting of wild and domesticated cowpea varieties (Lattanzio et al., 1997).

Specific *Vigna vexillata* genotypes show resistance against *Maruca vitrata*, the most widespread among insect pests that damage cowpea seed pods and young sensitive plant parts, thus, decreasing overall yield by up to 60% (Yao et al., 2023). Wild cowpea genotypes, which were entirely resilient against *M. vitrata*, were accessions of *V. vexillata*, *V. macrosperma*, *V. oblongifolia*, and *V. reticulata* (Boukar et al., 2020). Trichomes of *V. vexillata* lines (TVNu-73 and TVNu-72) contribute resilience against *M. vitrata*; however, after trichomes are trimmed off, the insect's larvae grow, though not effectively (Boukar et al., 2020). The resistance mechanism against insect larvae has been shown independently in the presence or absence of trichomes. However, when trichomes were excised from the pods, the feeding damage score of adult pod-sucking pests was somewhat more significant, and the two *V. vexillata* lines show antixenosis and antibiosis mechanisms of insect resistance. There was complete immunity to *Clavigralla tomentosicollis* Stl. among all three varieties of *V. luteola*, 17 of *V. vexillata*, two of *V. macrosperma*, and three of *V. angustifolia*. The cowpea seed weevil, *C. maculatus* Fabricius, could not infest any out of 27 and six accessions of *V. vexillata*, and *V. luteola*, respectively (Dabire-Binso et al., 2010).

Striga and Alectra Resistance

The parasitic weeds *Vatke* and *Alectra vogelii* [Benth], belonging to the Scrophulariaceae family, invade the cowpea crop in the field. These root pathogens can reduce cowpea yield by up to 50% or even 83%–100% (Boukar et al., 2020). A single *Striga* plant produces about 90,000 seeds, and many can survive in the soil for at least 15–20 years (Gbèhounou and Adango, 2003). Although *Alectra* is prevalent in wet savanna regions like the Guinea savannah, *Striga*, the most lethal weed, occurs in the arid savannah agroecology, which is the major production center of cowpea. The best strategy to manage these parasites in the fields of farmers is to develop cultivars that are resistant to them. Farmers in several nations are cultivating enhanced cultivars that combine genes offering resistance to both parasites. The fact that several races of *Striga* are prevalent in various nations poses a significant obstacle to spreading resistant cowpea cultivars across these regions. Initially, five races were recognized (Burkina Faso has race 1, Mali has race 2, Nigeria and Niger have race 3, the Republic of Benin has race 4, and Cameroon has race 5) (Botanga and Timko, 2006). Later reports claimed Nigeria was home to three races (1, 3, and 5) (Singh and Emechebe, 1997). Since then, two previously unidentified cowpea parasite races—one each in Senegal (race 6) and the Republic of Benin (SG4z) have been described (Botanga and Timko, 2006), raising the total number of known races to seven. From two distinct cowpea lines, the *Striga*-resistant genes (Rsg-1, 2, and 3) were isolated (Fery and Singh, 1997). The genes Rsg-1 and Rsg-2, duplicates, originate from the cowpea line B301. In heavily *Striga*-infected fields at Minjibir (northern Nigeria), situated in the Sudan agroecology, various wild cowpea genotypes were assessed for their responses against the parasite, some of which showed significant degrees of resistance. TVNu-1535, TVNu-1537, TVNu1647, TVNu-1070, TVNu-1083, TVNu-585, and TVNu-491 from the cowpea species; *oblongifolia*, *ambacensis*, *reticulata*, and *parkeri*, are among the wild relatives of cowpea that have shown resistance to *Striga* (Mohemed et al., 2016). Of the studied wild relatives, only a single *Striga*-tolerant cowpea line, TVNu-1589, belongs to section Catiang. Seeds of *Striga* cannot traverse borders due to plant quarantine restrictions; however, breeders all over the West African sub-region may share resistant cowpea seed varieties and cross-compatible wild genotypes for testing against the various races found in their respective nations. Rav-1 and Rav-2, two redundant genes, control cowpea resistance to *Alectra* (Oyatomi et al., 2016). The resistant wild relatives of cowpea may harbor new sources of resistance genes, and if they are non-allelic to any documented dominant tolerant genes integrated into improved varieties, they could increase resistance levels. The tolerant genes found in these wild cowpea genotypes may provide additional sources of protection against novel *Striga* varieties that could emerge due to the climate crisis.

Nutritional Qualities

Eight wild *Vigna* species (*V. ambacensis, V. unguiculata* dekindtiana, *V. vexillata*, *V. luteola*, *V. vexillata* macrosperma, *V. oblongifolia*, *V. racemosa*, and *V. reticulata*) were analyzed in a study for different biochemical properties including proteins, starch digestibility, amino acid status, as well as compounds like phytic acid, cysteine proteinase inhibitors, and tannins (Marconi et al., 1997). The goal was to discover substances that would effectively enhance cowpea's nutritional value and resistance against pests. The grains of the *V. vexillata* plant contained a high amount of proteins, up to 29.3%, and high sulfur-containing amino acids and starch content, ranging from 64% to 75%; as a consequence, all of the accessions showed high biochemical scores. A significant difference was identified between the grains regarding the amounts of trypsin inhibitors, lectins, and tannins. In addition, *V. luteola* also possessed significant amounts of these compounds; however, relatively low levels of these biochemicals were observed in *V. ambacensis* and *V. reticulata*; the same is true with *V. unguiculata* dekindtiana, progenitor of domesticated cowpea. Notwithstanding the high protein levels in the grains of wild cowpea genotypes, researchers found that their digestibility was much lower than that of domesticated varieties (Boukar et al., 2020). It was also evaluated that protein content was marginally better in the wild cowpea genotypes compared to the domesticated lines.

Other than producing grains rich in protein, some wild species of *Vigna* are used in various ways. The roots of *V. lobatofolia* and *V. vexillata*, which provide up to 15% more protein than a potato and six times more than cassava, are consumed in some tribes (Dakora and Belane, 2019). Increased protein content in the grains is one feature shown by certain wild cowpea relatives that might be beneficial in improving the cowpea's nutritional value.

Drought Tolerance

Cowpea has a stronger ability to withstand drought than many other crops. However, drought, particularly from the seedling to the blooming stage, might still negatively influence its production. Breeding can enhance drought tolerance in improved cultivars. Several wild relatives may be sources of drought-tolerant genes due to their growth patterns and the environments from which these germplasm accessions are obtained. Most stenophylla and tenuis germplasm were collected from dry savannah agroecological areas with sandy soils (Yahaya et al., 2019). There is a likelihood that wild genotypes of cowpea from such arid areas could be significant sources of water-stress tolerant genes because they have developed tolerance for these conditions. Some wild cowpea varieties have perennial growth habits, which may contribute significantly to their drought tolerance potential, as such plants can retain their color and survive dry periods from one cropping season to the next.

Longevity

Improved cultivars of cowpea may be classified as very early (harvested in 60 days or less), early (harvested in 65–75 days), medium (harvested in 75 days or more), or late (harvested in more than 100 days) (Boukar et al., 2020). However, the majority of cultivars used by conventional farmers are slow-maturing varieties. These later varieties are day-length sensitive and tend to swiftly expand and cover the land. When grown as a single crop, farmers often prefer day-neutral, very early, or early maturing lines, whereas when used as an intercrop, farmers use dual-purpose or late-maturing cowpeas. In the arid savannah, farmers value cowpea fodder for its excellent nutritional qualities for livestock (Samireddypalle et al., 2017). Farmers may sell cowpea fodder to livestock owners even after pod-sucking pests and legume pod borers destroy a crop in the field. Cut-and-carry methods are well established in Asia and Australia, where yields of cowpea forage may approach 4 tons per hectare (Plazas et al., 2021). Farmers and herders may cultivate perennial relatives of cowpea that are cross-compatible with cowpea to provide year-round forage for their livestock. Several RILs from a biparental hybrid between domesticated and wild cowpea relatives with perennial growth habits were long-lived, maintaining their color and leafiness for over 700 days after being planted in pots (Lo et al., 2020). The advancement of the novel, perhaps higher-yielding varieties of cultivated cowpea may be facilitated by a better understanding of the genetic underpinnings of perenniality. Varieties with long-lasting green features might benefit migratory ranchers, who are primarily centered in dry savannah regions but travel to humid coastal areas of West Africa during the dry season.

MULTI-OMICS: COWPEA PRODUCTIVITY ENHANCEMENT THROUGH ADVANCED TECHNIQUES

In the modern world, with the concept of smart bioengineering procedures that can overcome enormous obstacles in improving agricultural output production, multi-omics biotechniques are generally the game-changer. These diverse biotechnological tools, which include proteomics, genomics, metabolomics, and transcriptomics, provide enormous opportunities for enhancing crop protection and productivity and assuring sources of nourishment that are free from known dangers to human health. Using multi-omics technologies has improved plant breeding and developed more robust and resistant cultivars.

GENOMICS

Genomics, the study of an organism's entire DNA sequence, has revolutionized plant breeding by providing powerful tools for identifying and characterizing genes and other functional elements in the genome. In wild cowpea relatives, genomics research has yielded valuable insights into the genetic makeup and functional traits of these wild species, which can be incorporated for cowpea improvement (Boukar et al., 2020). One primary focus of genomics research in wild cowpea relatives is the identification of genes responsible for resistance to biotic stresses, such as diseases and pests. By sequencing the genomes of wild cowpea species and comparing them to the cultivated cowpea genome, researchers have identified genes associated with resistance to specific pathogens or pests. These genes can be introgressed into cultivated cowpea through traditional breeding or genetic engineering to develop resistant cultivars, reducing the reliance on chemical pesticides and increasing the sustainability of cowpea production (Boukar et al., 2019). Genomics has also been used to study the genetic basis of abiotic stress tolerance in wild cowpea relatives. Drought and heat stress are significant constraints to cowpea production, and wild relatives that thrive in harsh environments may carry genes conferring tolerance to these stresses. By identifying genes associated with drought or heat tolerance in wild cowpea species, researchers can develop cowpea cultivars that can withstand adverse climatic conditions, ensuring stable production even in the face of changing climates (Kapazoglou et al., 2023). Furthermore, genomics research has shed light on the genetic diversity and population structure of wild cowpea relatives. This information is crucial for conserving and utilizing these valuable genetic resources. Conservation strategies can be designed to protect the genetic diversity of wild cowpea species, and breeding programs can target specific wild populations or accessions that possess desired traits for cowpea improvement (Xiong et al., 2016). Thus, genomics research in wild cowpea relatives holds great promise for cowpea improvement by providing insights into the genetic makeup, functional traits, and diversity of these wild species. The knowledge gained from genomics research can be harnessed to develop cowpea cultivars with enhanced resistance to biotic and abiotic stresses, ultimately leading to improved cowpea production and food security in regions where cowpea is a critical crop. Moreover, genomics-based strategies can facilitate the conservation and sustainable utilization of cowpea wild relatives (CWRs), ensuring their preservation for future generations of plant breeders and farmers.

Transcriptomics

Transcriptomics, the study of gene expression at the transcript level, is a critical approach to understanding genetic diversity and the potential for improvement in CWRs. CWRs in diverse ecological niches hold genetic traits that could be exploited for improving cultivated cowpeas. Transcriptomics analysis of wild cowpea relatives involves characterizing the expression patterns of thousands of genes in response to different environmental conditions, stresses, and developmental stages. A transcriptome or the complete set of RNA (Ribose nucleic acid) molecules produced by a cell or tissue, can be determined using high-throughput RNA sequencing (RNA-seq) (Chen et al., 2017; Kang et al., 2023). One of the critical objectives of transcriptomics analysis in wild cowpea relatives is to identify differentially expressed genes that show significant changes in expression levels compared to cultivated cowpeas. These differentially expressed genes may be associated with important agronomic traits such as plant growth and development, nutrient uptake, disease resistance, and stress tolerance. By comparing the transcriptomes of wild cowpea relatives with those of cultivated cowpeas, researchers can identify candidate genes that may confer desirable traits to cultivated cowpeas. Transcriptomics analysis also provides insights into the molecular mechanisms underlying the unique traits of wild cowpea relatives. For example, it can reveal the regulatory networks and signaling pathways involved in stress responses, metabolic processes, and other biological processes. Understanding these molecular mechanisms can help researchers identify critical genes and pathways that can be targeted for the genetic improvement of cowpeas through breeding or genetic engineering approaches (Kang et al., 2023).

Moreover, transcriptomics analysis of wild cowpea relatives can aid in identifying functional genetic variants, such as structural variants and SNPs, which are variations in the DNA sequence that can affect gene expression and function. These genetic variants act as molecular markers for breeding programs, allowing for the selection of cowpea cultivars with desirable traits (Raizada and Souframanien, 2019). Furthermore, transcriptomics analysis can provide valuable information on gene regulatory networks, epigenetic modifications, and non-coding RNAs (RNA molecules not coding for proteins but essential in gene regulation). These factors can influence the expression of genes and contribute to the phenotypic diversity observed in wild cowpea relatives (Chen et al., 2016). Thus, transcriptomics analysis of wild cowpea relatives is a powerful method for understanding the genetic diversity and potential of these wild relatives for cowpea improvement. It provides insights into gene expression patterns, molecular mechanisms, functional genetic variants, and other factors that can be leveraged for developing improved cowpea cultivars with enhanced agronomic traits.

Proteomics

Proteomics, an essential tool for crop improvement and plant breeding, can help study wild cowpea's proteome and identify desirable trait-controlling proteins such as disease resistance, yield, and nutrient content. One of the critical areas of research in wild cowpea proteomics is identifying proteins associated with resistance to biotic and abiotic stresses. For example, a study identified several proteins upregulated in wild cowpea leaves in response to drought stress, including heat shock proteins, antioxidant enzymes, and proteins involved in energy metabolism. These proteins may be essential for improving drought tolerance in cowpeas (Kumar et al., 2022). Another area of research is the identification of proteins associated with yield and nutrient content. For example, a study identified several proteins associated with seed development and nutrient storage, including legumins, vicilins, and albumins. These proteins may be necessary to improve yield and nutrient content in cowpeas (Antonets et al., 2020). Proteomics can also be used to identify proteins associated with disease resistance in cowpeas. For example, a study identified several proteins upregulated in wild cowpea leaves in response to infection by the fungus *Colletotrichum destructivum*. These proteins may improve cowpea's resistance to this and other fungal diseases (Amadioha and Enyiukwu, 2019). In addition to identifying proteins associated with desirable traits, proteomics can also be used to understand the mechanisms underlying these traits. For example, a study identified several proteins involved in the biosynthesis of flavonoids, which are essential for plant defense and have antioxidant properties. Understanding the mechanisms underlying the biosynthesis of flavonoids may lead to developing new strategies for improving cowpea resistance to biotic and abiotic stresses (Deng et al., 2022). Overall, proteomics is an essential tool for enhancing cowpeas. By identifying proteins associated with desirable traits, proteomics can help breeders develop new varieties of cowpeas that are more resistant to disease, have higher yield, and have better nutritional value.

Metabolomics

Metabolomics is a powerful tool increasingly used in plant breeding and crop improvement. It involves the comprehensive analysis of metabolites in a biological sample, such as plant tissue, to understand the biochemical pathways and cellular processes underlying the observed phenotype. In the case of wild cowpea, metabolomics can be used to identify specific metabolites associated with traits of interest, such as resistance to pests and diseases, drought tolerance, or high nutritional content. By comparing the metabolite profiles of wild cowpea with those of domesticated varieties, researchers can identify potential targets for breeding or genetic engineering that may improve the performance of cowpea crops (Shahid et al., 2023). For example, a metabolomics study of wild cowpeas could identify metabolites involved in the biosynthesis of flavonoids, which are known to play a role in plant defense against fungal and bacterial pathogens. Identifying specific proteins involved in flavonoid biosynthesis and quantifying flavonoid levels in wild cowpea tissues could provide valuable information for cowpea improvement. By selecting cowpea varieties with higher levels of flavonoids, breeders could potentially develop crops more resistant to fungal and bacterial diseases (Falcone Ferreyra et al., 2012).

Similarly, metabolomics could identify metabolites associated with drought tolerance or nutritional quality in wild cowpeas. For example, identifying specific amino acids or other metabolites associated with drought tolerance could help breeders select cowpea varieties that perform better in arid or semi-arid regions. Alternatively, identifying specific metabolites associated with high protein or micronutrient content could help breeders develop cowpea varieties with improved nutritional quality (Goufo et al., 2017). Thus, metabolomics can provide valuable insights into the biochemical pathways and cellular processes that underlie essential traits in wild cowpeas. By identifying specific metabolites associated with traits of interest, metabolomics can inform breeding and genetic engineering efforts to improve the performance and nutritional quality of cowpea crops.

CONCLUSION AND PERSPECTIVE

Cultivating cowpea crops has faced numerous challenges due to biotic and abiotic stressors. However, breeders and scientists have identified wild cowpea relatives that resist these stressors and possess beneficial genes for drought tolerance and insect pest resistance. Despite the cross-incompatibility among various wild relatives, certain cross-compatible wild cowpea relatives have been used as sources of resistance genes. The production of grain-type cowpeas must be enhanced to address global concerns such as the pandemic, climate change, and food insecurity. This can be achieved by employing advanced biotechnological tools such as multi-omics techniques, genetic engineering, and precision agricultural practices. However, further research is required to fill gaps in knowledge and guarantee success. Coordinated initiatives are necessary to enable stakeholders to embrace sustainable biotechnological techniques, and an integrative and holistic approach to systematic biology can improve the cowpea crop. Cowpea-producing marginal communities have used synthetic microbial consortia to develop trait-specific cultivars customized to their niche habitats. With these tools and initiatives, it is possible to break the yield plateau and contribute to the sustainable development goals of reducing poverty, ending hunger, and addressing malnutrition.

REFERENCES

Abdou, S., Aken'Ova, M., Fatokun, C., & Alabi, O. (2013). Screening for resistance to cowpea aphids (Aphis craccivora KOCH) in wild and cultivated cowpea (Vigna unguiculata L. Walp) accessions. *International Journal of Environmental Science and Technology*, 2(4), 611–621.

Amadioha, A., & Enyiukwu, D. (2019). Alterations of biochemical composition of leaf and stem of cowpea (Vigna unguiculata L. Walp.) by Colletotrichum destructivum O'Gara in Nigeria. *Journal of Experimental Agriculture International*, 32(2), 1–7.

Antonets, K. S., Belousov, M. V., Sulatskaya, A. I., Belousova, M. E., Kosolapova, A. O., Sulatsky, M. I., et al. (2020). Accumulation of storage proteins in plant seeds is mediated by amyloid formation. *PLoS Biology*, 18(7), e3000564, https://doi.org/10.1371/journal.pbio.3000564.

Ba, F. S., Pasquet, R. S., & Gepts, P. (2004). Genetic diversity in cowpea [Vigna unguiculata (L.) Walp.] as revealed by RAPD markers. *Genetic Resources and Crop Evolution*, 51, 539–550.

Bata, H., Singh, B., Singh, S., & Ladeinde, T. (1987). Inheritance of resistance to aphid in Cowpea 1. Crop science, 27(5), 892-894.

Baudoin, J., & Marechal, R. (1989). *Taxonomy and Evolution of the Genus Vigna*. International Symposium on Mungbean, Thailand.

Botanga, C., & Timko, M. (2006). Phenetic relationships among different races of Striga gesnerioides (Willd.) Vatke from West Africa. *Genome/National Research Council Canada = Génome/Conseil National de Recherches Canada*, 49, 1351–1365. https://doi.org/10.1139/g06-086.

Boukar, O., Belko, N., Chamarthi, S., Togola, A., Batieno, J., Owusu, E., et al. (2019). Cowpea (Vigna unguiculata): genetics, genomics and breeding. *Plant Breeding*, 138(4), 415–424. https://doi.org/10.1111/pbr.12589.

Boukar, O., Abberton, M., Oyatomi, O., Togola, A., Tripathi, L., & Fatokun, C. (2020). Introgression breeding in cowpea [Vigna unguiculata (L.) Walp.]. *Frontiers in Plant Science*, 11, 567425. https://doi.org/10.3389/fpls.2020.567425.

Chen, H., Chen, X., Tian, J., Yang, Y., Liu, Z., Hao, X., et al. (2016). Development of gene-based SSR markers in rice bean (Vigna umbellata L.) based on transcriptome data. *PLoS One*, 11(3), e0151040.

Chen, H., Wang, L., Liu, X., Hu, L., Wang, S., & Cheng, X. (2017). De novo transcriptomic analysis of cowpea (Vigna unguiculata L. Walp.) for genic SSR marker development. *BMC Genetics*, 18(1), 65. https://doi.org/10.1186/s12863-017-0531-5.

Chheda, H. R., & Fatokun, C. A. (1982). Numerical analysis of variation patterns in okra (Abelmoschus esculentus [L.] Moench). *Botanical Gazette*, 143(2), 253–261.

Coulibaly, O., & Lowenberg-DeBoer, J. (2002). The economics of cowpea in West Africa. Paper presented at the Challenges and opportunities for enhancing sustainable cowpea production, Proceedings of the World Cowpea Conference III International Institute of Tropical Agriculture (IITA), Ibadan Nigeria.

D'Andrea, A. C., Kahlheber, S., Logan, A. L., & Watson, D. J. (2007). Early domesticated cowpea (Vigna unguiculata) from Central Ghana. *Antiquity*, 81(313), 686–698.

Dabire-Binso, C. L., Ba, N. M., Sanon, A., Drabo, I., & Bi, K. F. (2010). Resistance mechanism to the pod-sucking bug Clavigralla tomentosicollis (Hemiptera: Coreidae) in the cowpea IT86D-716 variety. *International Journal of Tropical Insect Science*, 30(4), 192–199.

Dakora, F. D., & Belane, A. K. (2019). Evaluation of protein and micronutrient levels in edible cowpea (Vigna Unguiculata L. Walp.) leaves and seeds. [original research]. *Frontiers in Sustainable Food Systems*, 3, 70. https://doi.org/10.3389/fsufs.2019.00070.

Deng, X., Shang, H., Chen, J., Wu, J., Wang, T., Wang, Y., et al. (2022). Metabolomics combined with proteomics provide a novel interpretation of the changes in flavonoid glycosides during white tea processing. *Foods*, 11(9), 1226. https://doi.org/10.3390/foods11091226.

Falcone Ferreyra, M. L., Rius, S. P., & Casati, P. (2012). Flavonoids: biosynthesis, biological functions, and biotechnological applications. *Frontiers in Plant Science*, 3, 222.

Gbèhounou, G., & Adango, E. (2003). Trap crops of Striga hermonthica: in vitro identification and effectiveness in situ. *Crop Protection*, 22(2), 395–404.

Gogile, A., Andargie, M., & Muthuswamy, M. (2013). Screening selected genotypes of cowpea [Vigna unguiculata (L.) Walp.] for salt tolerance during seedling growth stage. *Pakistan Journal of Biological Sciences: PJBS*, 16, 671–679. https://doi.org/10.3923/pjbs.2013.671.679.

Goufo, P., Moutinho-Pereira, J. M., Jorge, T. F., Correia, C. M., Oliveira, M. R., Rosa, E. A., et al. (2017). Cowpea (Vigna unguiculata L. Walp.) metabolomics: osmoprotection as a physiological strategy for drought stress resistance and improved yield. *Frontiers in Plant Science*, 8, 586.

Herniter, I. A., Muñoz-Amatriaín, M., & Close, T. J. (2020). Genetic, textual, and archeological evidence of the historical global spread of cowpea (Vigna unguiculata [L.] Walp.). *Legume Science*, 2(4), e57. https://doi.org/10.1002/leg3.57.

Huynh, B. L., Close, T. J., Roberts, P. A., Hu, Z., Wanamaker, S., Lucas, M. R., et al. (2013). Gene pools and the genetic architecture of domesticated cowpea. *The Plant Genome*, 6(3), plantgenome2013.2003.0005.

Kang, B. H., Kim, W. J., Chowdhury, S., Moon, C. Y., Kang, S., Kim, S. H., et al. (2023). Transcriptome analysis of differentially expressed genes associated with salt stress in cowpea (Vigna unguiculata L.) during the early vegetative stage. *International Journal of Molecular Sciences*, 24(5), 4762. https://doi.org/10.3390/ijms24054762.

Kapazoglou, A., Gerakari, M., Lazaridi, E., Kleftogianni, K., Sarri, E., Tani, E., et al. (2023). Crop wild relatives: a valuable source of tolerance to various abiotic stresses. *Plants*, 12(2), 328.

Kedisso, E. G., Barro, N., Chimphepo, L., Elagib, T., Gidado, R., Mbabazi, R., et al. (2022). Crop biotechnology and smallholder farmers in Africa. *Genetically Modified Plants and Beyond*, 15, 107–127.

Kumar, P., Singh, J., Kaur, G., Adunola, P. M., Biswas, A., Bazzer, S., et al. (2022). OMICS in fodder crops: applications, challenges, and prospects. *Current Issues in Molecular Biology*, 44(11), 5440–5473.

Lattanzio, V., Cardinali, A., Linsalata, V., Perrino, P., & Ng, N. (1997). Flavonoid HPLC fingerprints of wild Vigna species. In: B.B. Singh, D.R. Mohan Raji and K.E. Dashiel (eds.), *Advances in Cowpea Research*. Ibadan, Nigeria: IITA, (pp. 66–74).

Li, C. D., Fatokun, C. A., Ubi, B., Singh, B. B., & Scoles, G. J. (2001). Determining genetic similarities and relationships among cowpea breeding lines and cultivars by microsatellite markers. *Crop Science*, 41(1), 189–197.

Lo, S., Muñoz-Amatriaín, M., Boukar, O., Herniter, I., Cisse, N., Guo, Y.-N., et al. (2018). Identification of QTL controlling domestication-related traits in cowpea (Vigna unguiculata L. Walp). *Scientific Reports*, 8(1), 6261.

Lo, S., Fatokun, C., Boukar, O., Gepts, P., Close, T. J., & Muñoz-Amatriaín, M. (2020). Identification of QTL for perenniality and floral scent in cowpea (Vigna unguiculata [L.] Walp.). *PLoS One*, 15(4), e0229167.

Mahalakshmi, V., Ng, Q., Lawson, M., & Ortiz, R. (2007). Cowpea [Vigna unguiculata (L.) Walp.] core collection defined by geographical, agronomical and botanical descriptors. *Plant Genetic Resources*, 5(3), 113–119.

Marconi, E., Ruggeri, S., & Carnovale, E. (1997). Chemical evaluation of wild under-exploited Vigna spp. seeds. *Food Chemistry*, 59(2), 203–212. https://doi.org/10.1016/S0308-8146(96)00172-0.

Menancio-Hautea, D., Fatokun, C., Kumar, L., Danesh, D., & Young, N. (1993). Comparative genome analysis of mungbean (Vigna radiata L. Wilczek) and cowpea (V. unguiculata L. Walpers) using RFLP mapping data. *Theoretical and Applied Genetics*, 86, 797–810.

Mohammed, B., Ishiyaku, M., Abdullahi, U., & Katung, M. (2014). Response of transgenic Bt cowpea lines and their hybrids under field conditions. *Journal of Plant Breeding and Crop Science*, 6(8), 91–96.

Mohemed, N., Charnikhova, T., Bakker, E. J., van Ast, A., Babiker, A. G., & Bouwmeester, H. J. (2016). Evaluation of field resistance to Striga hermonthica (Del.) Benth. in Sorghum bicolor (L.) Moench. The relationship with strigolactones. *Pest Management Science*, 72(11), 2082–2090. https://doi.org/10.1002/ps.4426.

Oyatomi, O., Fatokun, C., Boukar, O., Abberton, M., & Ilori, C. (2016). Screening Wild Vigna Species and Cowpea (Vigna unguiculata) Landraces for Sources of Resistance to Striga gesnerioides. In: N. Maxted, M. Ehsan Dulloo and B.V. Ford-Lloyd (eds.), *Enhancing Crop Genepool Use: Capturing Wild Relative and Landrace Diversity for Crop Improvement*. Wallingford: CABI International, (pp. 27–31).

Padulosi, S. (1997). Origin, taxonomy, and morphology of Vigna unguiculata (L.) Walp. In: B.B. Singh, D.R. Mohan Raji and K.E. Dashiel (eds.), *Advances in Cowpea Research*. Ibadan, Nigeria: IITA, (pp. 1–12).

Padulosi, S., & Ng, N. (1990). Wild Vigna species in Africa: their collection and potential utilization. *Cowpea Genetic Resources*, 1990, 58–77.

Pasquet, R. (1996). Wild cowpea (Vigna unguiculata) evolution. In: B. Pickerskill and J. Lock (eds.), *Advances in Legume Systematics, Volume 8*. Kew: Royal Botanic Gardens, (pp. 95–100).

Plazas, M., Herniter, I. A., Cannon, S. B., Fatokun, C., Boukar, O., Abberton, M., et al. (2021). Introgression breeding in cowpea. *Introgression Breeding in Cultivated Plants*, 11, 567425.

Popelka, J. C., Gollasch, S., Moore, A., Molvig, L., & Higgins, T. J. (2006). Genetic transformation of cowpea (Vigna unguiculata L.) and stable transmission of the transgenes to progeny. *Plant Cell Reports*, 25(4), 304–312. https://doi.org/10.1007/s00299-005-0053-x.

Raizada, A., & Souframanien, J. (2019). Transcriptome sequencing, de novo assembly, characterisation of wild accession of blackgram (Vigna mungo var. silvestris) as a rich resource for development of molecular markers and validation of SNPs by high resolution melting (HRM) analysis. *BMC Plant Biology*, 19(1), 358. https://doi.org/10.1186/s12870-019-1954-0.

Rawal, K., Rachie, K., & Franckowiak, J. (1976). Reduction in seed size in crosses between wild and cultivated cowpeas. *Journal of Heredity*, 67(4), 253–254.

Samireddypalle, A., Boukar, O., Grings, E., Fatokun, C. A., Kodukula, P., Devulapalli, R., et al. (2017). Cowpea and groundnut haulms fodder trading and its lessons for multidimensional cowpea improvement for mixed crop livestock systems in West Africa. *Frontiers in Plant Science*, 8, 30.

Shahid, M., Singh, U. B., & Khan, M. S. (2023). Metabolomics-based mechanistic insights into revealing the adverse effects of pesticides on plants: an interactive review. *Metabolites*, 13(2), 246. https://doi.org/10.3390/metabo13020246.

Singh, B., & Emechebe, A. (1997). Advances in research on cowpea Striga and Alectra. In: B.B. Singh, D.R. Mohan Raji and K.E. Dashiel (eds.), *Advances in Cowpea Research*. Ibadan, Nigeria: IITA, (pp. 215–224).

Singh, S., & Jackai, L. (1988). Screening techniques for host plant resistance to cowpea insect pests. *Tropical Grain Legum Bull*, 35, 2–18.

Smartt, J. (1985). Evolution of grain legumes. III. Pulses in the genus Vigna. *Experimental Agriculture*, 21(2), 87–100.

Songsri, P., Jogloy, S., Holbrook, C., Kesmala, T., Vorasoot, N., Akkasaeng, C., et al. (2009). Association of root, specific leaf area and SPAD chlorophyll meter reading to water use efficiency of peanut under different available soil water. *Agricultural Water Management*, 96(5), 790–798.

Souleymane, A., Aken'Ova, M., Fatokun, C., & Alabi, O. (2013). Screening for resistance to cowpea aphid (Aphis craccivora Koch) in wild and cultivated cowpea (Vigna unguiculata L. Walp.) accessions. *International Journal of Environmental Science and Technology*, 2, 611–621.

Timko, M. P., Ehlers, J. D., & Roberts, P. A. (2007). Cowpea. In: C. Kole, (eds.), *Pulses, Sugar and tuber crops*. Berlin, Heidelberg: Springer, (pp. 49–67).

Togola, A., Boukar, O., Belko, N., Chamarthi, S. K., Fatokun, C., Tamo, M., et al. (2017). Host plant resistance to insect pests of cowpea (Vigna unguiculata L. Walp.): achievements and future prospects. *Euphytica*, 213(11), 239. https://doi.org/10.1007/s10681-017-2030-1.

Togola, A., Boukar, O., Servent, A., Chamarthi, S., Tamò, M., & Fatokun, C. (2020). Identification of sources of resistance in cowpea mini core accessions to Aphis craccivora Koch (Homoptera: Aphididae) and their biochemical characterization. *Euphytica*, 216(6), 88. https://doi.org/10.1007/s10681-020-02619-5.

Van Roekel, R., & Purcell, L. (2016). Understanding and increasing soybean yields. *In Proceedings of the Integrated Crop Management Conference, 2016 (Vol. 4).* Iowa State Univpuersity, IA.

Verdcourt, B. (1970). Studies in the Leguminosae-Papilionoideae for the'Flora of Tropical East Africa': IV. Kew Bulletin, 507-569.

Wani, S. H., Choudhary, M., Barmukh, R., Bagaria, P. K., Samantara, K., Razzaq, A., et al. (2022). Molecular mechanisms, genetic mapping, and genome editing for insect pest resistance in field crops. *Theoretical and Applied Genetics*, 135(11), 3875–3895. https://doi.org/10.1007/s00122-022-04060-9.

Xiong, H., Shi, A., Mou, B., Qin, J., Motes, D., Lu, W., et al. (2016). Genetic diversity and population structure of cowpea (Vigna unguiculata L. Walp). *PLoS One*, 11(8), e0160941.

Yahaya, D., Denwar, N., & Blair, M. W. (2019). Effects of moisture deficit on the yield of cowpea genotypes in the Guinea Savannah of Northern Ghana. *Agricultural Sciences*, 10(4), 577–595.

Yang, G., Zuo, Q., Liu, R., Yin, C., Shi, J., Hui, F., et al. (2016). Characteristics of P accumulation and distribution at the maturity stage of Brassica napus varieties with different phosphorus use efficiency for grain production. *Journal of Plant Nutrition*, 39(13), 1958–1970.

Yao, T., Xu, Y., Jiang, H., Chen, X., Liu, X., Chen, H., et al. (2023). Evaluating, Screening and Selecting Yardlong Bean [Vigna unguiculata subsp. sesquipedalis (L.) Verdc.] for Resistance to Common Cutworm (Spodoptera litura Fabricius). *Agronomy*, 13(2), 502.

8 Crop Wild Relatives of Pea (*Pisum sativum*) for Designing Future Climate-Resilient Cultivars

R. Beena, P.R. Nithya, and Roshni Vijayan

INTRODUCTION

The garden pea (*Pisum sativum* L.) belongs to the Fabeae tribe, which encompasses five genera, including notable grain legumes such as *Lathyrus* (grass pea), *Lens* (lentils), *Pisum* (peas), and *Vicia* (vetches). Two commonly recognized species within the genus *Pisum* are *P. fulvum* Sibth. & Sm. and *P. sativum* L. (Burstin et al., 2020), the latter further divided into two subspecies: the domesticated pea (subsp. *sativum*) and the wild form (subsp. *elatius* M. Bieb.) (Smýkal et al., 2012). Wild pea populations (*Pisum sativum* subsp. *elatius*) are found scattered throughout the Mediterranean basin, while *P. fulvum* is restricted to the Middle East. Despite the existence of approximately 98,000 accessions of worldwide pea germplasm, only a small fraction (less than 1%) represents wild pea varieties (Smýkal et al., 2013).

Pea is one of the major food legumes that can grow in different regions, and it ranks fourth in world food legume production next to soybean, peanut, and dry bean (Vidal-Valverde et al., 2003). Pea (*Pisum sativum* L.) is a self-pollinating annual crop consumed as a fresh vegetable (immature whole pods or seeds), as seed grain for food or livestock feed, or as a processed food ingredient. The pea vine is also used as fodder. Recently, pea protein has emerged as a favored and in-demand protein source. Furthermore, peas are becoming popular in the pet food market as they are grain-free and an excellent source of essential amino acids required by cats and dogs. The demand for pea protein is already high and is expected to soar in the coming years. To meet this growing demand, there is a need to double the current productivity from 1% to 2%.

Nutritionally, pea seeds are considered to have about 21%–33% protein and 56%–74% carbohydrate, with average iron, selenium, zinc, and molybdenum contents of about 97, 42, 41, and 12 ppm, respectively (Parihar et al., 2016, 2021). Therefore, it serves as an important ingredient in providing nutritional security for resource-poor people in developing countries (McCouch et al., 2016). *P. sativum* is an herbaceous annual, with a climbing hollow stem growing up to 2–3 m long. Leaves are alternate,

DOI: 10.1201/9781003434535-8

pinnately compound, and consist of 2–3 pairs of 1.5–8 cm long large leaf-like stipules. Flowers have five green fused sepals and five white to reddish-purple petals of different sizes. The fruit grows into a pod, 2.5–10 cm long, that often has a rough inner membrane. The pod is a seed container composed of two sealed valves and splits along the seam that connects the two valves. Seeds are round, smooth, and green.

The genetic diversity of a species is the outcome of cumulative mutation, recombination, and selection of individuals by the environment. Earlier, the selection was done from wild plants carrying promising traits for cultivation, resulting in locally adapted landraces that had lost many allele combinations that were disadvantageous to the farmer such as dehiscent pods and thick test as, and gained useful ones, such as increased seed size (Zohary and Hopf, 1973).

Wild varieties of pea (*Pisum* spp.) encompass a diverse range of genetic resources, holding great potential for crop improvement programs (Smykal et al., 2021). These wild relatives possess unique traits, including biotic resistance, that can be harnessed to enhance the resilience and productivity of cultivated pea varieties. Wild pea species exhibit a wide array of genetic variations and adaptations, enabling them to withstand diverse environmental conditions and biotic stresses. They have evolved mechanisms of resistance against pathogens, insects, and nematodes, making them valuable sources of genetic traits for biotic resistance in cultivated peas. These wild pea germplasm resources serve as an invaluable pool of genetic diversity that can be exploited through breeding and molecular techniques to develop improved pea varieties with enhanced resistance to biotic stresses (Cowling et al., 2017; Renzi et al., 2022).

Tolerance/resistance plays a crucial role in crop improvement by protecting plants against pests and pathogens, thereby enhancing yield potential, reducing dependence on chemical pesticides, and ensuring food security (Savary et al., 2019). Incorporation of resistance traits against major pests and diseases into cultivated varieties can mitigate the losses caused by these biotic factors and contribute to stable and sufficient food production. This approach also promotes sustainable agriculture by minimizing the environmental impact of chemical interventions (Pratap et al., 2020). Crop improvement programs with a focus on the development of biotic resistance have the potential to address the challenges of global food security and promote sustainable agricultural practices (Wolfe et al., 2008).

Centre of origin and taxonomy: The origin of *Pisum* spp. is in Southwestern Asia including Afghanistan, India, and Pakistan, and then spreads to the subtropic and tropic regions. The vernacular names of *P. sativum* include Chinese pea, edible pod pea, field pea, garden pea, green pea, honey pea, sugar pea, and sweet pea (English); ertjie (Afrikaans); katar (Bengali); ervilha (Brazil); jia wan dou (Chinese); doperwten (Dutch); petit pois (French); erbse (German); kacang ercis (Indonesian); endo (Japanese); sandaek (Khmer); kacang manis (Malay); ervilha (Portuguese); gorach (Russian); aroeja (Spanish); spritart (Swiss); thua lan tao (Thai); bezelye (Turkish); ropox (Ukrainian); and dau hoa lan (Vietnamese)(Lim, 2012). The taxonomy of *P. sativum* is as follows: the Kingdom (Plantae); Subkingdom (Viridiplantae); Infrakingdom (Streptophyta); Superdivision (Embryophyta); Division (Tracheophyta); Subdivision (Spermatophytina); Class (Magnoliopsida); Superorder (Rosanae); Order (Fabales); Family (Fabaceae); Genus (Pisum); and Species (*P. sativum*) (ITIS, 2016) (Table 8.1).

TABLE 8.1
***Pisum* Taxons Currently Listed Under GRIN Global Taxonomy**

Species, Authority	Subspecies, Authority
Pisum sativum L.	
	Pisum sativum L. subsp. *asiaticum* Govorov
	Pisum sativum L. subsp. *elatius* (M. Bieb.) Asch. & Graebn.
	Pisum sativum L. subsp. *elatius* var. *pumilio* Meikle
	Pisum sativum L. subsp. *elatius* var. *brevipedunculatum* P. H. Davis & Meikle
	Pisum sativum L. subsp. *jomardii* (Schrank) Kosterin
	Pisum sativum L. subsp. *sativum*
	Pisum sativum L. subsp. *sativum* var. *arvense* (L.) Poir.
	Pisum sativum L. subsp. *transcaucasicum* Govorov
Pisum abyssinicum A. Braun	
Pisum fulvum Sm.	

CYTOLOGY

Chromosome architecture and behavior: The Pisum chromosomes are ideally large and easily identifiable under a light microscope. Because of their well-defined structure, Pisum chromosomes found their place in the genetic studies from Mendel to current epigenetic research (Reid and Ross, 2011; Mathe and Vagas, 2013), providing insights into chromosome morphology, chromatin organization, and segregation mechanisms in plants. The unique characteristics of the Pisum chromosome structure, especially in the centromere region make them ideal for segregation studies. The foundation of genetics was based on Mendel's studies on Pisum, confirming the unique features of the extended primary constriction in Pisum chromosomes. These constrictions show histone phosphorylation patterns similar to holocentric chromosomes, which play a significant role in chromatid cohesion and accurate chromosome segregation (Neumann et al., 2016).

Somatic cell division: Our fundamental understanding of mitotic principles of inheritance, involving Mendel's crossbreeding experiments with peas, marks the first notable study. It is still remarkably interesting to note that the same *Pisum* is being studied for growth under spaceflight condition and stress (Yurkevich et al., 2018). The different stages of mitosis in *P. sativum* like condensation, separation alignment, and movement of chromosomes are being studied by researchers, paving the way to our understanding of the overall process of mitosis (Walczak et al., 2010).

Singh et al. (2017) studied the genetic variability in 32 pea genotypes using multivariate analysis of yield components. Genotyping of 3,020 germplasm collections revealed three groups like landraces, cultivars, and wild species. Later nested STRUCTURE analysis divided them into 14 sub-groups, many of which correlate with taxonomic sub-divisions of Pisum, domestication-related phenotypic traits, and/or restricted geographical locations (Jing et al., 2010). Mglinets et al. (2022) identified Psat0s797g0160 of the reference pea genome encoding for the seed cotyledon albumin SCA gene.

IDEAL SEVEN TRAITS OF MENDEL AND THEIR FUNCTIONS

Mendel's Trait	Phenotype	Mendel's Symbol	Chromosome/ Linkage group	Gene Function	Reference
Seed shape	Round/ Wrinkled	*R/r*	Chr3LGV	Starch branching enzyme1	Bhattacharyya et al. (1990)
Stem length	Tall/Dwarf	*LE/le*	Chr5LGIII	GA 3-oxidase1	Martin et al. (1997); Lester et al. (1997)
Cotyledon color	Yellow/ Green	*I/i*	Chr2LGI	Stay-green gene	Sato et al. (2007)
Seed coat/ flower color	Purple/ White	*A/a*	Chr6LGII	bHLH transcription factor	Hellens et al. (2010)
Pod color	Green/ Yellow	*GP/gp*	Chr3LGV	Encodes a protein involved in the differential development of bundle sheath and mesophyll cell chloroplasts/3' exoribonucleases family	Barth and Conklin (2003); Shirasawa et al. (2021)
Pod form	Inflated/ Constricted	*V/v*	Chr5LGIII	Cell wall remodeling	Opassiri et al. (2006)
Position of flowers	Axial/ Terminal	*FA/fa*	Chr4LGIV	Controls shoot and floral meristem size, and contributes to establishing and maintaining floral meristem identity	Clark et al. (1993, 1997)

SOURCES OF WILD RELATIVES OF *PISUM SATIVUM*/ GERMPLASM DIVERSITY

Crop wild relatives (CWRs) are commonly considered a potential resource of novel genes/alleles for crop improvement. The wild relatives of *Pisum sativum* (common pea) can be found in various regions worldwide. They are an important source of germplasm diversity for breeding programs aimed at improving cultivated peas. Here are some common sources of wild relatives of *Pisum sativum* and germplasm diversity.

BIOTIC STRESSES

Pea crops (*P. sativum*) are susceptible to various biotic stresses that can significantly impact their growth and yield. The major biotic stresses affecting pea crops include pathogens, insects, and nematodes. Pathogens such as fungal pathogens (e.g., powdery mildew, *Ascochyta* blight), bacterial pathogens (e.g., bacterial blight), and viral pathogens (e.g., pea seed-borne mosaic virus) can cause devastating diseases, leading to yield losses and reduced quality of harvested peas. Insects also pose a significant threat to pea crops, with common pests including pea aphids, pea weevils, and caterpillars, which can feed on foliage, flowers, and pods, causing direct damage and reducing crop productivity (Li et al., 2021). Additionally, nematodes such as root-knot nematodes and cyst nematodes can attack the roots of pea plants, leading to stunted growth, nutrient deficiencies, and yield reduction. Effective management strategies to combat these biotic stresses are crucial to ensure the success and sustainability of pea crop production.

BIOTIC RESISTANCE TRAITS IN WILD PEA SPECIES

The yield potential of cultivated pea, *P. sativum* is greatly affected by its low resistance to biotic stress and other environmental conditions. The success in developing varieties resistant to biotic stresses depends on the availability of good sources of resistance, an accurate scoring method, and the inheritance of resistance. While single-gene resistances can be easily introduced through backcrossing, introducing quantitative resistance governed by multiple minor genes/quantitative trait loci (QTLs) is challenging. Most resistant varieties are based on major genes, with fewer successes achieved in polygenic resistances. The wild germplasm of *P. sativum* has been recognized as a valuable resource for biotic resistance due to its diverse genetic makeup and adaptations to various environmental conditions. Varied levels of resistance to different diseases, pests, and nematodes were identified from wild pea species such as *P. abyssinicum*, *P. fulvum* and *P. sativum* ssp. *elatius*. By tapping into these genetic reservoirs of wild pea relatives, new resistance traits can be introduced to enhance the overall resilience of peas against wild stresses.

DISEASE RESISTANCE IN WILD PEAS

Several wild pea accessions possess natural resistance to fungal pathogens that commonly affect cultivated peas. For example, *P. fulvum* and *P. sativum* ssp. *elatius* exhibit resistance against powdery mildew caused by *E. pisi* (Pavan et al., 2011) and pea seed-borne mosaic virus (PSbMV) (Nayidu et al., 2014).

P. abyssinicum, an Ethiopian native, showed race non-specific resistance or partial resistance to pea bacterial blight caused by *P. syringaepv. pisi* (Taylor et al., 1994). Extensive screening of a large pool of wild germplasm identified high levels of resistance in *P. fulvum* against rust disease caused by *U. pisi* (Barilli et al. 2018). Barilli et al. (2010) integrated a high-density linkage map of *P. fulvum* and identified QTLs responsible for resistance to rust diseases. Other wild pea species, including *P. abyssinicum* and *P. sativum* ssp. *elatius*, exhibit resistance against *Ascochyta* blight (Burstin et al., 2015; Smýkal et al., 2018). Wild pea germplasm also holds the potential for resistance against bacterial pathogens that affect pea crops. For instance, wild pea accessions such as *P. sativum* ssp. *elatius* exhibit resistance against bacterial blight caused by *P. syringae* pv. *pisi* (Singh et al., 2016). These resistant traits can be valuable for developing pea cultivars with enhanced resistance to diseases.

INSECT RESISTANCE IN WILD PEAS

Wild pea species possess natural resistance to various insect pests, providing opportunities for developing insect-resistant pea varieties. Wild pea accessions, such as *P. fulvum* and *P. sativum* ssp. *elatius*, exhibit resistance to pea aphids and pea weevils (Mansouripour and Mousavi, 2015; Smith et al., 2019). These resistant traits can be introgressed into cultivated pea varieties to enhance their resistance to insect pests. Through extensive research involving moderately resistant and resistant accessions of *P. fulvum*, a wild species from the eastern Mediterranean and Near East areas where pea domestication originated, sources of antibiosis resistance were identified (Clement et al., 2000). This knowledge enables the transfer of weevil resistance traits into pea cultivars, enhancing their ability to resist infestations by the pea weevil *B. pisorum* (Clement et al., 2002). As different parts and tissues of *P. fulvum* express resistance, it may be feasible to develop a multi-tiered defense mechanism against *B. pisorum* in a single pea cultivar to provide long-term stability in weevil resistance, contributing to the improved protection of pea crops. Seed resistance in *P. fulvum* to pea weevil controlled by at least three major recessive alleles, pwrl, pwr2, and pwr3, and complete susceptibility by three major dominant alleles, PWR1, PWR2, and PWR3, could be transferred effectively to cultivated peas through hybridization and repeated backcrossing (Byrne et al., 2008).

NEMATODE RESISTANCE IN WILD PEAS

Wild pea species offer potential sources of resistance against nematode pests that affect pea crops. Wild pea accessions, including *P. fulvum* and *P. sativum* ssp. *elatius*, exhibit resistance to root-knot nematodes and cyst nematodes (Dalmadi et al., 2012; Kharrat et al., 2017). These wild relatives provide potential sources for breeding nematode-resistant pea cultivars.

Pea wild relatives are valuable sources of genetic diversity for improving abiotic stress resistance in cultivated peas (*Pisum sativum*). Here are some examples of how wild relatives can contribute to enhancing abiotic stress tolerance:

1. **Drought resistance**: Wild relatives, such as *Pisum fulvum* and *Pisum elatius*, possess traits that enable them to withstand drought conditions. These traits include deep root systems, efficient water-use efficiency, and mechanisms for osmotic adjustment. Introducing genes from these wild relatives into cultivated pea varieties can enhance drought tolerance.
2. **Heat tolerance**: Wild relatives adapted to warmer climates, such as *Pisum sativum* subsp. *asiaticum*, often possess natural heat tolerance mechanisms. They may exhibit traits such as improved thermotolerance, increased antioxidant capacity, and heat shock protein production, which can help cultivated peas withstand high temperatures.
3. **Cold tolerance**: Wild relatives like *Pisum sativum* subsp. *elatius*, which are native to higher altitudes, often exhibit improved cold tolerance compared to cultivated varieties. They possess genetic traits such as increased freezing tolerance, cold acclimation capacity, and the ability to regulate ice formation, which can be introduced into cultivated peas to enhance their cold tolerance.
4. **Salinity tolerance**: Salinity is a major abiotic stress that affects crop productivity. Some wild relatives of pea, including *Pisum fulvum and Pisum*

elatius, possess natural salt tolerance mechanisms. They may have traits such as efficient ion exclusion, osmotic adjustment, and enhanced antioxidant activity that can help cultivated peas tolerate saline conditions.

5. **Aluminum tolerance**: Aluminum toxicity in acidic soils is a significant constraint for pea cultivation. Wild relatives, such as *Pisum sativum* subsp. *elatius*, often exhibit natural tolerance to aluminum toxicity. They possess genes and mechanisms that allow them to detoxify or exclude aluminum ions, which can be introduced into cultivated peas to improve their aluminum tolerance.

By utilizing wild relatives with inherent abiotic stress tolerance, breeders can incorporate these desirable traits into cultivated pea varieties through targeted breeding programs, thereby enhancing their resilience and productivity under challenging environmental conditions (Kaloo, 1993). However, it is important to note that introgressing these traits requires careful selection, backcrossing, and screening to retain desirable agronomic characteristics and eliminate undesirable traits from the wild relatives.

1. ***Pisum fulvum***: This wild relative of *Pisum sativum* is found in the eastern Mediterranean region, particularly in Israel, Lebanon, Syria, and Jordan. It is known for its resistance to several diseases, such as powdery mildew (Wand et al., 2021) and pea enation mosaic virus (Mahmond, 2021). *P. fulvum* is resistant to pea weevil infestation, and resistance alleles of pea weevil from *P. fulvum* have been identified (Clement et al., 2002; Byrne et al., 2008). Fondevilla et al. (2007) recognized *P. fulvum*as the source of resistance for powdery mildew.
2. ***Pisum elatius***: Also known as the field pea or wild field pea, *Pisum elatius* is native to the Mediterranean region, including areas in Europe, North Africa, and the Middle East (Kosterin et al., 2010). It is closely related to Pisum sativum and is a valuable source of genetic diversity for traits like tolerance to abiotic stress and resistance to pests and diseases.
3. ***Pisum abyssinicum***: This wild relative of *Pisum sativum* is found in Ethiopia and other parts of East Africa. It possesses traits such as resistance to fusarium wilt and is being explored for its potential in breeding programs (Weller et al., 2002).
4. ***Pisum sativum* subsp. *asiaticum***: Native to the Himalayan region and parts of China, this wild subspecies of Pisum sativum is known for its tolerance to high-altitude conditions. It offers potential for improving cold tolerance and adaptation to mountainous environments.
5. ***Pisum sativum* subsp. *transcaucasicum***: This wild subspecies is found in the Transcaucasus region, which includes countries like Georgia, Armenia, and Azerbaijan. It is adapted to diverse climatic conditions and is a valuable resource for broadening the genetic base of cultivated peas.
6. ***Pisum sativum* subsp. *elatius***: This subspecies is found in parts of Europe, including the Balkans and the Carpathian Mountains. It is considered one of the primary wild ancestors of *Pisum sativum* and contributes to the genetic diversity of cultivated peas.

WILD RELATIVES OF PEA USED IN PLANT BREEDING FOR ABIOTIC, BIOTIC RESISTANCE, AND QUALITY TRAITS

Trait	Wild sp	Reference
Yield improvement	*Pisum fulvum*	Prescott-Allen and Prescott-Allen (1988)
Ascochyta Blight Resistance	*Pisum fulvum*	Wroth (1998)
	Pisum sativum subsp. *elatius var. pumilio*	Ladizinsky and Abbo (2015)
		Gurung et al. (2002)
	Lathyrus clymenum	Gurung et al. (2002)
	Lathyrus ochrus	Gurung et al. (2002)
	Lathyrus sativus	Fondevilla et al. (2005)
	Pisum fulvum	Ellis (2011)
	Pisum sativum subsp. *elatius*	Ellis (2011)
	P. sativum ssp. syriacum	Fondevilla et al. (2022)
Pea Bacterial Blight Resistance	*Pisum abyssinicum*	Elvira-Recuenco (2000)
Pea Weevil Resistance	*Pisum fulvum*	Warkentin et al. (2015)
Powdery Mildew Resistance	*Pisum fulvum*	Fondevilla et al. (2007)
Rust Resistance	*Pisum fulvum*	Barilli et al. (2009)
Drought Tolerance	*Lathyrus gloeospermus*	Schaefer et al. (2012)
Frost Tolerance	*Pisum abyssinicum*	Elvira-Recuenco et al. (2003)
Leaf spot		
Broomrape Resistance	*Pisum abyssinicum*	Ellis (2011)
	Pisum fulvum	Ellis (2011)
	Pisum sativum subsp. *elatius*	Ellis (2011)
	Pisum sativum subsp. elatius var. pumilio	Ladizinsky and Abbo (2015)
Bruchid Resistance	*Pisum fulvum*	Ellis (2011)
Powdery Mildew Resistance	*Lathyrus cicera*	Patto et al. (2007)
	Pisum fulvum	Ellis (2011)
Seed lipoxygenase levels	*Pisum fulvum*	Forster et al. (1999)

LIMITATIONS AND POTENTIAL OF UTILIZING WILD RELATIVES OF PEA

Utilizing wild relatives of pea (*Pisum sativum*) in breeding programs has both limitations and potential. Here are some of the main limitations and potential benefits associated with the use of wild relatives:

LIMITATIONS

1. **Genetic compatibility**: Wild relatives and cultivated peas may have genetic barriers, such as differences in ploidy level or chromosomal structure, which can hinder successful hybridization and gene transfer.
2. **Undesirable traits**: Wild relatives may carry undesirable traits, such as bitterness, increased seed dormancy, or smaller seed size, which can be challenging to eliminate through breeding.

3. **Domestication traits**: Wild relatives may lack certain traits that have been selected and improved through domestication, such as increased yield, uniformity, or specific agronomic characteristics.
4. **Time-consuming and costly**: Utilizing wild relatives in breeding programs requires extensive resources, including time, labor, and funding, to effectively screen and evaluate large populations for desired traits.
5. **Regulatory considerations**: Accessing and utilizing genetic resources from wild relatives may require compliance with international and national regulations related to access and benefit-sharing.

POTENTIAL

1. **Genetic diversity**: Wild relatives offer a vast reservoir of genetic diversity, including resistance to diseases, pests, and abiotic stresses. This diversity can be harnessed to enhance the resilience and adaptability of cultivated peas to various environmental conditions.
2. **Trait introgression**: Through targeted breeding efforts, genes from wild relatives can be introgressed into cultivated pea varieties to confer specific desirable traits, such as disease resistance, tolerance to drought or heat, or improved nutritional content.
3. **Improvement of agronomic traits**: Wild relatives can contribute genes for traits like root architecture, nutrient uptake efficiency, or flowering time regulation, which can enhance the overall performance and productivity of cultivated peas.
4. **Future adaptability**: The genetic diversity present in wild relatives can provide a source of adaptive traits that may be valuable in the face of changing climatic conditions or emerging pests and diseases.
5. **Reservoir of novel genes**: Wild relatives may harbor unique genes or gene combinations that can provide novel traits or biochemical compounds with potential applications in food, pharmaceutical, or industrial sectors.

It is important to note that utilizing wild relatives in breeding programs requires careful evaluation, selection, and backcrossing to ensure the successful incorporation of desired traits while minimizing the introduction of undesirable characteristics. Collaboration between breeders, geneticists, and conservationists is crucial to effectively harness the potential of wild relatives of peas while addressing the limitations associated with their utilization.

UTILIZATION OF PEA FOR NUTRIENT USE EFFICIENCY

Pea (*Pisum sativum*) can be utilized to enhance nutrient use efficiency, which refers to the plant's ability to acquire, assimilate, and utilize nutrients effectively for growth and development. Here are some ways in which peas can contribute to nutrient use efficiency:

1. **Nitrogen fixation**: Pea has a unique ability to form symbiotic associations with nitrogen-fixing bacteria called rhizobia. These bacteria colonize the roots of pea plants and convert atmospheric nitrogen into a plant-usable

form, ammonium. This symbiotic relationship allows peas to obtain a significant portion of their nitrogen requirements without relying solely on external nitrogen fertilizers, thus improving nitrogen use efficiency.

2. **Phosphorus uptake and utilization**: Pea has developed adaptive mechanisms to efficiently acquire phosphorus from the soil, particularly in phosphorus-deficient conditions. It possesses extensive root systems with root hairs that enhance phosphorus uptake, as well as enzymes and transporters that facilitate phosphorus assimilation and utilization within the plant. These traits contribute to improved phosphorus use efficiency.
3. **Micronutrient availability**: Pea exhibits traits that enhance the availability and uptake of essential micronutrients, such as iron and zinc. For example, some pea cultivars have been bred to have enhanced iron and zinc concentrations in their seeds, which can contribute to addressing micronutrient deficiencies in human diets. Peas were naturally rich in Fe (46–54 mg kg^{-1}), Zn (39–63 mg kg^{-1}), and Mg (1,350–1,427 mg kg^{-1}). A single serving of field pea could provide 28%–68% of the recommended daily allowance (RDA) for Fe, 36%–78% of the RDA for Zn, and 34%–46% of the RDA for Mg. Field pea is not a good source of Ca (622–1,219 mg kg^{-1}; 6%–12% of RDA). In addition, these field peas are naturally low in PA (4.9–7.1 mg g^{-1} of PA or 1.4–2 mg g^{-1} of phytic-P) despite very high total P concentrations (3.5–5 mg g^{-1}) (Amarakoon et al., 2012).
4. **Nutrient cycling and organic matter**: Pea can contribute to nutrient cycling and organic matter accumulation in agricultural systems. When pea residues are incorporated into the soil after harvest, they can improve soil fertility by adding organic matter, increasing nutrient availability, and enhancing the soil's nutrient-holding capacity.
5. **Interactions with mycorrhizal fungi**: Pea can form symbiotic associations with mycorrhizal fungi, which can enhance nutrient uptake, particularly for phosphorus and micronutrients. Mycorrhizal fungi extend the root system and increase the nutrient-absorbing surface area, improving nutrient acquisition efficiency.
6. **Breeding for nutrient use efficiency**: Through breeding programs, pea varieties can be selected and developed for improved nutrient use efficiency. Traits such as enhanced nutrient uptake, assimilation, and utilization can be targeted for improvement, along with symbiotic interactions and root characteristics, facilitating nutrient acquisition.

Efforts to enhance nutrient use efficiency in peas can contribute to sustainable agriculture by reducing fertilizer inputs, minimizing nutrient losses to the environment, and improving crop productivity. Breeding, agronomic practices, and crop management strategies aimed at optimizing nutrient use efficiency in peas can have positive environmental and economic impacts while ensuring food security.

OTHER TRAITS FROM PEA FOR BREEDING PROGRAMS

1. **Seed traits**: Wild relatives can offer diverse seed traits, such as different seed coat colors, shapes, and sizes. These traits can be utilized for aesthetic purposes, specialty markets, or specific culinary preferences.

2. **Flowering time and photoperiod sensitivity:** Wild relatives may exhibit different flowering times and photoperiod sensitivities than cultivated peas. By introgressing genes from wild relatives, breeders can develop pea varieties with altered flowering times, allowing adaptation to different geographical regions or specific production systems.
3. **Tolerance to environmental stresses**: Apart from abiotic stresses, wild relatives of peas may possess tolerance or adaptation to other environmental stresses such as high altitude, low fertility, heavy metals, or toxic compounds. Incorporating these traits into cultivated peas can expand their range of adaptation and resilience to various challenging environments.
4. **Nutritional traits**: Wild relatives may contain unique nutritional profiles, such as higher levels of specific vitamins, antioxidants, or phytochemicals. These traits can be valuable for improving the nutritional quality of cultivated pea varieties and enhancing their potential health benefits.

Utilizing these diverse traits from wild relatives through targeted breeding efforts can help improve the overall performance, adaptability, and sustainability of cultivated peas while preserving and utilizing the genetic resources available in wild populations.

SPEED BREEDING STRATEGIES ADOPTED IN PEA

Speed breeding strategies have been adopted in pea (*Pisum sativum*) to accelerate the breeding process and reduce the time required to develop new pea varieties. These strategies aim to shorten the generation time, increase the number of generations per year, and expedite the selection process. Here are some commonly used speed breeding strategies in peas:

1. **Controlled environments**: Pea plants are grown in controlled environments, such as growth chambers or greenhouses, where environmental conditions like temperature, humidity, and photoperiod can be manipulated and optimized for rapid growth and development. This allows researchers to create ideal conditions for pea plants, enabling them to grow faster and complete their life cycle in a shorter time.
2. **Extended photoperiod**: Pea is a long-day plant, meaning it requires a specific minimum day length to initiate flowering. By extending the photoperiod using artificial lighting, breeders can induce earlier and more synchronized flowering in pea plants, reducing the time required for plants to reach reproductive stages.
3. **High-density planting**: Pea plants are densely packed in trays or containers to maximize space utilization and increase the number of plants per unit area. This approach allows breeders to screen a large number of plants simultaneously, enabling faster evaluation of traits and selection of desired individuals.
4. **Early seedling production**: Pre-germinated seeds or young seedlings can be used to establish plants, reducing the time required for seed germination

and early growth stages. This approach allows researchers to skip the initial growth phase and accelerate the development of plants for subsequent evaluations and selections.

5. **Application of growth regulators**: Growth regulators, such as gibberellic acid, can be used to promote rapid growth and development in pea plants. By applying specific concentrations of growth regulators at appropriate growth stages, researchers can stimulate plant growth and hasten the overall development process.
6. **Marker-assisted selection**: Genetic markers associated with target traits can be used to identify desirable individuals at early stages of plant development. This approach allows breeders to select plants based on genetic markers linked to specific traits of interest, enabling faster and more efficient selection processes.
7. **Tissue culture and micropropagation**: Tissue culture techniques can be employed to rapidly multiply desirable pea genotypes. Through techniques like shoot multiplication and somatic embryogenesis, large numbers of identical plants can be produced in a short period, facilitating the propagation and distribution of selected varieties.

Cazzola et al. (2020) compared three systems: *in vitro, in vivo – in vitro,* and *in vivo* for speed breeding in pea. Greater efficiency (51%–95%) and lower cost were found for *in vivo* systems. It consisted of a hydroponic system, with a 22-h photoperiod supplied by fluorescent T5 tubes, a temperature of 20 ± 2°C, flurprimidol antigiberelin, and early grain harvest. Rapid generation technology used 0.6 µM flurprimidol, 266 plants per square meter, a 20-hour photoperiod, 21°C/16°C light/dark, a hydroponic system with vermiculite substrate, scheduled fertilizer application, and 500 µM m^{-2} s^{-1} light intensity using T5 fluorescent bulbs. This approach was 30–45 days per generation faster than conventional single seed descent (SSD) methods (Mobini and Warkentin, 2016).

By combining these speed breeding strategies, researchers can significantly reduce the breeding cycle time in peas, allowing for the development of new varieties more quickly. These strategies have the potential to accelerate the genetic improvement of pea crops, enhance breeding efficiency, and address the challenges posed by changing agricultural needs and environmental conditions.

RESEARCH RELATED TO GENOME EDITING IN PEA

Increasing yield potential and reliability is a complex challenge that will benefit from the adoption of novel strategies, such as genomics-assisted selection (marker-assisted selection or genomic selection) and precision breeding (gene editing).

Genome editing techniques, such as clustered regularly interspaced short palindromic repeats and CRISPR-associated protein 9 (CRISPR-Cas9), have gained significant attention in the field of plant research, including in peas (*Pisum sativum*). While research on genome editing in peas is still relatively limited compared to other crops, there have been some notable studies and advancements. Here are a few examples of research related to genome editing in peas:

1. **Targeted gene knockout**: Researchers have successfully used CRISPR-Cas9 to generate targeted gene knockouts in peas. For instance, Li et al. (2023) demonstrated the use of CRISPR-Cas9 to disrupt the *PsPDS* gene, which is involved in carotenoid biosynthesis, resulting in an albino phenotype in pea seedlings. They used the CRISPR/Cas9 system to edit the *pea phytoene desaturase* (*PsPDS*) gene, causing albinism, by Agrobacterium-mediated genetic transformation. This was the first report of a successful generation of gene-edited pea plants by this route. This study showcased the potential of genome editing for functional gene analysis and trait modification in peas.
2. **Pest resistance**: CRISPR-Cas9 technology is a very efficient functional analysis tool and has been developed in several insects to edit their genomes through the injection of eggs with guide RNAs targeting coding sequences of genes of interest. Le Trionnaire et al. (2019) used CRISPR-Cas9 mutagenesis to create and maintain stably edited aphid lineages. Stylin-01 edited stable lineages have been successfully generated, and colonies are now well-established.
3. **Disease resistance**: Genome editing holds promise for improving disease resistance in peas. Pavan et al. (2011) described the obtainment of a novel *er1* resistant line by experimental mutagenesis with the alkylating agent diethyl sulfate. This line was found to carry a single nucleotide polymorphism in the *PsMLO1* gene sequence, predicted to result in premature termination of translation and a non-functional protein. A cleaved amplified polymorphic sequence (CAPS) marker was developed on the mutation site and shown to be fully cosegregating with resistance in F_2 individuals. Sequencing of PsMLO1 from three powdery mildew-resistant cultivars also revealed the presence of loss-of-function mutations. Taken together, results reported in this study strongly indicate the identity between er1 and *mlo* resistances and are expected to be of great breeding importance for the development of resistant cultivars via marker-assisted selection. CRISPR-Cas9 can be used to generate targeted mutations in the PsMLO1 gene, which is associated with susceptibility to powdery mildew. The mutated plants showed reduced susceptibility to powdery mildew, suggesting the potential for developing powdery mildew-resistant pea varieties through genome editing.

It's worth noting that while these studies demonstrate the potential of genome editing in peas, further research is needed to optimize and validate the efficiency, specificity, and long-term stability of genome-edited traits in pea plants. Additionally, as genome editing technologies continue to advance, more research is expected to emerge, providing new insights and applications for genome editing in peas and contributing to the genetic improvement of this important crop.

The garden pea (*Pisum sativum* L.) played an integral role as the model organism in Mendel's groundbreaking discovery of the laws of inheritance in 1866, a major cornerstone of modern plant genetics. Even before Mendel's experiments, the pea had already been subject to experimental investigations (Smykal, 2014; Smykal et al., 2016). Despite the close evolutionary connections among the major leguminous crops, there are considerable disparities in genome size, fundamental chromosome number, ploidy level, reproductive characteristics, etc. For a comprehensive framework for legumes, two species namely *Medicago truncatula* and *Lotus japonicus*, were chosen as models for exploring legume biology and genomics (Cook, 1999; Sato et al., 2008).

The preservation of vast pea germplasm in various germplasm collections institutions, such as French National Institute for Agriculture, Food and Environemnt (NIAR, France) (8,839), ATFCC, Australia (7,432), N. I. Vavilov all Russian science research Institute of Plant Industry's fruit collection (NIVRIPR, Russia) (6,790), International Center for Agricultural Research in the Dry Areas ICARDA, Syria (6,105), exhibits huge diversity characterized by various morpho-agronomic traits, geographical distributions, etc. (Kreplak et al., 2019). The infusion of genetic traits from the diverse pea germplasm helps improve genetic diversity. The process starts with the pre-selection of desired traits through observable characteristics or Marker Assisted Selection (MAS). The core emphasis pursued by breeders and geneticists involves improving resistance to biotic and abiotic stresses, refining quality attributes, and optimizing yield attributes.

Even though various leguminous crops have been extensively researched for their genetic framework, the genome size remains a major constraint for thorough exploration. Among all cultivated legumes, soybean is the pre-eminent and economically pivotal leguminous crop whose genome has been available since 2010, followed by the common bean in 2014 and the garden pea in 2019 (Schmutz et al., 2010, 2014; Kreplak et al., 2019). Subsequently, a large number of leguminous species have undergone sequencing. These genomic sequences have been augmented by an extensive array of genomic assets, encompassing resources such as tools for conducting genome-wide association studies, diversity panels, and online germplasm databases (Bauchet et al., 2019). These resources have eased the expansion of molecular breeding endeavors in legumes. The possession of established reference genomes holds pivotal importance in propelling the progression of genomic mapping methodologies (Kaloumenos et al., 2019), either through conventional biparental populations or through association mapping panels.

Possessing comprehensive genomic data encompassing diversity across a substantial and varied collection of accessions, coupled with the accumulation of descriptive phenotypic data, equips the researchers for genome-wide association studies and genomic selection. This will finally lead to the identification of candidate genes/loci that regulate the traits under study and offer markers with practical applications in plant breeding (Tafesse et al., 2020; Annicchiarico et al., 2020).

Understanding the molecular mechanisms behind resistance to biotic and abiotic stresses points to a new area in peas. This exploration enhances the comprehension of evolutionary dynamics between pathogens and their host organisms and streamlines breeding methodologies. With the substantial variability exhibited by pathogens and their ability to overcome single-gene or allelic resistance, it is crucial to identify the allelic diversity within specific genes as in the case of powdery mildew resistance in peas (Sun et al., 2019).

The influence of climate change is already being felt across various crops, peas being no exception. It is imperative to understand the mechanisms of stress tolerance,

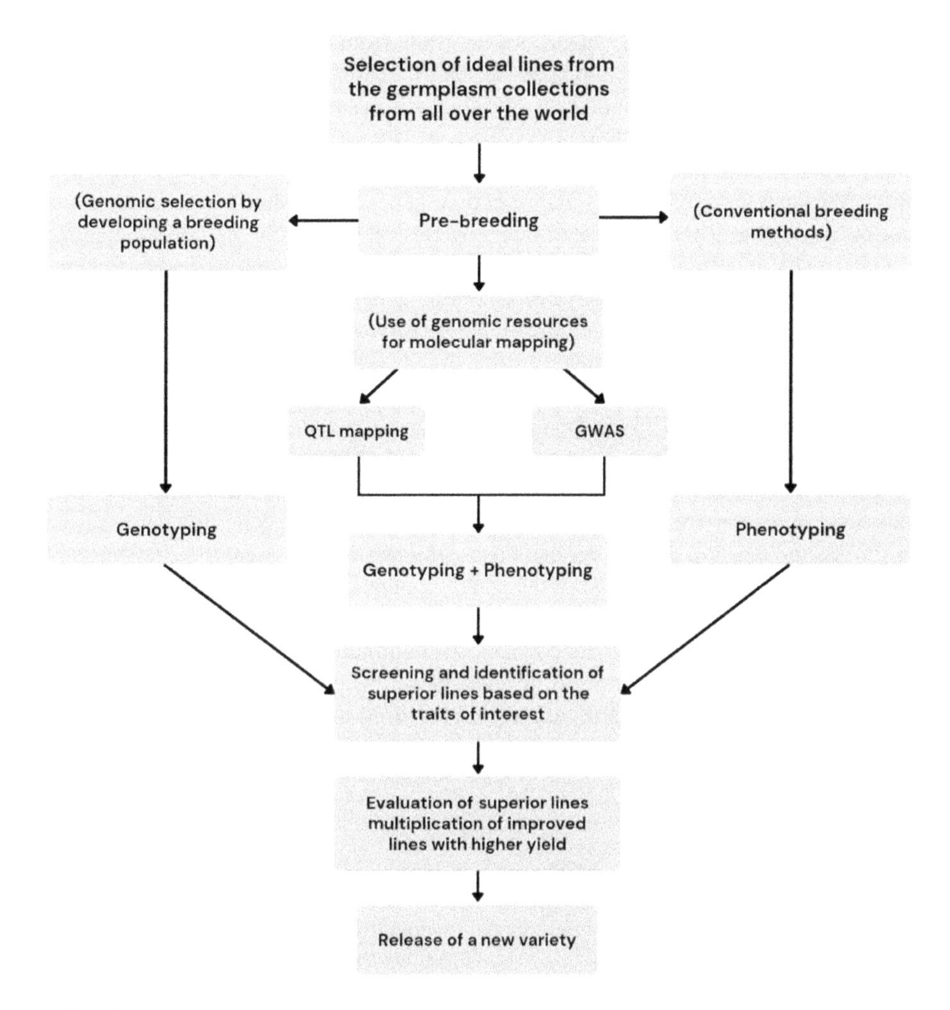

FIGURE 8.1 Breeding strategies in peas.

avoidance, and resistance in peas to mitigate the ensuing aftereffects. According to Kumar et al. (2019), there is a critical need for developing legume crops resilient to changing climates, especially for tropical and sub-tropical regions struggling with limited water and soil resources. Among the various breeding methodologies, selective breeding needs to integrate conventional methods with new high-throughput technologies of genomics, with special emphasis on speed breeding to help develop improved climate-resilient varieties in garden peas (Figure 8.1).

CONCLUSION

CWRs of pea (*Pisum sativum*) play a crucial role in designing future climate-resilient cultivars. These wild relatives possess valuable genetic diversity and adaptive traits that can be utilized to enhance the resilience of cultivated peas to changing environmental conditions. By incorporating traits from wild relatives into breeding programs, researchers can develop pea varieties that are better equipped to withstand abiotic stresses, such as drought, heat, and soil salinity, as well as resist pests and diseases. The genetic diversity present in CWRs provides a rich resource for breeders to improve traits, such as abiotic stress tolerance, nutrient use efficiency, and yield stability. Through targeted crosses and selection, genes responsible for desirable traits can be introgressed into cultivated pea varieties, allowing for the development of climate-resilient cultivars. Additionally, CWRs can offer valuable traits beyond abiotic stress resistance. They may possess unique genes related to disease resistance, insect resistance, seed traits, and nutritional quality, which can further enhance the adaptability, productivity, and market value of cultivated peas. Conservation and utilization of CWRs are essential to ensure the availability of genetic resources for future breeding efforts. Efforts should be made to collect, preserve, and characterize diverse wild pea populations to understand their genetic potential and identify valuable traits for breeding programs. Collaboration among breeders, geneticists, conservationists, and farmers is crucial to effectively utilize the genetic diversity present in CWRs and develop climate-resilient pea cultivars that can meet the challenges of a changing climate. By harnessing the genetic potential of CWRs of peas, breeders can contribute to sustainable agriculture, food security, and the resilience of farming systems in the face of climate change. The continued exploration and utilization of CWRs will play a vital role in developing future pea cultivars that can thrive in diverse environmental conditions while providing nutritious food for a growing global population.

REFERENCES

Amarakoon, D., Thavarajah, D., McPhee, K., & Pushparajah Thavarajah, P. (2012). Iron-, zinc-, and magnesium-rich field peas (Pisum sativum L.) with naturally low phytic acid: a potential food-based solution to global micronutrient malnutrition. *Journal of Food Composition and Analysis*, 27(1), 8–13. https://doi.org/10.1016/j.jfca.2012.05.007.

Annicchiarico, P., Nazzicari, N., Laouar, M., Thami-Alami, I., Romani, M., & Pecetti, L. (2020). Development and proof-of-concept application of genome-enabled selection for pea grain yield under severe terminal drought. *International Journal of Molecular Sciences*, 21, 2414.

Barilli, E., Sillero, J. C., Moral, A., & Rubiales, D. (2009). Characterization of resistance response of pea (Pisum spp.) against rust (Uromyces pisi). *Plant Breeding*, 128(6), 665–670.

Barilli, E., Satovic, Z., Rubiales, D., & Torres, A. M. (2010). Mapping of quantitative trait loci controlling partial resistance against rust incited by Uromyces pisi (Pers.) Wint. in a Pisum fulvum L. intraspecific cross. *Euphytica*, 175, 151–159.

Barilli, E., Cobos, M. J., Carrillo, E., Kilian, A., Carling, J, & Rubiales, D. (2018). A high-density integrated DArTseq SNP-based genetic map of *Pisum fulvum* and identification of QTLs controlling rust resistance. *Frontiers in Plant Science*, 9, 167. https://doi.org/10.3389/fpls.2018.00167.

Barilli, E., et al. (2018). The use of marker-assisted selection to improve common bean. *Frontiers in Plant Science*, 9, 995.

Barth, C., & Conklin, P. L. (2003). The lower cell density of leaf parenchyma in the *Arabidopsis thaliana* mutant lcd1-1 is associated with increased sensitivity to ozone and virulent *Pseudomonas syringae*. *Plant Journal*, 35, 206–218.

Bauchet, G. J., Bett, K. E., Cameron, C. T., Campbell, J. D., Cannon, E. K., Cannon, S. B., Carlson, J. W., Chan, A., Cleary, A., Close, T. J., et al. (2019). The future of legume genetic data resources: challenges, opportunities, and priorities. *Legume Science*, 1, e16.

Bhattacharyya, M. K., Smith, A. M., Ellis, T. H., Hedley, C., & Martin, C. (1990). The wrinkled-seed character of pea described by mendel is caused by a transposon-like insertion in a gene encoding starch-branching enzyme. *Cell*, 60(1), 115–122. https://doi.org/10.1016/0092-8674(90)90721-p. PMID: 2153053.

Burstin, J., et al. (2015). Ecogeographic analysis of the worldwide pea collection: Revisiting its classification. *Crop Science*, 55(5), 1939–1951.

Burstin, J., Kreplak, J., Macas, J., & Lichtenzveig, J. (2020). *Pisum sativum* (Pea). *Trends in Genetics*, 36(4), 312–313. https://doi.org/10.1016/j.tig.2019.12.009.PMID:31959367.

Byrne, O. M., Hardie, D. C., Khan, T. N., Speijers, J., Yan, G., Byrne, O. M., Hardie, D. C., Khan, T. N., Speijers, J., & Yan, G. (2008). Genetic analysis of pod and seed resistance to pea weevil in a Pisum sativum × P. fulvum interspecific cross. *Australian Journal of Agricultural and Resource*, 59, 854–862. https://doi.org/10.1071/AR07353.

Cazzola, F., Bermejo, C. J., Guindon, M. F., Cointry, E. (2020). Speed breeding in pea (Pisum sativum L.), an efficient and simple system to accelerate breeding programs. *Euphytica* 216, 178. https://doi.org/10.1007/s10681-020-02715-6.

Clark, S. E., Running, M. P., & Meyerowitz, E. M. (1993). CLAVATA1, a regulator of meristem and flower development in arabidopsis. *Development*, 119(2), 397–418. https://doi.org/10.1242/dev.119.2.397. PMID: 8287795.

Clark, S. E., Williams, R. W., & Meyerowitz, E. M. (1997). The CLAVATA1Gene encodes a putative receptor kinase that controls shoot and floral meristem size in arabidopsis. *Cell*, 89(4), 575–585. https://doi.org/10.1016/s0092-8674(00)80239-1.

Clement, S. L., Wightman, J. A., Hardie, D. C., Bailey, P., Baker, G., & McDonald, G. (2000). Opportunities for integrated management of insect pests of grain legume. In: R. Knight (ed.) *Linking and Tissues Expressing Resistance, It may be Possible to Research and Marketing Opportunities for Pulses in the 21st Century*. Kluwer, Dordrecht, The Netherlands, pp. 467–480.

Clement, S. L., Hardie, D. C., & Elberson, L. R. (2002). Variation among accessions of *Pisum fulvum* for resistance to pea weevil. *Crop Science*, 42, 2167–2173. https://doi.org/10.2135/cropsci2002.2167.

Cook, D. R. (1999). *Medicago truncatula*-a model in the making! *Current Opinion in Plant Biology*, 2, 301–304.

Cowling, W. A., Li, L., Siddique, K. H. M., Henryon, M., Berg, P., Banks, R. G., & Kinghorn, B. P. (2017). Evolving gene banks: improving diverse populations of crop and exotic germplasm with optimal contribution selection. *Journal of Experimental Botany*, 68, 1927–1939.

Dalmadi, Á., et al. (2012). Evaluation of root-knot nematode (Meloidogyne spp.) resistance in pea (Pisumsativum L.) lines and wild relatives. *Euphytica*, 185(3), 479–486.

Ellis, T. H. N. (2011). Pisum. In: Chittaranjan Kole (ed.) Wild Crop Relatives: Genomic and Breeding Resources. Springer, Berlin Heidelberg, pp. 237–248.

Elvira-Recuenco, M. (2000). Sustainable control of pea bacterial blight: approaches for durable genetic resistance and biocontrol by endophytic bacteria. PhD thesis, Wageningen University, The Netherlands.

Elvira-Recuenco, M., Bevan, J. R., & Taylor, J. D. (2003). Differential responses to pea bacterial blight in stems, leaves and pods under glasshouse and field conditions. *European Journal of Plant Pathology*, 109(6), 555–564.

Fondevilla, S., Avila, C. M., Cubero, J. I., & Rubiales, D. (2005). Response to Mycosphaerella pinodes in a germplasm collection of Pisum spp. *Plant Breeding*, 124, 313–315.

Fondevilla, S., Carver, T. L. W., Moreno, M. T., & Rubiales, D. (2007). Identification and characterization of sources of resistance to *Erysiphe pisi* Syd. in Pisum spp. *Plant Breeding*, 126, 113–119.

Fondevilla, S., Torres, A. M., Moreno, M. T., & Rubiales, D. (2007). Identification of a new gene for resistance to powdery mildew in *Pisumfulvum*, a wild relative of pea. *Breeding Science*, 57, 181–184. https://doi.org/10.1270/jsbbs.57.181.

Fondevilla, S., Krezdorn, N., Rubiales, D., Rotter, B., & Winter, P. (2022). Bulked segregant transcriptome analysis in pea identifies key expression markers for resistance to Peyronellaea pinodes. *Scientific Reports*, 12(1), 18159.

Forster, C., North, H., Afzal, N., Domoney, C., Hornostaj, A., Robinson, D. S., & Casey, R. (1999). Molecular analysis of a null mutant for pea (Pisum sativum L.) seed lipoxygenase-2. *Plant Molecular Biology*, 39(6), 1209–1220.

Gurung, A., Pang, E., & Taylor, P. (2002). Examination of Pisum and Lathyrus species as sources of ascochyta blight resistance for field pea (Pisum sativum). *Australasian Plant Pathology*, 31, 41–45.

Hellens, R. P., Moreau, C., Lin-Wang, K., Schwinn, K. E., Thomson, S. J., et al. (2010). Identification of mendel's white flower character. *PLoS One*, 5(10), e13230. https://doi.org/10.1371/journal.pone.0013230.

Integrated Taxonomic Information System (ITIS). (2016). Pisum sativum L. Taxonomic Serial No.: 26867. Geological Survey, VA, USA.

Jing, R., Vershinin, A., Grzebyta, J., Shaw, P., Smýkal, P., Marshall, D., Ambrose, M. J., Ellis, T. N., & Flavell, A. J. (2010). Transcriptome sequencing for high throughput SNP development and genetic mapping in Pea. *BMC Evolutionary Biology*, 10, 44. https://www.biomedcentral.com/1471-2148/10/44.

Kalloo, G. (1993). *Genetic Improvement of Vegetable Crops||Pea*. Pergamon Press, Oxford, New York, pp. 409–425. https://doi.org/10.1016/b978-0-08-040826-2.50033-3.

Kaloumenos, N., et al. (2019). GWAS on temperate and pea: unraveling the polygenic control of a complex trait. *Frontiers in Plant Science*, 10, 1421.

Kharrat, M., et al. (2017). Genetic diversity and population structure in Pisumsativum L. *G3: Genes, Genomes, Genetics*, 7(2), 541–553.

Kosterin, O. E., Zaytseva, O. O., Bogdanova, V. S., & Ambrose, M. J. (2010). New data on three molecular markers from different cellular genomes in Mediterranean accessions reveal new insights into phylogeography of Pisum sativum L. subsp elatius (Bieb.) Schmalh. *Genetic Resources and Crop Evolution*, 57, 733–739.

Kreplak, J., Madoui, M. A., & Cápal, P. et al. (2019). A reference genome for pea provides insight into legume genome evolution. *Nature Genetics*, 51, 1411–1422. https://doi.org/10.1038/s41588-019-0480-1.

Kumar, J., Choudhary, A. K., Gupta, D. S., & Kumar, S. (2019). Towards exploitation of adaptive traits for climate resilient smart pulses. *International Journal of Molecular Sciences*, 20, 2971.

Ladizinsky, G., & Abbo, S. (2015). The pisum genus. In: *The Search for Wild Relatives of Cool Season Legumes*. Springer International Publishing, Berlin Heidelberg, pp. 55–69.

Lester, D. R., Ross, J. J., Davies, P. J., & Reid, J. B. (1997). Mendel's stem length gene (Le) encodes a gibberellin 3 beta-hydroxylase. *The PlantCell*, 9(8), 1435–1443. https://doi.org/10.1105/tpc.9.8.1435.

Le Trionnairea, G., Tanguya, S., Hudaverdiana, S., Gleonneca, F., Richarda, G., Cayrolb, B., Monsionb, B., Pichonb, E., Deshouxb, M., Websterb, C., Uzestb, M., Herpinc, A., & Tagub, D. (2019). An integrated protocol for targeted mutagenesis with CRISPR-Cas9 system in the pea aphid. *Insect Biochemistry and Molecular Biology*, 110, 34–44. https://doi.org/10.1016/j.ibmb.2019.04.016.

Li, X., et al. (2021). Sustainable management of insect pests in pulse crops: Challenges and opportunities. *Journal of Pest Science*, 94, 121–142.

Li, G., Liu, R., Xu, R., Varshney, R. K., Ding, H., Li, M., Yan, X., Huang, S., Li, J., Wang, D., Ji, Y., Wang, C., He, J., Luo, Y., Gao, S., Wei, P., Zong, X., & Tao Yang, T. (2023). Development of an Agrobacterium-mediated CRISPR/Cas9 system in pea (*Pisum sativum* L.). *The Crop Journal*, 11, 132–139. https://doi.org/10.1016/j.cj.2022.04.011.

Lim, T. (2012). *Edible Medicinal and Non-Medicinal Plants. Vol. 2. Fruits*. Springer, Netherlands.

Mahmoud, G. A. E. (2021). Biotic stress to legumes: fungal diseases as major biotic stress factor. *Sustainable Agriculture Reviews 51: Legume Agriculture and Biotechnology*, 2, 181–212.

Mansouripour, S. M., & Mousavi, S. A. (2015). Screening pea (Pisumsativum L.) genotypes for resistance to pea weevil, Bruchuspisorum L. *Journal of Applied Entomology*, 139(5), 378–385.

Martin, D. N., Proebsting, W. M., & Hedden, P. (1997). Mendel's dwarfing gene: cDNAs from the Le alleles and function of the expressed proteins. *Proceedings of the National Academy of Sciences of the United States of America*, 94(16), 8907–8911. https://doi.org/10.1073/pnas.94.16.8907. PMID: 9238076; PMCID: PMC23192.

Máthé, C., & Vasas, G. (2013). Microcystin-LR and cylindrospermopsin induced alterations in chromatin organization of plant cells. *Marine Drugs*, 11(10), 3689–3717. https://doi.org/10.3390/md11103689.

McCouch, S. R., et al. (2016). Agriculture: feeding the future. *Nature*, 540(7631), 30–32.

Mglinets, A. V., Bogdanova, V. S., & Kosterin, O. E. (2022). Identification of the gene coding for seed cotyledon albumin SCA in the pea (Pisum L.) genome. *Vavilovskii Zhurnal Genet Selektsii*, 26(4), 359–364. https://doi.org10.18699/VJGB-22-43.

Mobini, S. H., & Warkentin, T. D. (2016). A simple and efficient method of in vivo rapid generation technology in pea (Pisum sativum L.). *Vitro Cellular and Developmental Biology - Plant*, 52(5), 530–536. https://doi.org/10.1007/s11627-016-9772-7.

Nayidu, N. K., et al. (2014). Genetic and molecular analysis of resistance to pea seed-borne mosaic virus (PSbMV) in the Pisum collection. *Journal of General Plant Pathology*, 80(1), 26–39.

Neumann, P., Schubert, V., Fuková, I., Manning, J. E., Houben, A., & Macas, J. (2016). Epigenetic histone marks of extended meta-polycentric centromeres of lathyrus and pisum chromosomes. *Frontiers in Plant Science*, 7, 174512. https://doi.org/10.3389/fpls.2016.00234.

Opassiri, R., Pomthong, B., Onkoksoong, T., Akiyama, T., Esen, A., & Ketudat Cairns, J. (2006). Analysis of rice glycosyl hydrolase family 1 and expression of Os4bglu12 β-glucosidase. *BMC Plant Biology*, 6(1), 33. https://doi.org/10.1186/1471-2229-6-33.

Parihar, A. K., Bohra, A., & Dixit, G. P. (2016). Nutritional benefits of winter pulses with special emphasis on peas and rajmash. In: U. Singh, C. S. Praharaj, S. S. Singh, and N. P. Singh (eds) *Biofortification of Food Crops*, Springer, New Delhi, pp. 61–71. https://doi.org/10.1007/978-81-322-2716-8_6.

Parihar, A. K., Dixit, G. P., Singh, U., Singh, A. K., Kumar, N., & Gupta, S. (2021). Potential of field pea as a nutritionally rich food legume crop. In: D. S. Gupta, S. Gupta, and J. Kumar (eds) *Breeding for Enhanced Nutrition and Bio-Active Compounds in Food Legumes*, Springer, Cham, pp. 47–82. https://doi.org/10.1007/978-3-030-59215-8_3.

Patto, M. V., Fernndez-Aparicio, M., Moral, A., & Rubiales, D. (2007). Resistance reaction to powdery mildew (Erysiphe pisi) in a germplasm collection of Lathyrus cicera from Iberian origin. *Genetic Resources and Crop Evolution*, 54(7), 1517–1521.

Pavan, S., Schiavulli, A., Appiano, M., Marcotrigiano, A. R., Cillo, F. F., Visser, R. G. F., Yuling Bai, Y., Lotti, C., & Ricciardi, L. (2011). Pea powdery mildew er1 resistance is associated to loss-of-function mutations at a MLO homologous locus. *Theoretical and Applied Genetics*, 123, 1425–1431. https://doi.org/10.1007/s00122-011-1677-6.

Pavan, S., et al. (2011). Characterization of resistance genes to powdery mildew in pea (Pisumsativum L.). *ActaHorticulturae*, 907, 195–199.

Pratap, A., Douglas, C., Prajapati, U., Kumari, G., War, A. R., Tomar, R., Pandey, A. K., & Dubey, S. (2020). Breeding progress and future challenges: biotic stresses. *The Mungbean Genome*, 2020, 55–80.

Prescott-Allen, R., & Prescott-Allen, C. (1988). *Genes From the Wild: Using Wild Genetic Resources for Food and Raw Materials*. 2nd edn. International Institute for Environment and Development/Earthscan Publications, London.

Reid, J. B., & Ross, J. J. (2011). Mendel's genes: toward a full molecular characterization. *Genetics*, 189(1): 3–10. https://doi.org/10.1534/genetics.111.132118.

Renzi, J. P., Coyne, C. J., Berger, J., von Wettberg, E., Nelson, M., Ureta, S., Hernández, F., Smýkal, P., & Brus, J. (2022). How could the use of crop wild relatives in breeding increase the adaptation of crops to marginal environments?. *Frontiers in Plant Science*, 13, 886162.

Sato, Y., Morita, R., Nishimura, M., Yamaguchi, H., & Kusaba, M. (2007). Mendel's green cotyledon gene encodes a positive regulator of the chlorophyll-degrading pathway. *Proceedings of the National Academy of Sciences of the United States of America*, 104(35), 14169–14174. https://doi.org/10.1073/pnas.0705521104.

Sato, S., Nakamura, Y., Kaneko, T., Asamizu, E., Kato, T., Nakao, M., Sasamoto, S., Watanabe, A., Ono, A., Kawashima, K., et al. (2008). Genome structure of the legume, *Lotus japonicus*. *DNA Research*, 15, 227–239.

Savary, S., Willocquet, L., Pethybridge, S. J., Esker, P., McRoberts, N., & Nelson, A. (2019). The global burden of pathogens and pests on major food crops. *Nature Ecology & Evolution*, 3(3), 430–439.

Schaefer, H., Hechenleitner, P., Santos-Guerra, A., de Sequeira, M. M., Pennington, R. T., Kenicer, G., & Carine, M. A. (2012). Systematics, biogeography, and character evolution of the legume tribe Fabeae with special focus on the middle-Atlantic island lineages. *BMC Evolutionary Biology*, 12(1), 1.

Schmutz, J., Cannon, S. B., Schlueter, J., Ma, J., Mitros, T., Nelson, W., Hyten, D. L., Song, Q., Thelen, J. J., Cheng, J., et al. (2010). Genome sequence of the palaeopolyploid soybean. *Nature*, 465, 120.

Schmutz, J., McClean, P. E., Mamidi, S., Wu, G. A., Cannon, S. B., Grimwood, J., Jenkins, J., Shu, S., Song, Q., Chavarro, C., et al. (2014). A reference genome for common bean and genome-wide analysis of dual domestications. *Nature Genetics*, 46, 707–713.

Shirasawa, K., Sasaki, K., Hirakawa, H., & Isobe, S. (2021). Genomic region associated with pod color variation in pea (*Pisum sativum*). *G3 Genes|Genomes|Genetics*, 11(5), jkab081. https://doi.org/10.1093/g3journal/jkab081.

Singh, A. K., Rai, R., Singh, B. D., Chand, R., and Srivastava, C. P. (2015). Validation of SSR markers associated with rust (Uromyces fabae) resistance in pea (Pisum sativum L.). *Physiology and Molecular Biology of Plants* 21, 243–247. doi: 10.1007/s12298-015- 0280-8.

Singh, S. K., Pathak, R., Choudhary, V. (2016). Plant growth-promoting rhizobacteria-mediated acquired systemic resistance in plants against pests and diseases. *Microbial-mediated Induced Systemic Resistance in Plants*, 125–134.

Singh, S. R., Ahmed, N., Singh, D. B., Srivastva, K. K., Singh, R. K., & Abid, M. (2017). Genetic variability determination in garden pea (*Pisum sativum* L sub sp. Hortense Asch. and Graebn.) by using the multivariate analysis. *Legume Research*, 40(3), 416–422.

Smith, L., et al. (2019). Exploring the genetic basis of pea weevil resistance in wild pea relatives. *Frontiers in Plant Science*, 10, 523.

Smykal, P. (2014). Pea (*Pisum sativum* L.) in Biology prior and after Mendel's Discovery. *Czech Journal of Genetics and Plant Breeding*, 50, 52–64.

Smýkal, P., Aubert, G., Burstin, J., Coyne, C. J., Ellis, N. T., Flavell, A. J., Ford, R., Hýbl, M., Macas, J., Neumann, P., McPhee, K. E., Redden, R. J., Rubiales, D., Weller, J. L., & Warkentin, T. D. (2012). Pea (*Pisum sativum* L.) in the Genomic Era. *Agronomy*, 2(2), 74–115. https://doi.org/10.3390/agronomy2020074.

Smykal, P., Varshney, R. K., Singh, V. K., Coyne, C. J., Domoney, C., Kejnovsky, E., & Warkentin, T. (2016). From mendel's discovery on pea to today's plant genetics and breeding. *Journal of Applied Genetics*, 129, 2267–2280.

Smýkal, P., et al. (2018). Genetic diversity and population structure of pea (Pisum sativum L.) varieties derived from combined retrotransposon, microsatellite and morphological marker analysis. *Theoretical and Applied Genetics*, 131(11), 2411–2428.

Smýkal, P., et al. (2021). Pea (Pisum sativum L.). In: *Handbook of Legumes of World Economic Importance*, Springer, Cham, pp. 135–206.

Sun, S., Deng, D., Duan, C., Zong, X., Xu, D., He, Y., & Zhu, Z. (2019). Two Novel er1 alleles conferring powdery mildew (Erysiphe pisi) resistance identified in a worldwide collection of pea (Pisum sativum L.) germplasms. *International Journal of Molecular Sciences*, 20, 5071.

Tafesse, E. G., Gali, K. K., Lachagari, V. B. R., Bueckert, R., & Warkentin, T. D. (2020). Genome-wide association mapping for heat stress responsive traits in field pea. *International Journal of Molecular Sciences*, 21, 2043.

Taylor, J. D., Roberts, S. J., & Schmit, J. (1994). Screening for resistance to pea bacterial blight {Pseudomonas syringaepv. pisi). In: M. Lemaittre, S. Freignoun, K. Rudolph, and J. G. Swings, (eds) *Proceedings of the 8th International Conference of Plant Pathogenic Bacteria*, CABI Compendium, Paris, pp. 1027.

Vidal-Valverde, C., Frias, J., Hernandex, A., Martin-Alvarez, P., Sierra, I., Rodriquez, C., et al. (2003). Assessment of nutritional compounds and antinutritional factors in pea (Pisum sativum) seeds. *Journal of the Science of Food and Agriculture*, 83, 298–306

Walczak, C. E., Cai, S., & Khodjakov, A. (2010). Mechanisms of chromosome behaviour during mitosis. *Nature Reviews. Molecular Cell Biology*, 11(2), 91. https://doi.org/10.1038/nrm2832.

Wang, L., et al. (2021). Proteomics analysis reveals the molecular basis of resistance to powdery mildew in pea (Pisumsativum L.). *Journal of Proteomics*, 231, 104005.

Warkentin, T. D., Smykal, A. P. P., Coyne, C. J., Weeden, N., Domoney, C., Bing, D. J., & McPhee, K. E. (2015). Pea. In: De Ron, A. M. (ed.) *Grain Legumes*. Springer, Cham, pp. 37–83.

Weller, J. L., et al. (2002). Genetic mapping of resistance to Fusarium wilt race 2 in pea (Pisumsativum L.). *Theoretical and Applied Genetics*, 105(5), 809–814.

Wolfe, M. S., Baresel, J. P., Desclaux, D., Isabelle, G., Hoad, S., Kovacs, G., Löschenberger, F., Miedaner, T., Østergård, H., & Lammerts van Bueren, E. T. (2008). Developments in breeding cereals for organic agriculture. *Euphytica*, 163, 323–346.

Wroth, J. M. (1998). Possible role for wild genotypes of Pisum spp. to enhance ascochyta blight resistance in pea. *Australian Journal of Experimental Agriculture*, 38(5), 469–479.

Yurkevich, O. Y., Samatadze, T. E., Levinskikh, M. A., Zoshchuk, S. A., Signalova, O. B., Surzhikov, S. A., Sychev, V. N., Amosova, A. V., & Muravenko, O. V. (2018.) Molecular cytogenetics of Pisum sativum L. grown under spaceflight-related stress. *BioMed Research International*, 2018, 4549294. https://doi.org/10.1155/2018/4549294.

Zohary, D., & Hopf, M. (1973). Domestication of pulses in the old world. *Science*, 182, 887–894.

9 An Ethnobotanical Review of Tuberous Legumes

Mackenzie K. Laverick and
Eric J. Bishop von Wettberg

Tuberous legumes are unique among the Fabaceae due to their elongated or enlarged roots that are often edible, their high content of protein and starches, and their potential to act as natural bio-drills to break up compacted soils (Stai et al., 2020). Like potatoes, tuberous legumes are highly productive food crops with yields anywhere from 4 to 20 times those of wheat or corn (Mann, 2011). Since the plant does not have to carry the weight of the tuber, it can grow much larger. Legume tubers can have a size and biomass similar to that of potatoes but have a much higher protein content (~12%), more micronutrients, and often other edible parts like pods or seeds. They also have promising soil remediation qualities like symbiotic nitrogen fixation and bio-drilling. Like potatoes, the tubers can be harvested anytime during the end of the growing season but unlike potatoes, many other parts of the plant including the beans or seeds, flowers, and shoots are also edible. There are about six major grain legumes: soybean, peanut, dry bean, chickpea, pigeon pea, and cowpea. However, most other cultivated legumes are only grown on a small scale for feeding families, especially in areas with limited resources or harsh conditions, and are thus vital parts of many communities (Yang et al., 2018). There are many minor grain legumes, such as winged beans, that have a lot of agricultural potential as tuber crops (Hymowitz and Boyd, 1977).

Humans and plants have had a complex relationship from agriculture to culture and beyond for thousands of years. Indigenous and local communities have used plants as "food, medicine, shelter, decoration, construction, and clothing", giving certain plants huge cultural value (Kumar et al., 2021). Ethnobotany bridges the disciplines of social and environmental sciences, as the intersection of plants and people provides knowledge that can help us face global challenges such as ending poverty, achieving zero hunger, improving nutrition, boosting health, promoting sustainable agriculture, combating climate change, and more (Kumar et al., 2021). With the rise of food insecurity, climate change's adverse effects on the environment, and issues in agriculture, there is an increased demand for more suitable alternatives. At the forefront of this are wild foods that are locally grown, adapted to local environmental conditions, resilient, and beneficial to the human diet (Kumar et al., 2021).

DOI: 10.1201/9781003434535-9

Ethnobotanists already have a significant amount of insight into plants with rich cultural history that are nutritious and can help promote biodiversity and aid in soil remediation.

In Vermont, a state in the Northern United States just south of the Canadian province of Quebec and the city of Montreal, many diverse ethnic groups have chosen or been forced to leave their traditional homes. Most of the larger groups include people from Asia, such as Tibetans, Burmese, and Laotians. There is also a substantial group of Africans, such as Burundi, Somali, and Sudanese, as well as even Europeans from Bosnia and Herzegovina (Vermont Folklife Center, 2021). Additionally, the majority of undocumented migrant workers on Vermont farms hail from Mexico and Central America (Wakefield, 2019). Many of these groups are at a higher risk for food insecurity and may miss some traditional foods from their home regions. Here, we take a global perspective on tuber-forming legumes to find species potentially suitable for the small-scale agriculture of these groups.

This chapter explores ten tuberous legumes with potential as new crops for a more diverse and sustainable food system by helping increase nutrition, biodiversity, and soil remediation. The plants are: *Apios americana* (groundnut, hopniss), *Glycine tomentella* (wild soybean), *Lathyrus linifolius* (everlasting pea), *Melilotus alba* (white sweet clover), *Pachyrhizus tuberosus* (Jicama), *Phaseolus coccineus* (scarlet runner bean), *Psophocarpus tetragonolobus* (winged bean), *Sphenostylis stenocarpa* (African yam bean), *Vigna lanceolata* (bush carrot), and *Vigna vexilata* (zombi pea). General information on each of the legumes in the study is listed below in Table 9.1. These legumes hail from countries worldwide with varied cultural histories, and many have ancient roots in food and medicine, giving them a unique ethnobotanical background. Several legumes, such as *Phaseolus coccineus*, *Sphenostylis stenocarpa* (African yam bean), *Vigna lanceolata*, and *Apios americana*, are known to form tuberous roots that serve as storage organs for starches and other carbohydrates (Saxon, 1981), and a few are cultivated commercially, including *Pachyrhizus tuberosus* (Jicama).

What are the ethnobotanical characteristics of each crop including their places of origin, optimal habitats, morphological characteristics, cultural uses, and edibility rating? How well can these plants grow in Vermont and in what ways can they aid in soil remediation?

MATERIALS AND METHODS

SOURCING ETHNOBOTANICAL LITERATURE

Sources were collected from online databases including AGRICOLA, ProQuest, JSTOR, and Google Scholar, using keywords such as [ethnobotany], [food], [culture], [growth], [crop], [domestication], [history], [origins], [tuber], [tuberous], [legumes], and [edible], in combination with the plant's scientific names.

CULTIVATION IN VERMONT

We conducted a trial study in the summer of 2020 to calculate growth rates and a harvest index. The scarlet runner beans were started in the UVM Greenhouse and

TABLE 9.1
A Chart of Promising Tuberous Legumes and Basic Background Information about Each

Scientific Name	Common Name	Origin	Edible Parts	Uses
Apios americana	Groundnut, hopniss	North America	Tuber, seeds, shoots, flowers	Food
Glycine tomentella	Wild soybean	Asia	Tuber, seeds	Food, Medicine
Lathyrus linifolius	Bitter-vetch	Europe	Tuber	Food, Medicine, appetite suppressant
Melilotus alba	White sweet clover	Europe	Leaves for medicinal tea	Livestock forage crop
Pachyrhizus tuberosus	Jicama	Mesoamerica	Tuber	Food
Phaseolus coccineus	Scarlet runner bean	Mesoamerica	Dry seeds, immature pods, flowers	Food, Medicine
Psophocarpus tetragonolobus	Winged bean	Asia	Pods, leaves, shoots, tuber, seeds	Food
Sphenostylis stenocarpa	African yam bean	Sub-Saharan Africa	Tuber	Food
Vigna lanceolata	Bush carrot or pencil yam	Australia	Tuber	Food
Vigna vexillata	Zombi pea	Africa/Indonesia	Tuber, seeds	Food

later transplanted to the UVM Horticulture Farm, where the study was conducted. The rest of the seeds were directly planted into the soil on June 16, 2020. The plants were grown in a high tunnel, and data was collected on plant growth habits, associated pollinators, rhizobia presence, tuber size, and estimated yields. The plants were harvested on November 6, 2020, after roughly 20 weeks or 143 days.

Most of the seeds were obtained from GRIN (seed appendix for order list), and the *Phaseolus* seeds were ordered from Johnny's Seeds (https://www.johnnyseeds.com/). Many *Apios* tubers were bought from a Louisiana company called Interwoven Permaculture Farm, and some were collected from local wild populations.

We also planted a second batch of crops for a follow-up study for taste test analysis in 2021 after sprouting the plants in the greenhouse during the spring with seeds leftover from the previous year. The plants were started in the UVM Greenhouse as seedlings on April 10, 2021, and transplanted to the UVM Horticulture Farm on 04/30/2021, where they were grown until harvest on August 31, 2021 (Figures 9.1 and 9.2).

Data Collection

When harvested, measurements were taken on shoot height, root length, shoot biomass, root biomass, presence of a tuber, and the presence of rhizobia, including nodule count. Other characteristics such as the shape and size of bean pods, tubers,

FIGURE 9.1 From top left to right and down. (a) *Apios americana* flowers (b) *Glycine tomentella* flowers (c) *Lathyrus linifolius* winged rachis (d) *Melilotus alba* flowers (e) Massive *Pachyrhizus tuberosus* leaves (f) *Phaseolus coccineus* flowers (g) *Psophocarpus tetragonolobus* flowers (h) *Sphenostylis stenocarpa* flowers (i) *Vigna lanceolata* flowers.

FIGURE 9.2 (a) *Vigna vexillata* flowers (b) *Lathyrus linifolius* tuber and nodules (c) Young *Melilotus alba* roots and nodules (d) *Psophocarpus tetragonolobus* tetrapod bean (e) *Phaseolus coccineus* beans (f) Bag of *Apios americana* tubers (g) Hoop house at the UVM Horticulture Farm (h) Massive *Pachyrhizus tuberosus* plants pre-harvest.

and flowers were also noted. Overall health and growth of the plants were observed including a comparison of the number of plants that survived versus seeds planted (germination rate).

Root Growth Analysis

WinRhizo was used to scan roots and determine root length, number of lateral roots, and count nodules (Regent Co., Quebec). RStudio was used to analyze the data including various analysis of variance (ANOVA) tests to determine if tuber size impacted other aspects of the plant's growth and health (based on biomass), and if rhizobia presence affected any of these characteristics. ANOVA tests were also run on lateral root count versus tuber presence/size and overall root biomass to see if the presence and extent of lateral roots were positively or inversely related to tuber size.

Edibility Tests

To test the edibility of the tubers and other parts of the plant, taste tests were conducted, and recipes were developed. A good portion of the plants had recipes listed online, which were followed when relevant. If none were found, we derived cooking instructions from ethnobotanical literature sources and followed them accordingly. After harvest, the tubers and other edible parts of each species of legumes were prepared, and a blind taste test was conducted to determine the edibility and tastiness of the plants. The panel consisted of ten people of varying ages and backgrounds in an attempt to reflect the taste preferences of the Vermont community and determine the viability of these crops in the food market. All the recipes were vegan and gluten-free to ensure everyone in the taste test was able to participate. The study was approved by the IRB under the number STUDY00001606.

The participants followed the guidelines for sensory evaluation from UVM Extension's article by Roy Desrochers, Heather Darby, and Sara Ziegler titled "Evaluating the Sensory Characteristics of Organic Grass Fed Milk" from November 2020. The analysis observed mouthfeel, balance, fullness, intensity of flavor and aroma, and the five basic tastes. Participants filled out a form ranking the level of smell, intensity of flavor, fullness of flavor, and likability. They were asked to describe the flavor, and balance of the dish, and whether or not they would make this dish again for themselves. The board included several people from across the United States, including the Northeast, East, Midwest, and Western regions of the country. There were also participants originally from Turkey, Ghana, and Iran.

RESULTS

Ethnobotanical Review

Apios americana: It is also known as groundnut or hopniss and is a perennial legume that is naturally found across North America with a history entwined with various Native American cultures as well as settler cultures (Turner and Aderkas, 2012).

The plant grows in a range of habitats but favors wet soils near bodies of water. Plants can be either diploid or triploid (diploid being more common in the south and lower Midwest while triploid is more common in the north) and reproduce sexually or asexually from tuber sprouting (Belamkar et al., 2020). This plant was an early cultivar, as the tuber could be harvested at any time of the year, providing nutrients such as proline and complex starches (Kalberer et al., 2020). The taste is often described as a "mix of boiled peanut and Irish potatoes", and due to the tubers' low levels of reducing sugars, they are claimed to be excellent in the form of fries or chips (Belamkar et al., 2015). The tubers can also be dried and powdered for use in cooking as a starch, similar to potato or corn starch (Belamkar et al., 2015). The seeds, shoots, and flowers of this plant are also edible. *Apios* is a promising crop not only for its nutritional value but also for its ease of cooking, taste, long shelf life when refrigerated (over a year), and wide adaptation to geographic climates and conditions (Belamkar et al., 2015). One of the few cons with this plant is the growth form. *Apios* tubers are root or stolon tubers that grow at nodes from a central "mother" tuber and continue to spread outwards. This growth pattern can make it difficult to grow for harvest due to the extensive sprouting, space needed for the roots to grow, and trellising needed for the shoots. The tubers can be small when young, so for maximum yield they should be harvested in late fall, which unfortunately takes a fair amount of labor to dig as the tubers grow wherever they please and can stretch for feet in various directions (Belamkar et al., 2020).

Apios americana is one of five species in the genus *Apios* (Li et al., 2014). There is some evidence that Price's groundnut, *Apios priceana,* the only other North American species, may also have been used as a wild-collected food crop and has potential for further cropping and domestication (Blackmon and Reynolds, 1986). The rest of the genus is East Asian. Interestingly, although we are not aware of reports (in English at least) of the use of the native *Apios* in East Asia, there is the importation of *Apios americana* for use as flour (Kalberer et al., 2020).

Glycine tomentella: It is a wild member of the soybean genus (*Glycine*) with a tetraploid ($2n=80$) genotype (Bell, 2011). Soybean is a well-known, economically important legume grown for feed, oil, and various soy food products. The genetic base of cultivated soybean is very narrow, with the genus originating in East Asia, and many ancient cultivars are now on the edge of extinction as farmers focus on growing the more commercial varieties. There are 16 perennial species of soybean, with *Glycine max* being the most commonly cultivated species, domesticated from the wild *Glycine soja* (Singh and Hymowitz, 1999). The seeds of *Glycine tomentella* are promising as they have various compounds like phytoestrogens that may have several health benefits besides being rich in protein (Bell, 2011). Most soybean seeds contain roughly 40% proline and 20% oil. *G. tomentella* is commonly found on sandy soil and has smaller leaflets than some other members of the genus. Some breeding attempts have been made between *G. max* and *G. tomentella*, resulting in fertile diploid lines that were more resistant to pests and pathogens (Singh and Hymowitz, 1999). Growing this and other wild species of soybean is important to maintain genetic diversity within this genus to prevent species loss and maintain an adaptable gene pool. We are not aware of the use of the tubers of *G. tomentella*, although this may be due to a lack of investigation by ethnobotanists.

Lathyrus linifolius: It is also known as everlasting pea or bitter vetch, is a tuberous legume from Europe whose roots were previously used as food during times of famine. It has also been used for medicinal purposes in areas such as Scotland, where the tubers were used to aid with chest ailments, offset inebriation, relieve flatulence, and add a licorice-like flavor to various drinks (Smykal, 2020). The tubers can also be ground into flour and fermented to make beer or distilled into spirits (Dello Jacovol et al., 2019). *Lathyrus* tubers have been noted to increase feelings of satiation, which some think could be promising as an appetite suppressant or to boost energy. The compound likely responsible for this is trans-anethole (Smykal, 2020). *Lathyrus linifolius* is also effective at fixing nitrogen in low-nutrient soils and supports various moth and butterfly species which help promote insect biodiversity. While originally from the Mediterranean, this plant is now found naturalized or in some cases, planted intentionally across the globe. It can often be seen in grazed or ungrazed grasslands at low altitudes in low pH (4–7) and nutrient-poor soils. In a study by Dello Jacovol et al. (2019), a comparative phylogenetic characterization using the nodA gene sequence revealed that all "rhizobia isolated from *L. linifolius* segregated with *Rhizobium pisi,* alongside *R. leguminosarum, R. lentis, R. bangladeshense, R. binae*, and *R. laguerreae* and a strain from *L. sativus* (grass pea)". *Lathyrus linifolius* is also cultivated for aesthetic purposes due to its aromatic flowers. However, it has not been domesticated with loss of seed dormancy or indehiscence, like the crop grasspea, *Lathyrus sativus.*

***Melilotus alba* or white sweet clover**: It is a biennial plant that produces a taproot with fibrous secondary roots and often nodulates with bacterial symbionts. *Melilotus* originated in Eurasia as a foraging plant and is currently being used in Australia and South America to help restore salt-affected roadsides and saline soils for agriculture (Wolf et al., 2004). Since the 1900s, sweet clovers have been planted to help stabilize eroded soils, fix nitrogen, and serve as a forage crop for livestock (Gucker, 2009). Agricultural lands typically have low C:N ratios, and in these poor soils, non-native species that are good nitrogen fixers can easily exploit the area and spatially outcompete native species (Wolf et al., 2004). *Melilotus alba* and *M. officinalis* (white and yellow sweet clover) are such plants. These biennials have taken over a good portion of the Midwest due to their competitive ability to colonize disturbed areas, produce large amounts of seeds, grow early in the season, produce extensive root systems, and fix nitrogen almost immediately (Wolf et al., 2004). This is in part due to symbiotic relationships with *Rhizobium* bacteria that aid in nitrogen fixation within depleted soils. The same study by Wolf et al. found that these species may alter soil nutrient and moisture levels, and they have a concerning habit of spreading beyond their planted areas and outcompeting native plants. While promising for soil restoration, *Melilotus* plantings should be controlled outside of their native range to prevent invasion. Like other forage legumes, it is not domesticated in the sense that it has seed shattering and seed dormancy (e.g., von Wettberg et al., 2019).

Pachyrhizus tuberosus: It is also known as jicama and is a semi-domesticated tropical tuberous legume originally from the Americas. This legume has been cultivated in South America and East Asia and can already be found in many grocery stores in the United States and the Western world. There are several other semi-domesticated species within this genus (Rodriguez-Navarro et al., 2020), though *P. tuberosus* is the most popular, with production beyond its native range in the Philippines and Southeast Asia. Several *Pachyrhizus* species are native to South America, specifically the Andean region, and were previously cultivated by indigenous cultures (pre-Columbian) for food or medicinal uses (Rodriguez-Navarro et al., 2020). This legume is notable for its large root that can be eaten raw, cooked, or even turned into juice (Sorensen, 1997). Eaten as a vegetable, *Pachyrhizus* provides vitamin C, starch, sugar, and fiber. The roots are used to make West African foods like gari, can be made into biodegradable plastics for industrial applications, used as a root starch to be a gluten-free flour or food thickener, and they have a higher protein content than similar tuberous roots like sweet potatoes and yams (Rodriguez-Navarro et al., 2020). *P. tuberosus* is able to fix nitrogen and grows quickly. The roots are also rich in iron, magnesium, and zinc. Unfortunately, the seeds are not edible due to rotenone, a chemical compound that acts as a natural insecticide (Agaba et al., 2016).

As a semi-domesticated species that has received less attention than many other cultivated legumes, it is not clear how its distribution in the Andes is driven by humans versus natural processes. Furthermore, determining differences among wild and semi-cultivated populations is simply challenging. Further work is needed to examine genetic patterns, tuber characteristics, and ecological factors in this group to better understand the balance between human selection and natural processes, crafting variation in this group.

Phaseolus coccineus: It is commonly known as scarlet runner bean and is a perennial legume native to Mesoamerica. It is widely cultivated as a garden crop or in small-scale agriculture for its dry seeds and immature pods, both of which are edible (Guerra-Garcia et al., 2017). As a domesticate, it may be part of a Vavilovian series with common bean, due to its capacity to tolerate more humid conditions (von Wettberg et al., 2020b). In the United States, the large-seeded scarlet runner is the most common, with seeds that are normally purple or variegated, though white varieties exist as well (Kaplan, 1965). Cultivation of this bean by indigenous peoples is greatest in the humid uplands of Chiapas and the cooler regions of Guatemala above 1,800 m. The beans are often planted with corn in fields or treated as perennials in house gardens. Corn and beans together make up a large portion of the human diet and are quite nutritious. The joint cultivation of these plants began roughly 7,000 years ago in Mesoamerica, and in the Aztec Empire, maize and beans were often used in annual tributes to the emperor. Notably, it was Moctezuma Xocoyotzin II, the last Aztec Emperor, who required around 371 towns to pay tribute to him and took record of it, noting 21 bins of beans which was estimated to be around 230,000 bushels or 5,000 tons a year (Kaplan, 1965). The beans were mixed but likely *P. vulgaris, P. lunatus, P. coccineus*, and *P. acutifolius* were most common. There is also indication of domesticated runner beans from Coxcatalan Cave in Tehucana valley at least 2,200 years ago, but the wild variety could have been used before then (Kaplan, 1965). Due to globalization, the diet of the majority of Mexicans has shifted away from traditional foods to industrialized food products, which has increased chronic diseases within the community while also causing a recognition and consumption of non-industrialized foods with local/regional identities (Manzanero-Medina et al., 2020). Differences in zinc, copper, and magnesium levels can damage metabolism, growth, and the immune system, while iron deficiency causes anemia. "Quelites" are young plants, shoots, and flowers of edible vegetables that are eaten for their vitamins, minerals, and proteins (Manzanero-Medina et al., 2020). The flowers are also edible and thought to help relieve respiratory, gastric, and epidermal issues due to the vitamins, minerals, and other compounds found within them (Teliban, 2016). The flowers of *P. coccineus* have healthy protein and fiber contents, as well as zinc, calcium, and magnesium, and consuming them fresh or fried can help people meet their daily mineral requirements (Manzanero-Medina et al., 2020). Many *Phaseolus* species also have high levels of lysine and tryptophan, two essential amino acids used to build proteins (Kaplan, 1965).

As an open-pollinated species, cultivated scarlet runner bean has substantial introgression from its wild relatives (Guerra-Garcia et al., 2022). As a close relative of the more widely grown common bean, *P. vulgaris*, both cultivated and wild forms of the scarlet runner bean are useful for breeding efforts (e.g., Porch et al., 2013). The species has cold tolerance and disease resistance traits that are likely useful for both its improvement and that of other *Phaseolus* species (Schwember et al., 2017).

***Psophocarpus tetragonolobus* or winged bean**: It is a tuberous legume likely native to Papua New Guinea or Southeast Asia (Lepcha et al., 2017). The legume is notable for its quadrangular pods which are ridged, giving the appearance of 'wings'. The entire plant is edible, from the leaves to the seeds, flowers, and tubers which contain a high percentage of protein and other micronutrients (Abberton, 2020). The seed's protein content is 30%–45%, (comparable to soybean) and high in amino acids as well as micronutrients like vitamin A, C, calcium, and iron. However, there are some anti-nutrient factors such as trypsin inhibitors which require thorough soaking, rinsing, and cooking to make the older beans palatable (Yang et al., 2018). The seeds can be roasted and boiled like peanuts and the tubers are boiled, steamed, baked, fried, or roasted. *P. tetragonolobus* is nicknamed the "one species supermarket" due to its versatility as a crop. Currently, the plant is widely grown in Papua New Guinea and other locations in Asia where the young pods, flowers, leaves, and shoots are eaten raw or cooked while unripe seeds are used in soups or curries, and mature seeds are consumed after roasting. Medicinally, *P. tetragonolobus* leaves are used to treat smallpox and the roots have been used in a poultice to cure vertigo in Malaya (Hymowitz and Boyd, 1977). *P. tetragonolobus* is a self-pollinator, but it can outcross with the help of carpenter bees (Tanzi et al., 2019). *P. tetragonolobus* can be grown at a range of elevations (sea level to above 2,000 m) in hot, humid climates (Tanzi et al., 2019). Estimated yields include 10 t/ha green pods, 2 t/ha mature seeds, and 11 t/ha tubers, and research suggests that while *P. tetragonolobus* has higher water requirements for optimal growth (compared to other legumes) it can be grown without irrigation with slightly lower yields than if regularly watered (Tanzi et al. 2019). In other areas, the winged bean is grown as a forage crop or for soil remediation. *P. tetragonolobus* readily nodulates with a broad spectrum of rhizobia strains to fix nitrogen and thus can survive on a wide range of soil types from poor acidic clay and loam to sandy, swamp peats and heavy clay soils. It is also very salt-tolerant and drought-resistant (Tanzi et al., 2019).

As a tropical legume that has received relatively little attention, large gaps exist in our knowledge of its domestication (e.g., Lepcha et al., 2017). The genus occurs across Southeast and tropical South Asia and has been spread by people across the humid tropics from Madagascar to South Florida (Lepcha et al., 2017). The World Vegetable Center, the international genebank with winged beans within its mandate, has only 285 accessions, of which only 13 are wild. Consequently, as with many legumes (e.g., Coyne et al., 2020), increasing wild collections should be a priority, despite the challenges imposed by the Nagoya Protocol.

Sphenostylis stenocarpa: It is also known as African yam bean and is an annual legume from sub-Saharan Africa. Some indigenous names include "mpempo", "pempo", and "mfuyu" from Bas-Zaire and Menkao. *S. stenocarpa* is often considered the most economically important species of the genus due to its tuberous roots, which can be used as a food crop, as well as the seeds that are full of protein, carbohydrates, and vitamins (Abberton, 2020). The *Sphenostylis* genus is made up of seven species from Southern Africa, four of which have documented human uses. *S. stenocarpa* is foraged in central and east Africa, while it is cultivated in West Africa for its seeds and tubers. In West Africa, *Sphenostylis* is cultivated mainly for seeds rather than tubers, although in Nigeria they are grown annually as the rootstocks return after the dry season (Potter, 1992). The tuber contains up to 67% starch (dry) and is used to combat malnutrition (Malumba et al., 2016). The tubers are high in protein and minerals such as potassium, iron, and magnesium, resembling sweet potatoes, and are often boiled and eaten plain or added to stews (Abberton, 2020). The seeds are eaten boiled after soaking for a few hours or used in stews. They can also be ground into a seasoned paste and wrapped in banana leaves then boiled or eaten fried which gives the eater a valuable helping of protein and amino acids. However, the seeds contain some anti-nutrients such as trypsin inhibitors, tannins, and oxalates which can cause diarrhea and other health problems related to digestion unless they are cooked for 4–6 hours, which limits the food use of the seeds (Azeke et al., 2005). These legumes normally are found in dry forests, forested or open savannas, and the cultivated versions have a few morphologically distinct characteristics from the wild species. Some of these cultivar traits include delayed dehiscence of pods, greater pod, seed, and tuber size, and seeds are glamorous (wild seeds have a black webbed surface) (Potter, 1992). The cultivated varieties are likely semi-domesticated, lacking pod indehiscence or loss of seed dormancy. Cultivated lines are often grown with cassava and take around 8 months to mature. This plant is an important family or home garden crop, and some people in Africa consider *S. stenocarpa* an important income source. As with the other species here, relatively little is known of their domestication and wild diversity (Abberton, 2020).

Vinga lanceolata: It is commonly called pencil yam or bush carrot and is a native Australian tuberous legume whose rhizomes resemble those of a carrot and are often baked or eaten raw (Castelli and Mikic, 2019). *Vigna lanceolata* is an endemic Australian legume, and some other wild species in this genus are cultivated as cover crops and supplementary foods, including Bambara groundnut, cowpea, zombi pea, mungbean, adzuki bean, black gram, moth bean, rice bean, and creole bean (Nubankoh et al., 2015). These legumes contain proteins, carbohydrates, vitamins, and minerals that are important dietary staples in Asia, Africa, America, and Australia (Nubankoh et al., 2015). Bush carrot is wild harvested, and the extent to which human use has led to domestication is not clear from the few studies performed on it. Its distribution in Australia, a continent with a high diversity of other wild *Vigna* species, deserves greater exploration to find and preserve useful variation (e.g., Lawn, 2015).

Vinga vexilata: It is also known as zombi pea, growing naturally in Africa and Indonesia, and surprisingly, it is also found in deserts (Von-Wettberg, 2020a). This unusual distribution means it is almost certainly part of a complex of different taxa that are phenotypically similar. As with many of the other species here, greater attention is needed to find and characterize wild diversity in this taxonomic group(s). Nine *Vigna* species are considered as domesticated crops and are grown mainly for dry seeds or the pods, leaves, and stems fed by small farmers located in tropical and subtropical regions. *V. vexilata* often develops edible tubers that contain around 15% protein, which is roughly three times higher than that of potato or yam and six times higher than that of cassava (Dachapak et al., 2017). *V. vexilata* is a fast ground-covering legume that has a generative life cycle of less than 4 months and can grow in poor, acidic, and rocky soils. Tuberous roots and nodules develop early on, making it a good potential forage or green manure and erosion control plant (Karuniawan et al., 2006) (Table 9.2).

Growth Results Form Planting in Vermont

All of the plants were able to grow in Vermont, but with varying levels of harvestable biomass. The number of seeds planted versus the number of plants that survived revealed the germination success rates of the plants, as listed below in Table 9.3. Three species had a 100% germination rate: *Apios, Pachyrhizus,* and *Phaseolus.* Unfortunately, due to COVID-19 complications, we were not able to plant according to our original schedule, which may have affected yields a bit, as the summer was also unusually hot.

Data collected from the 2020 planting revealed the average growth results, and a harvest index was calculated by examining the percentage of edible biomass from the total yields by species. The species with the highest average edible biomass (for human consumption) were *Psophocarpus, Pachyrhizus,* and *Apios* (Table 9.4).

Plant Health and Correlation Variables

WinRhizo was used to gather data by scanning roots and measuring the length of the primary root, secondary roots, etc., as well as counting nodules and estimating the total root area. RStudio was used to compute the impact of root area and length to see if there was any correlation between tuber size and root length (see Table 9.5). Correlation statistics were calculated via ANOVA tests, and p-values under 0.05 were considered significant. Other calculations included testing the correlation between root length and volume (*Pachyrhizus* was too large to put into WinRHIZO, so water displacement measurements were used to calculate root volume). Correlation between nodule presence was tested against the shoot height and root length. Tuber presence was tested against root length and root biomass, as well as shoot biomass and height, to see if there was any significant correlation.

TABLE 9.2
Literature Review Summary Table

Scientific Name	Morphology	Reproduction	Habitat	Cultural	Challenges
Apios americana	Twining perennial vine, pinnate leaves 5-7 leaflets, tuberous "beaded" roots, pink-purple panicle flowers	Mostly asexually via stoloniferous growth. Sexually, insect-pollinated	Wet, sandy soils near bodies of water, cold hardy	Used primarily as a food crop	Difficulties in planting and harvesting due to extensive sprouting and horizontal growth patterns.
Glycine tomentella	Trifoliate leaves, spreading perennial. Small purple flowers, pointy legume pods.	Sexually, insect-pollinated.	Sandy soil, variety of habitats, sun, and humid areas	Used as forage, seeds for food, or medicinal tea	Seeds can take years to mature in abundance, which makes harvesting difficult
Lathyrus linifolius	Twining perennial with tendrils and winged rachis.	Sexually, insect-pollinated	Low altitudes, soil generalist, grassy areas	Biodiversity promoters, medicinal teas, food, and drink additive, soil remediation	Smaller plant, lower nutritional value
Melilotus alba	Shrubby herbaceous biennial, small white panicle flowers, trifoliate leaves	Sexually, insect-pollinated	Generalist, will grow on almost anything	Medicinal tea, soil remediation, and cover crop, important for wildlife and maintaining biodiversity	Invasive
Pachyrhizus tuberosus	Large trifoliate leaves, white and lilac flowers, vine	Sexually, pollination, or asexually from regrowth from tubers	Hot humid climates, grows well on acidic soils and low to high elevations	Food, (tuber), can be made into biodegradable products	Seeds and green biomaterials can be toxic to humans

(*Continued*)

TABLE 9.2 (*Continued*)
Literature Review Summary Table

Scientific Name	Morphology	Reproduction	Habitat	Cultural	Challenges
Phaseolus coccineus	Trifoliate leaves, climbing vine, brilliant scarlet flowers, legume pod	Sexually, insect- and hummingbird-pollinated	Generalist can tolerate humidity and high elevations	Food, medicine, ornamental	Debate on whether or not the root is edible, contains some toxic compounds
Psophocarpus tetragonolobus	Trifoliate leaves, climbing vine, brilliant blue flowers. Tetrahedral legume pod.	Sexually, insect-pollinated, or self-pollinated	Generalist, wide range of soils, tropical climates, low to high elevations	Food, medicinal, soil remediation, green manure, ornamental	Prefers tropical humid and hot weather, which Vermont is not
Sphenostylis stenocarpa	Trifoliate leaves, dark purple almost black flowers with white speckles, thin legume pods that explode with heat	Sexually, insect-pollinated (wasps and ants primarily)	Dry forests, open savannas, often in rocky areas and from low to high elevations	Food	Not much data available
Vigna lanceolata	Trifoliate leaves, yellow flowers	Sexually, insect-pollinated	Dry climates, desserts	Food	Not much data is available
Vigna vexillate	More angular trifoliate leaves, yellow to pink flowers, perennial climbing vine	Sexually, insect-pollinated	Dessert grasslands, tropics, subtropics, low to high elevations	Food, erosion control, green manure, soil remediation	Not much data is available

TABLE 9.3
Germination Rates

Scientific Name	Planted	Final	Germination
Apios americana	9	9	100%
Glycine tomentella	20	4	20%
Lathyrus linifolius	15	4	27%
Melilotus alba	16	2	13%
Pachyrhizus tuberosus	20	20	100%
Phaseolus coccineus	45	45	100%
Psophocarpus tetragonolobus	9	6	67%
Sphenostylis stenocarpa	13	11	85%
Vigna lanceolata	15	11	73%
Vigna vexilata	11	6	55%

There was a correlation between root length and root area with significant p-values (less than 0.05) in *Apios, Glycine, Phaseolus, Psophocarpus, Sphenostylis, Vigna lanceolata.* Root length and volume were calculated using WinRhizo as another formatting for root size, also showed significance in *Apios, Glycine, Lathyrus, Phaseolus, Vigna lanceolata, and vexilata.* Tuber presence and root length were correlated in *Glycine, Phaseolus, Sphenostylis.* Tubers and root biomass were correlated in *Glycine, Lathyrus, Phaseolus, Psophocarpus, Sphenostylis,* and *Vigna lanceolata.* In summary (and as expected), larger roots are correlated with larger tubers.

Tuber presence was also tested in RStudio using ANOVA tests to determine if there was a correlation between general plant health and tubers. Correlation between tuber presence and plant height was significant in *Glycine, Phaseolus, Sphenostylis,* and *V. lanceolata.* Tubers and shoot biomass were significant in *Glycine* and *Sphenostylis.* Root width, another WinRhizo calculation, and plant height were significant in *Glycine, Phaseolus,* and *Vigna lanceolata.* In general, larger plants produce bigger tubers.

Due to a lack of nodules found in most species, potentially from not inoculating the plants, there was no significance found between nodule count and plant health.

Edibility

Recipe Development

Apios americana: The tubers can be eaten in any way that potatoes are. For the taste test, the lead authors separated the tubers from the roots and boiled them for around 30 minutes. Once boiled, a portion of the roots was roasted in the oven for 30 minutes at 400°F with olive oil and salt. We also made "mashed apios" using vegan butter (olive oil based), almond milk, salt, pepper, and oregano. Other cooking methods for the tubers include making "potato" fries and slicing them into disks and dehydrating them, and blending them into flour to use as a soup or to stew thickener, or make farinata (crisp bread). *Safety notes*: tubers can cause some digestion issues such

TABLE 9.4
Average Growth Results from Data Collected from the 2020 Planting

Species	Number of Seeds Planted	Average Number of Flowers/Pods	Average Biomass (Root)	Average Biomass (Shoot)	Average Height (Shoot)	Average Length (Roots)	Average Tubers per Plant	Average Edible Biomass
Apios americana	10	3*	109.4	29.9	48	42	102.4	77.7% (flowers, seeds, tubers)
Glycine tomentella	20	6.1	1.3	8.7	17.9	6.9	1	4.5% (seeds, tuber)
Lathyrus linifolius	15	0	0.5	0.5	7.6	6	1	54% (tubers)
Melilotus alba	15	25	6.6	97.6	47.1	8.4	1	78% (leaves, flowers for livestock)
Pachyrhizus tuberosus	20	0	135.6	24.4	38.9	7.2	1	80.3% (tuber)
Phaseolus coccineus	42	14.4	6.3	85.6	46.5	8.6	1	8.5% (flowers, pods)
Psophocarpus tetragonolobus	8	5*	18.6	40.3	57.8	9.8	1	100% (entire plant)
Sphenostylis stenocarpa	13	50*	2.8	13.4	29.9	12.3	3	12.7% (tuber)**
Vigna lanceolata	5	15.5	37.4	298.8	54.8	34.9	1	17.6% (tuber)
Vigna vexillata	11	5*	56.6	370.3	26.2	37.0	4	11.9% (seeds, tuber)

* Estimated (flowered/seeded before harvest date so not counted…and *Sphenostylis* had too many to keep track and the seeds method of dispersal was via exploding dehiscence, which made them difficult to gather).

** Seeds were not collected.

TABLE 9.5
Correlation Stats

P-values	*Apios*	*Glycine*	*Lathyrus*	*Pachy-rhizus*	*Phaseolus*	*Psopho-carpus*	*Spheno-stylis*	*V. Lanceolata*	*V. Vexilata*
Root Length: Area	0.01743744	0.000913073	0.4381274	NA	$2.94275*10^{-12}$	0.0311662	0.03116626	0.008379944	0.187676
Root Length: Volume	0.01293362	0.000771980	0.00166443	NA	$5.555391*10^{-8}$	0.8540202	0.8540202	0.000151813	0.00117048
Nodule: Height/ Length	NA	NA	NA	NA	0.26383/0.887398	NA	NA	NA	NA
Tuber: Height	0.6120884	0.008997582	0.5448798	0.82853	0.02894518	0.572311	0.0125171	0.00120393	0.654301
Tuber: Length	0.09058131	0.0175274	0.269703	0.842424	0.00267416	0.106187	0.02736985	0.0551455	0.053534
Tuber: Root Biomass	0.053739	0.00794789	0.0386319	0.7099008	0.0026741	0.00590893	0.00844022	0.00191971	0.9665906
Tuber: Shoot Biomass	0.6120884	0.0016563	0.5083517	0.775488	0.411383	0.0856680	0.01654247	0.812719	0.416865
Root Width: Height	0.1764745	0.0215845	0.9759697	NA	$1.778739*10^{-15}$	0.0545762	0.0545762	0.0003560844	0.907307

[*] Data was N/A for *Melilotus* (too small of a sample size).

as gas, bloating, abdominal cramps, possible diarrhea, and vomiting if not cooked properly, so be sure to boil the tubers for 30 minutes before consuming.

Glycine tomentella: Not enough plant material from the 2021 planting was available to cook. Normally the seeds are cooked like any other soybeans, and the root can be used in tea.

Lathyrus linifolius: Not enough plant material from the 2021 planting was available to cook. The tubers are normally used as an appetite suppressant, and the dried root can be roasted and used as a drink flavoring, adding a licorice taste.

***Melilotus alba*:** The young leaves can be eaten raw (before the plant blossoms) and are bitter and aromatic which can add a nice flavoring to salads. Dried leaves and flowers can be used to make sweet clover tea, which we tried for the taste test by combining a handful of sweet clover (leaves and flowers), 3 tbsp sugar, 1 L water, and a dash of vanilla. We stirred everything together and let the tea infuse overnight in the refrigerator. Apparently, the seeds can also be used as a spice but should be used sparingly because excess consumption can make you throw up! *Medicinal uses*: Sweet clover is a traditional external poultice for external and internal tissue inflammation. The tea has been used in many countries as a stomach soother, but the presence of coumarin can combine inappropriately with prescription drugs, resulting in digestive issues.

Pachyrhizus tuberosus: There are so many recipes for jicama! We made "jicaletas" by washing and slicing the tuber into spears or sticks. Then we squeezed lime on top of the spears and sprinkled a dash of chili con limon (cayenne pepper). There are many other ways to use jicama, including in a slaw with cabbage, carrots, and avocado. The roots can be added to a stir fry with broccoli, garlic, ginger, scallions, and toasted sesame seeds. The tuber can easily be turned into fries, added to a taco salad, and much more.

Phaseolus coccineus: The bean pods can be eaten raw when young or cooked like string beans when older. Dry beans should be soaked before cooking, while fresh beans do not require soaking before cooking for 30 minutes. We soaked and cooked the dry beans before preparing a batch of roasted beans using olive oil and salt. We used the remaining cooked beans to make hummus with tahini, cumin, paprika, lemon, salt, and a dash of turmeric.

Psophocarpus tetragonolobus: No plants were grown in the 2021 planting season, but back in 2020, we were able to harvest a few pods. We made a stir-fry dish using one teaspoon of toasted sesame oil, one tablespoon of soy sauce, one large diced tomato, half a teaspoon of Kosher salt, one cup of winged beans (cut into bite-size lengths), three cloves garlic (sliced thinly), and a dash of oil. We cooked everything in a pan over medium heat until the food was nicely roasted.

Sphenostylis stenocarpa: The tubers are edible and can be roasted or boiled like any potato. We chopped and cooked the roots with oil and salt in the oven at 400°F for 30 minutes. The tubers can also be ground into flour and used to make biscuits and other baked goods. Seeds are edible when cooked thoroughly but none were collected for the taste test.

Vigna lanceolata: Not enough plant material from the 2021 planting was available to cook. The tuber can be eaten raw or cooked like a potato.

Vigna vexilata: The tuber can be eaten raw or cooked. We peeled the skin off the tubers and roasted them in the oven for 30 minutes at 400°F with a little bit of olive oil and salt. None were collected for the taste test, but young leaves and pods can be cooked and eaten as a vegetable (Figure 9.3).

Sensory Evaluation

After preparing all the food, ten people from the Plant and Soil Science/Plant Biology department at the University of Vermont were recruited to assist with the sensory analysis. Afterward, we analyzed the data from the participants and averaged the results for a more cohesive table (Table 9.6).

DISCUSSION

Ethnobotanical Importance

It is important to maintain a diverse agricultural system not only to preserve biodiversity for the health of the planet but also to maintain a rich ethnobotanical ecosystem. Difficulties accessing data for ethnobotanical research did arise, as most of these plants have been studied minimally—if at all—in the past. Research on *Vigna lanceolata* and *Vigna vexilata* was particularly scarce. Based on as thorough a literature review as we could conduct, we did find that many of these plants have historical importance as food crops or medicinal plants for many indigenous cultures and even for medieval Europeans as well. Several research articles pointed to issues with the globalization of agriculture and the commercialization of food products. The distinct trend away from native plants to more global cash crops such as corn, wheat, potatoes, and soybeans uses a significant amount of resources and overall does not comprise a completely nutritious diet. Most of the tuberous legumes in this study are high in proteins and rich in other micronutrients, especially those with edible flowers and beans/seeds. Many of the plants also have medicinal uses, though We did not find any research suggesting ceremonial uses, which leads us to believe these plants are primarily used for nutrition. Some future movements could be reaching out globally to indigenous ethnobotanists in the respective home areas of the plants and interviewing them for greater insight into the plant's history, current uses, and meanings to the people.

Viability of Crops in Vermont

Due to difficulties with COVID-19 at the start of the summer (2020), the growth period of the plants was much shorter than initially planned, which may skew the growth data and final harvest yields. Nevertheless, all of the plants were able to be grown in Vermont! The plants with the highest germination rates were *Apios, Pachyrhizus,* and *Phaseolus*, and the lowest germination rates were *Glycine, Lathyrus*, and *Melilotus*. The tallest average plants were *Sphenostylis, Vigna, Pachyrhizus*, and *Psophocarpus*, with *Psophocarpus* coming in first in the 2020

FIGURE 9.3 (a) *Vigna vexillata* tubers (raw and cooked) (b) Mashed and cooked *Apios* tubers and beans (c) *Phaseolus* hummus and cooked beans (d) Raw Sphenostylis roots (e) Boiled and roasted *Apios* tubers (f) Cooking the *Phaseolus* beans (g) Raw *Pachyrhizus* "Jicaleta" slices and cooked *Sphenostylis* tubers (h) Array of taste test samples.

TABLE 9.6
Average Sensory Evaluation Results

Average Results	Level of Smell (0-7, No Smell to Strong Smell)	How Would You Describe the Flavor?	Intensity of Flavor? (0–10, Not Noticeable to Very Intense)	Fullness of Flavor? (Thin, Moderate, Full)	How would You Describe the Mouthfeel?	Was the Dish Balanced? (Y/N)	Did You Like This? (0–10 Terrible to Great)	Would You Make This for Yourself? (Y/N)
Boiled *Apios* (roots)	Slight to moderate (3.8)	Earthy, potato, nutty, like peanuts and yucca	3.7	Thin	Potato-like, more gritty	No, more flavor was needed	5.1	Yes
Roasted *Apios* (roots)	Moderate (4.9)	Nutty, potato-like, earthy	5.8	Full	Creamy, like potatoes	Yes	7.4	Yes
Mashed *Apios* (roots)	Moderate to Strong (5.6)	Nutty, rich, warm	7.5	Full	Smooth, creamy	Yes	9	Yes
Melilotus (flowers) tea	Slight to moderate (4.0)	Vanilla, sweet, honey	6.9	Moderate	Smooth, almost creamy, would be better strained	Yes	8.3	Yes
Raw *Pachyrhizus* (roots)	Slight to moderate (3.9)	Watery, slightly bitter, vegetable taste similar to celery	3.6	Moderate	Crunchy, watery, starchy	Yes	6	Yes
Cooked *Phaseolus* (beans)	Moderate (5.4)	Nutty, bean-like	6.4	Moderate to full	Bean-like, slightly chewy	Yes	6.9	Yes
Phaseolus (beans) hummus	Moderate to strong (5.6)	Nutty, smooth, aromatic	7.7	Full	Smooth, bean-like	Yes	8.1	Yes
Roasted *Sphenostylis* (roots)	Slight to moderate (3.8)	Rooty, earthy	3.6	Thin to moderate	Fibrous, stringy, hard to chew	No	2.3	No
Raw *Vigna* (roots)	Slight (3.3)	Earthy, carrot, and citrus notes	3.4	Thin to moderate	Raw, fibrous, stringy	No	2.1	No
Roasted *Vigna* (roots)	Slight (2.6)	Nutty, earthy, tough	3	Thin to moderate	Stringy, hard to chew	No	1.4	No

summer study. Many of the plants grew quite tall, as was expected from twining legumes, so support systems are highly recommended, such as trellises or a wall for the plants to grow on. A note from the 2021 planting, the *Pachyrhizus* plants grew very large (over 6 ft!) which they did not do the previous year. The plants with the largest average shoot biomass were *Vigna vexillata* at 370 g per plant. However, this was mostly older leaves, and supposedly only the young leaves are edible. Looking further, *Psophocarpus* had the next highest edible shoot biomass at around 40 g per plant. The lowest on the scale was *Lathyrus linifolius* with only 0.5 g per plant on average. However, *Sphenostylis* produced a significant number of seeds that we were not able to quantify as they were in continuous production throughout the summer and also exploded as their form of dispersal. In reality, *Sphenostylis* was probably the most productive. However, as a largely wild taxon, neodomestication via selection for pod indehiscence (Ogutcen et al., 2020) may be necessary simply to adequately propagate seeds.

The largest average root biomass, which was also edible, was *Pachyrhizus* with 135 g per plant and *Apios* coming in second at 109 g per plant. An interesting note is the number of tubers on each *Apios* plant almost exactly matched the weight of the overall root biomass, meaning one tuber averaged around 1 g after 20 weeks of growth. The longest average roots were *Apios*, *Sphenostylis,* and *Vigna* species, while the *Pachyrhizus* plants had very short, large roots. The highest average seed/pod production was around 25 flowers per plant in *Melilotus* and around 14 per plant for *Phaseolus*. Not as many plants as expected nodulated, though no commercial inoculant was added so it may not be that surprising. The plants that did have nodules were *Apios, Phaseolus*, and *Melilotus*.

Edibility

Some interesting notes on cooking the plants: we found that *Apios* produces a slight latex that coats the pots when boiling, making it very sticky like glue and hard to remove from the pot. However, once cooked, the tubers act almost identically to potatoes (with a hint of more yucca-like flavor and texture), making it a very versatile and delicious crop. The hummus and mashed *Apios* were our favorites, but as with most recipes, spice is life, so flavor your foods.

Recommendations

Based on this data overall, the best viable and marketable plants for Vermont's climate, whether for home growers or commercial consumers, would be *Apios americana, Phaseolus coccineus*, *Psophocarpus tetragonolobus*, and *Pachyrhizus* tuberosus, based on their ethnobotanical backgrounds, germination rates, harvest index, and taste test results. The least recommended plants from this study would be *Melilotus, Glycine*, and *Lathyrus* due to lower germination and growth rates and harvest index. Another note of concern is that while *Melilotus* may be a good forage or soil remediation crop, it can be an aggressive invasive, so it may be better not to plant more of it. The *Vigna* species and the *Sphenostylis* plants grew fairly well but ranked low on the edibility scale and moderately on the harvest index.

Future Movements

Further investigation into bio-drilling and soil remediation with a deeper analysis of soil compaction and nitrogen levels in the soil would be very interesting. We were also unable to test the nitrogen-fixing capabilities of the plants but did notice some rhizobia associations in the form of nodules on many of the plants' roots. No inoculation was used on the plants, so all of the rhizobia formed already existed within the soils at the farm. I did look into which bacteria these plants commonly nodulate with, and the majority of the literature points to *Bradyrhizobium* associations with *Pachyrhizus, Apios, Vigna,* and *Glycine. Glycine* is also associated with *Sinorhizobium fredii. Sphenostylis, Phaseolus, Psophocarpus*, and *Lathyrus* associate with *Rhizobium* and are normally more generalist in the species (Zakhia and La Judie, 2001).

CONCLUSIONS

We were able to successfully grow these ten promising tuberous legumes in Vermont and gain some insight into the estimated yields and growth habits of the plants. After a thorough literature review, we compiled some data surrounding the historical uses and cultural importance of each of the legume species. All of the plants in this study come from varied backgrounds, reaching across five continents. Most of the plants are primarily used as food for both humans and livestock; while several are also used medicinally. The edibility of the plants was tested via a sensory analysis, and results showed *Apios* and *Phaseolus* ranked the highest. Unfortunately, *Psophocarpus* was not grown during the 2021 planting, so no one in the taste test was able to sample it. However, two of the researchers were able to try the pods and leaves from the previous year and ranked them as highly palatable, with a taste very similar to green beans. There were several existing recipes found online, and due to the similarities between many of the tubers and beans to potatoes and commercial beans, these plants could easily be substituted in any recipes that use the more common vegetables. The only safety note would be to make sure that some of the plants, like *Apios* tubers and *Sphenostylis* seeds, are properly cooked long enough to ensure no harmful or irritating compounds remain before consuming them. Overall, based on access to seeds, growth habits, edible harvest index, and taste test results, the top recommended tuberous legumes to incorporate into the Vermont agriculture and gardening market would be *Apios americana* (American groundnut), *Pachyrhizus tuberosus* (jicama), *Phaseolus coccineus* (scarlet runner bean), and *Psophocarpus tetragonolobus* (winged bean).

ACKNOWLEDGMENTS

We thank Dan Tobin, Michael Sundue, Dave Barrington, Bailey Kretzler, and Gianna Sassi for their helpful comments on an earlier version of our manuscript. We would also like to thank the Crop Heritage Lab in the UVM Department of Plant and Soil Science for their assistance during the 2021 planting and taste test. Finally, we owe our substantial gratitude to the UVM Horticulture Farm and UVM Greenhouse for their assistance.

FUNDING

This project was funded with the support of the Hatch Multistate project (VT-H02811MS), "Improving Forage and Bioenergy Crops for Better Adaptation, resilience, and Flexibility" (MS# NE1710), and the New Roots for Restoration Biology Integration Institute (NSF BII, 2021).

CONFLICT OF INTEREST

The author reports no conflict of interest.

DATA AVAILABILITY

All data is included in the supporting information/appendix below.

SUPPORTING INFORMATION

Sensory evaluation background information-Section below describing sensory characteristics is taken from UVM Extension's Article by Roy Desrochers, Heather Darby, and Sara Ziegler on "Evaluating the Sensory Characteristics of Organic Grass Fed Milk" from November 2020.

"What are sensory characteristics?

Sensory characteristics are grouped into aroma and flavor. Aroma is what you smell when you sniff with your nose. Flavor includes everything you perceive when you put a product in your mouth, such as what you taste on your tongue, what you smell in your nose, and how it makes your mouth feel.

Your tongue detects five basic tastes using its taste buds: sweet, sour, salty, bitter and savory (umami). Everything else that you think you taste in your mouth is actually flavor compounds that travel from your mouth to your nose through the back nasal passage.

You can try this at home by simply holding your nose while you taste a food product. While holding your nose, you will be able to taste any of the five basic tastes present, but nothing else. When you let your nose go, you will taste (actually smell) the rest of the flavor.

Mouthfeel is exactly what the term implies, which is how your mouth feels during and after consuming a product. For example, your mouth might feel dry when eating bread, or you may pucker when you drink a sour lemonade. In the case of milk, you will often experience a fatty or creamy coating inside your mouth.

Total intensity of aroma: The overall intensity of smell (aroma) of the sample.

Balance: The harmony of the flavor. If individual flavors stick out and are easy to detect, the flavor is not balanced.

Fullness: Fullness is a measure of the complexity of flavor. If the product tastes simple and thin, it is not very full.

Total intensity of flavor: The overall intensity of taste of the sample.

Five basic tastes: The intensity of sweet, salty, sour, bitter and savory taste on the tongue.

Mouthfeel: The overall intensity of mouthfeel regardless of the type. This may be a mouth coating, dryness, salivation, etc.

Others: The overall intensity of other notes detected that do not belong in one of the previous categories".

REFERENCES

Abberton, M., T. T. Adegboyega, B. Faloye, R. Palawan, and O. Oyatomi. 2020. Winged bean, psophocarpus tetragonolobus. *Legume Perspectives Issue*. 13(19), 589–595.

Abberton, M., T. T. Adegboyega, B. Faloye, R. Palawan, and O. Oyatomi. 2020. African yam bean, Sphenostylis stenocarpa. *Legume Perspectives Issue*. 25(19), 209–210.

Agaba, R., P. Tukamuhabwa, P. Rubaihayo, S. Tumwegamire, A. Ssenyonjo, R. O. M. Mwanga, J. Ndirigwe, and W. J. Grüneberg. 2016. Genetic variability for yield and nutritional quality in yam bean (Pachyrhizus sp.). *HortScience*. 51(9), 1079–1086.

Azeke, M. A., B. Fretzdorff, H. Buening-Pfaue, W. Holzapfel, and T. Betsche. 2005. Nutritional value of African yambean (Sphenostylis stenocarpa L): improvement by lactic acid fermentation. *Journal of the Science of Food and Agriculture*. 85, 963–970.

Belamkar, V., A. Wenger, S. R. Kalberer, V. G. Bhattacharya, W. J. Blackmon, and S. B. Cannon. 2015. Evaluation of phenotypic varation in a collection of apios Americana: an edible tuberous legume. *Crop Science*. 55, 712–726.

Belamkar, V., J. Singh, S. Kalberer, and S. Cannon. 2020. Apios Americana: crop improvement and genomic characterization. *Legume Perspectives Issue*. 6, 34908.

Bell, L. W., R. G. Bennett, M. H. Ryan, and H. Clarke. 2011. The potential of herbaceous native Australian legumes as grain crops: a review. *Renewable Agriculture and Food Systems*. 26, 72–91.

Blackmon, W. J., and B. D. Reynolds. 1986. The crop potential of Apios americana-preliminary evaluations. *HortScience*, 21(6), 1334–1336.

Castelli, A., and A. Mikic. 2019. Vigna lanceolata in the fire-stick farming and the Australian aboriginal culture. *Ratarstvo i povrtarstvo*. 56(2), 56–64.

Coyne, C. J., S. Kumar, E. J. von Wettberg, E. Marques, J. D. Berger, R. J. Redden, T. N. Ellis, J. Brus, L. Zablatzká, and P. Smýkal. 2020. Potential and limits of exploitation of crop wild relatives for pea, lentil, and chickpea improvement. *Legume Science*. 2(2), e36.

Dachapak, S., P. Somata, S. Poonchaivilaisak, T. Yimram, and P. Srinives. 2017. Genetic diversity and structure of the zombie pea (Vigna vexilata (L.) A. Rich) gene pool based on SSR marker analysis. *Genetica*. 145(2), 189–200.

Dello Jacovo, E., et al. 2019. Towards a characterisation of the wild legume bitter vetch (Lathyrus linifolius L. (Reichard) Bässler): heteromorphic seed germination, root nodule structure and N-fixing rhizobial symbionts. *Plant biology (Stuttgart, Germany)*. 21(3), 523–532. https://doi.org/10.1111/plb.12902

Guerra-Garcia, A., M. Suarez-Atilano, A. Mastretta-Yanes, A. Delgado - Salinas, and D. Pinero. 2017. Domestication genomis fo the open-polliinated scartlet runner bean (Phaseolus coccineus L.). *Frontiers in Plant Science*. 8, 1891.

Guerra-García, A., I. C. Rojas-Barrera, J. Ross-Ibarra, R. Papa, and D. Piñero. 2022. The genomic signature of wild-to-crop introgression during the domestication of scarlet runner bean (Phaseolus coccineus L.). *Evolution Letters*. 6(4), 295–307.

Gucker, C. L. 2009. Melilotus alba, M. officinalis. In: *Fire Effects Information System*. U.S. Department of Agriculture, Forest Service, Rocky Mountain Research Station, Fire Sciences Laboratory (Producer). Available: https://www.fs.fed.us/database/feis/plants/forb/melspp/all.html

Hymowitz, T., and J. Boyd. 1977. Origin, ethnobotany and agricultural potential of the winged bean: psophocarpus tetragonolobus. *Economic Botany*. 31(2), 180–188.

Kalberer, S., J. S. Belamkar, and S. Cannon. 2020. Apios americana: natural history and ethnobotany. *Legume Perspectives Issue*. 19, 29–32.

Kaplan, L. 1965. Archeology and domestication in American Phaseolus (beans). *Economic Botany*. 19(4), 358–368.

Karuniawan, A., A. Island, P. R. Kale, J. Heinzmann, and W. J. Gruneberg. 2006. Vigna vexilata (L.) A. Rich. Cultivated as a root crop in Bali and Timor. *Genetic Resources and Crop Evolution*. 53, 213–217.

Kumar, A., S. Kumar, N. R. Kemal, and P. Singh. 2021. Role of traditional ethnobotanical knowledge and indigenous communities in achieving sustainable development goals. *Suitability*. 13, 3062.

Lawn, R. J. 2015. The Australian Vigna species: a case study in the collection and conservation of crop wild relatives. *Crop Wild Relatives and Climate Change*. 2015, 318–335.

Lepcha, P., A. N. Egan, J. J. Doyle, and N. Sathyanarayana. 2017. A review on current status and future prospects of winged bean (Psophocarpus tetragonolobus) in tropical agriculture. *Plant Foods for Human Nutrition*, 72, 225–235.

Li, J., J. Jiang, H. V. Stel, A. Homkes, J. Corajod, K. Brown, and Z. Chen. 2014. Phylogenetics and biogeography of Apios (Fabaceae) inferred from sequences of nuclear and plastid genes. *International Journal of Plant Sciences*. 175(7), 764–780.

Malumba, P., B. M. Denis, K. K. Joseph, L. Doran, S. Danthine, and F. Béra. 2016. Structural and physicochemical characterization of Sphenostylis stenocarpa (Hochst. ex A. Rich.) Harms tuber starch. *Food Chemistry*. 212, 305–312.

Mann, C. E. 2011. *1493: Uncovering the New World Columbus Created*. New York: Knopf.

Manzanero-Medina, G. I., M. A. Vásquez-Dávila, H. Lustre-Sánchez, and A. Pérez-Herrera. 2020. Ethnobotany of food plants (quelites) sold in two traditional markets of Oaxaca, Mexico. *South African Journal of Botany*. 130, 215–223.

Nubankoh, P., S. Pimtong, P. Somta, S. Dachapak, and P. Srinives. 2015. Genetic diversity and population structure of pencil yam (Vigna lanceolata) (Phaseoleae, Fabaceae), a wild herbaceous legume endemic to Australia, revealed by microsatellite markers. *Botany*. 93(3), 183–191.

Ogutcen, E., Pandey, A., Khan, M.K., Marques, E., Penmetsa, R.V., Kahraman, A. and Von Wettberg, E.J., 2018. Pod shattering: a homologous series of variation underlying domestication and an avenue for crop improvement. *Agronomy,* 8(8), p.137.

Potter, D. 1992. Economic botany of sphenostylis (leguminosae). *Economic Botany*. 46(3), 262–275.

Porch, T. G., J. S. Beaver, D. G. Debouck, S. Jackson, J. D. Kelly, and H. Dempewolf. 2013. Use of wild relatives and closely related species to adapt common bean to climate change. *Agronomy*. 3(2), 433–461.

Rodriguez-Navarro, D. N., M. Sorensen, and E. O. Leidi. 2020. Ahipa, Pachyrzhius aphis: a legume with edible tuberous roots. *Legume Perspectives Issue*. 19, 449–469.

Saxon, E. C. 1981. Tuberous legumes: preliminary evaluation of tropical australian and introduced species as fuel crops. *Economic Botany*. 35(2), 163–173.

Schwember, A. R., B. Carrasco, and P. Gepts. 2017. Unraveling agronomic and genetic aspects of runner bean (Phaseolus coccineus L.). *Field Crops Research*. 206, 86–94.

Singh, R. J., and T. Hymowitz. 1999. Soybean genetic resources and crop improvement. *Genome*. 42(4), 605–616.

Smykal, P. 2020. Laszlo Erdos "European tuberous Lathyrus species". *Legume Perspectives Issue*. 19, 36–38.

Sorensen, M., S. Doygaard, J. Estrella, L. Kristen, and P. Nielsen. 1997. Status of the South American tuberous legume Pachyrhizus tuberosus (Lam.) Spreng. *Biodiversity and Conservation*. 6, 1581–1625.

Stai J. S., E. J. von Wettberg, P. Smýkal, and S. B. Cannon. 2020. Which came first: the tuber or the vine? A taxonomic overview of underground storage in legumes. *Legume Perspectives*. 19, 5–7.

Tanzi, A. S., G. E. Eagleton, W. K. Ho, Q. N. Wong, S. Mayes, and F. Massawe. 2019. Winged bean (Psophocarpus tetragonolobus (L.) DC.) for food and nutritional security: synthesis of past research and future direction. *Planta*. 250(3), 911–931. https://doi.org/10.1007/s00425-019-03141-2

Teliban, G., M. Burducea, A. Lobiuc, E. Rosenhech, V. Stoleru, V. Onofrei, M.-M. Zamfirache, N. E. Munteanu. 2016. Phenolic contents and antioxidant activities in Paseolus coccineus L. flowers. Medicinal Plants-Present and Perspectives, Poster Presentation at Analele Stiintifice ale Universitatii.

Turner, N. J., and P. V. Aderkas. 2012. Sustained by First Nations: European newcomers use of Indigenous plant foods in tempterate North America. *Acta Societatis Botanicorum Poloniae*. 81(4), 78–87.

Vermont Farm to Plate Strategic Plan. "3.2 Farm Inputs: Soil" 2013. Accessed at https://www.vtfarmtoplate.com/assets/plan_sections/files/3.2_Farm%20Inputs_Soil_MAY%202013.pdf

Vermont Folklife Center. "Culture Groups" Updated 2021. Accessed at https://www.vermontfolklifecenter.org/culturegroups

von Wettberg, E. J., J. Ray-Mukherjee, K. Moriuchi, and S. S. Porter. 2019. Medicago truncatula as an ecological, evolutionary, and forage legume model: new directions forward. *The Model Legume Medicago Truncatula*. 38, 31–40.

Von Wettberg, E. B., K. Natio, A. Kur, and N. Ludidi. 2020a. The unusual biogeography of zombi pea, Vigna vexillata. *Legume Perspectives Issue*. 19, 8–9.

Von Wettberg, E., T. M. Davis, and P. Smykal. 2020b. Editorial: wild plants as source of new crops. *Frontiers in Plant Science*. 11, 591554.

Wakefield, J. 2019. New Book: Majority of Vermonts Migrant Farm Workers are Food Insecure. *UVM Today*. https://www.uvm.edu/news/story/new-book-majority-vermonts-migrant-farm-workers-are-food-insecure

Wolf, J. J., S. W. Beatty, and T. R. Seastedt. 2004. Soil characteristics of rocky mountain national park grasslands invaded by melilotus officinalis and M. Alba. *Journal of Biogeography*. 31(3), 415–424. https://www.jstor.org/stable/3554730

Yang, S., A. Grall, and M. A. Chapman. 2018. Origin and diversification of winged bean (Psophocarpus tetragonolobus (L.) DC.), a multipurpose underutilized legume. *American Journal of Botany*. 105(5), 888–897.

Zakhia, F., and P. de La Judie. 2001. Taxonomy of rhizobia. *Agronomie*. 48(5), 369–382.

10 Soybean Wild Relatives for Designing Future Climate-Resilient Cultivars

Manisha Saini, Subhash Chandra, N. Krishan Kumar Rathod, Manu Yadav, Arjun Sharma, and Akshay Talukdar

INTRODUCTION

Soybean [*Glycine max* (L.) Merr.] is an economically important oilseed crop containing oil (18%–20%), protein (36%–42%), and other vital nutrients needed for human health. In terms of animal nutrition, almost two-thirds of the world's protein meal demands are fulfilled by soybeans alone. With increased awareness, the demand for soybeans has been on the rise (OECD Food and Nations AOOTU, 2016). However, a consistent increment in the global population, which is expected to be around 10 billion by 2050, and a food deficit of more than 50% is predicted (Ranganathan et al., 2016). At the same time, consequences of climate change, rising weed prevalence, the occurrence of severe disease outbreaks, and drought and flood stresses (Raza et al., 2019), etc., are predicted to cause crop losses of billions of dollars in value globally (Gregory et al., 2009; Mittler and Blumwald, 2010). The yield of soybean is also at a current deficit of 1.2% per annum (Ray et al., 2013), even though output has more than doubled and yield per acre has improved by almost 40% over the past 25 years (1990–2015) (USDA, 2016). However, the current soybean yield potential is constrained by numerous factors including the limited genetic diversity in *G. max* making it difficult to develop soybean varieties with high yields. Besides, the narrow genetic diversity of *G. max* has made it harder to create soybean varieties with high levels of climatic resilience. Therefore, there is an urgent need to collect, characterize, and utilize the soybean wild relatives (SWRs), which are treasures of tolerance to various stresses, to widen the genetic base for designing and developing varieties of soybean resilient to the changing climatic variables.

DOI: 10.1201/9781003434535-10

SOYBEAN WILD RELATIVES (SWRs): EVOLUTION, DOMESTICATION, AND PROSPECTS UNDER CHANGING CLIMATIC SCENARIO

Evolution of Soybean and Soybean Wild Relatives (SWRs)

Soybean is an angiosperm plant that belongs to the order Fabales, family Fabaceae, and genus *Glycine*. The genus *Glycine* is divided into two subgenera, *Glycine and Soja*, that comprise around 28 species (Table 10.1) (Sherman-Broyles, 2014).

TABLE 10.1
Classification of Soybean and Soybean Wild Relatives (SWRs)

Genus	Sub-genus	Species	Authority	Chromosome Number (2*n*)
Glycine	Soja	*G. max*	Merr.	40
		G. soja	Sieb. & Zucc.	40
	Glycine	*G. canescens*	F. J. Hermann	40
		G. clandestine	Wendl.	40
		G. latrobeana	Benth.	40
		G. tabacina	Benth.	40, 80
		G. tomentella	Hayata	38, 40, 78, 80
		G. falcate	Benth.	40
		G. latifolia	(Benth) Newell & Hymowitz	40
		G. argyrea	Tindale	40
		G. cyrtoloba	Tindale	40
		G. curvata	Tindale	40
		G. arenaria	Tindale	40
		G. microphylla	(Benth.) Tindale	40
		G. albicans	Tindale & Craven	40
		G. hirticaulis	Tindale & Craven	40, 80
		G. lactovirens	Tindale & Craven	40
		G. dolichocarpa	Tateishi & Ohashi	40
		G. pindanica	Tindale & Craven	40
		G. stenophita	B. Pfeil & Tindale	40
		G. peratosa	B. E. Pfeil & Tindale	40
		G. rubiginosa	Tindale & B. E. Pfeil	40
		G. aphyonota	B. Pfeil	40
		G. pullenii	B. Pfeil, Tindale & Craven	40
		G. gracei	B. E. Pfeil & Craven	40
		G. montis-douglas	B. E. Pfeil, Tindale & Craven	40
		G. pescadrensis	Hayata	40
		G. syndetika	B. E. Pfeil & Craven	40

The sub-genus *Soja* comprises the annual plants, wild soybean (*G. soja* Siebold & Zuccarini), and cultivated soybean [*G. max* (L.) Merr.].

The cultivated soybean, *Glycine max*, descended from the wild soybean, *Glycine soja* (Siebold & Zucc.), is native to East Asia with a wide geographic range that stretches from East Russia to South China. The accessions of *G. soja* dehisce the pods in response to drying after maturity, scattering their seeds effectively, which is nearly absent in the cultivated soybean (Funatsuki et al., 2014). The cultivated soybean is erect compared to other species, which are either twining or climbing (William and Akiko, 1980) (Figure 10.1). Despite differing in numerous phenotypic traits, both *G. soja* and *G. max* have the same number of chromosomes ($2n=40$), display typical meiotic chromosomal pairing, and are cross-compatible (Carter et al., 2013). The sub-genus *Glycine* considered an Australian sub-genus (Sherman-Broyles et al., 2014) includes perennial types such as *G. canescens* and *G. tomentella*, known to be found in Australia and Papua New Guinea (Hymowitz, 1995). The species in this sub-genus shared a common ancestor around 5 million years ago (Innes et al., 2008). The perennials are predominantly autogamous plants having both cleistogamous and chasmogamous flowers (Sherman-Broyles et al., 2014).

The wild *Neonotonia wightii* is a perennial soybean included in a different genus and is known to have originated in Africa (Heuzé et al., 2015). Most of the species in the sub-genus *Glycine* contain $2n=40$ chromosomes, except for accessions of both *G. hirticaulis* and *G. tabacina*, which have diploid ($2n = 40$) and tetraploid ($2n = 80$) chromosome numbers. In contrast, the accessions of *G. tomentella* contain four cytotypes ($2n = 38, 40, 78$, or 80). The accessions of *G. max* do not readily cross with perennial species of the sub-genus *Glycine* due to crossability barriers that can be overcome by extensive hybridization, immature seed rescue, and cytogenetic manipulations (Singh et al., 2019). The cross-compatible accessions of *G. max* and *G. soja*, along with the accessions of *G. gracilis*, are included in the primary gene pool (GP-1), whereas the secondary gene pool (GP-2) has no accessions in soybeans. The tertiary gene pool (GP-3) consists of 26 wild species from the sub-genus *Glycine* (Chung and Singh, 2008).

(a)

(b)

FIGURE 10.1 Morphological variation in soybean species; (a) *G. soja* accession DC2008-1 and (b) *G. max* cv. DS9712.

Domestication of Soybean and SWRs

Around 3,000–5,000 years ago, the cultivated soybean is known to have been derived from its wild progenitor *G. soja* in East Asia (China, Korea, Japan, and part of Russia) (Larson et al., 2014) along the Yellow River or Huang-He valley in Central China (Li et al., 2010), resulting in the landraces of *G. max* and subsequently the modern (elite) cultivated soybean.

The domestication process of soybean can be studied under three overarching hypotheses: the single origin hypothesis, the multiple origin hypothesis, and the complex hypothesis (Sedivy et al., 2017). Line depictions of the hypotheses are given in Figure 10.2. According to the single origin hypothesis, cultivated soybean emerged from the wild soybean through a single domestication event in Central China. This widely accepted hypothesis is supported by several studies (Zhou et al., 2015b; Han et al., 2016; Guo et al., 2010; Dong et al., 2014; Chung et al., 2014). In contrast, the multiple origin hypothesis suggests that the domestication process continued over a long period of time through multiple events from *G. soja* to *G. max* between 5,000 and 9,000 years ago (Xu et al., 2002; Abe et al., 2003). The nuclear and chloroplast genomes of Korean and Japanese soybeans, which have significantly different gene pools, support the idea of independent domestication (Xu et al., 2002; Abe et al., 2003; Zhou et al., 2015b). The larger soybean seeds found in Korea and Japan, in contrast to those in China (Yellow River basin), indicate that there may have been multiple independent efforts to domesticate wild soybeans, either *G. soja* or the *G. soja/G. max* complex, at different locations in East Asia (Lee et al., 2011) (Figure 10.2).

According to the complex hypothesis, the G. soja/G. max complex first diverged before multiple domestication events (Kim et al., 2010b; Li et al., 2014). The whole genome comparison of one wild soybean ecotype to one soybean cultivar indicated that the age of the *G. soja/G. max* complex is 0.27 million years ago (Kim et al., 2010b), and a pangenome comparison of seven wild soybean ecotypes indicated it to be 0.8 million years ago (Li et al., 2014). The domestication might have occurred from an already diverged *G. soja/G. max* complex (Sedivy et al., 2017). This theory explains that the early-domesticated *G. soja* or (*G. soja* × *G. max*) complex was distributed from China to Korea and Japan and subsequently underwent varying degrees of domestication according to local needs. Thus, it is now widely accepted that *G. max* was created from the *G. soja or G. soja* × *G. max* complex through a long, slow, and complex domestication process involving countless independent efforts (Sedivy et al., 2017).

Prospects of SWRs Under Climate Change Regime

Over the decades of crop domestication and active breeding, gene or genetic erosion has been occurring at a modest but consistent rate (Schouten et al., 2019). Crop types have reportedly lost more than 75% of their genetic diversity during the past 100 years (FAO, 1999; Khoury et al., 2022). The wild soybean, *Glycine soja* (Siebold & Zucc.), which is included in gene pool 1 (GP1), has still retained a great deal of genetic diversity and holds great potential to improve its agriculturally important domesticated relative by supplying genes and gene families responsible

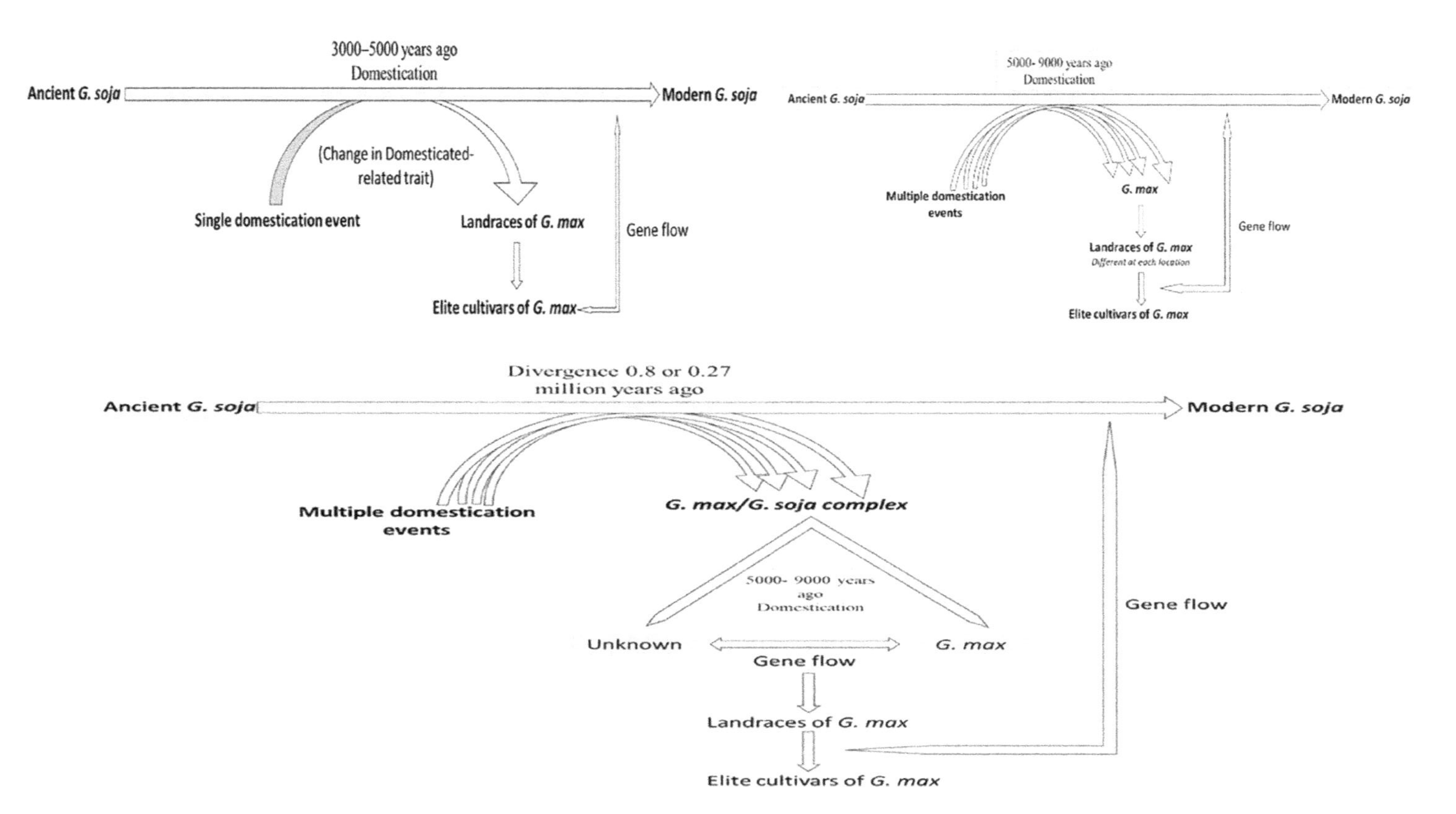

FIGURE 10.2 Depiction of single, multiple, and complex origin hypotheses of soybean.

for higher oil and protein contents, resistance to drought, high temperatures, disease, and insect pests (Perkins, 1995; Murphy, 2002). However, relatively little work has been done to exploit *G. soja* as a genetic reservoir for soybean improvement (Kim et al., 2011; Qi et al., 2014; Zhang et al., 2016). To develop soybean varieties that could cope with harsh climates due to global warming, limited water supplies, and rapidly evolving plant pests and pathogens, it is essential to explore the soybean wild relatives for effective genes or quantitative trait loci (QTLs). Several studies have characterized the genes/QTLs of domestication-related traits and provided the details of major abiotic and biotic stress tolerance-related genes in wild soybeans (Meyer and Purugganan, 2013; Murphy, 2002; Rathod et al., 2019; Manu et al., 2020). The introgression of several disease-resistance genes from wild relatives, and advances in breeding techniques, such as genome sequencing, pangenome construction, *de novo* domestication, etc., have been facilitating traits/gene selection from both closely and distantly related species. Activities of farming companies worldwide have demonstrated that utilization of wild relatives in breeding has resulted in large economic gains with a yield improvement of about 30%, accounting for an estimated global value of about $100 billion (Pimentel et al., 1997; Brozynska et al., 2016).

BREEDING APPROACHES FOR DESIGNING CLIMATE-RESILIENT SOYBEAN USING SWRS

In terms of commercial cultivation, soybean is relatively a new crop in India; however, its spread in terms of the area of cultivation is rapid. In the 1970s, soybean was introduced from the United States, and commercial cultivation was started. For higher yield and adaptation, the initial cultivars were developed through the process of selection from local cultivars (e.g. Bhat, Kalitur) or direct adaptation of introduced varieties (e.g., Bragg, Lee, Hardee, etc.) (Tiwari, 2014). Subsequently, mutation breeding was adopted and several varieties such as NRC 2, Birsa Soya, VLSoya 1, DS 9712, NRC 12, Aarti, TAMS 98-21, etc., were released (Tiwari, 2014). After 1990, major mega varieties such as JS 335, JS 9305, NRC 7, JS 9560, and JS 2034 were developed through hybridization and pedigree selection. The molecular breeding approach is now gaining momentum, and some varieties such as NRC127, NRC142, NRC-SL 2, etc., have been developed through it (Chandra et al., 2022a). The fact remains that India's share of the global soybean area and production is 10% and 4%–5%, respectively, indicating that the productivity of soybeans in India is still very low (~1.2 t/ha) compared to global productivity (~2.9 t/ha) (AMIS, FAO website). Little use of wild-type soybean in the breeding program is one of several factors contributing to low productivity in the country (Maranna et al., 2021). Ideally, a soybean cultivar should have determinate/semi-determinate growth habits, erect and non-lodging characteristics suitable for mechanical harvesting, a good long juvenile period, rapid seed fill nature, and a maturity duration of 95–100 days. However, resilience to climatic variables is the primary factor in the context of climate change. Here, we present the breeding approaches in the context of developing climate-resilient soybean cultivars using SWRs.

Pre-Breeding

Pre-breeding is an approach to generating breeding materials that can eventually be utilized for developing high-yielding cultivars with desirable attributes (Singh et al., 2019). The main objective of pre-breeding is to broaden the genetic base in cultivated genotypes through the identification and introgression of valuable traits from non-adapted genetic resources. For effective pre-breeding, the constitution and compatibility of chromosomes of the wild-type genotypes are very important. Cultivated soybean [*Glycine max* (L.) Merrill] and its wild ancestor *Glycine soja* are annuals belonging to the sub-genus *Glycine* and are cross-compatible. The sub-genus *Soja* possesses more than 25 wild perennial species. The transfer of traits from it to cultivated species needs special techniques like protoplast fusion and embryo rescue. Of late, *Glycine tomentella* and *Glycine tabacina* have been used for gene transfer to cultivated soybean for a number of useful traits (Singh, 2019). Transferring useful genes from *G. soja* accessions, Talukdar et al. (2013) have developed soybean genotypes resistant to yellow mosaic virus (YMV) disease. Similarly, using primary gene pool *Glycine soja* accessions, i.e., PI 593983, PI 407170, PI 549046, and tertiary gene pool (*Glycine microphylla*), lines with early maturity, rapid grain filling traits, and insect resistance have been developed (Annual Report, ICAR-IISR 2021). Such successes show the promise of pre-breeding in the development of climate-resilient soybeans.

Conventional Breeding

Pedigree selection after hybridization between elite parents and backcrossing breeding for introgression of target traits has been very much utilized in soybean improvement. These conventional breeding techniques have led to the development of more than 4,000 varieties of soybean worldwide for commercial cultivation. In India too, several mega varieties like JS 335, JS 95-60, JS 93-05, JS 20-34, etc., were bred through conventional strategies of pedigree selection. The soybean wild species *Glycine soja* is easily crossable with cultivated soybean (*Glycine max*); it has been used effectively for widening the genetic base and transferring of certain useful traits. In India, mungbean yellow mosaic virus (MYMV) resistance has been introgressed from *Glycine soja* to JS 335 (*Glycine max*) through backcrossing programs, resulting in high-yielding YMV resistant lines such as YMV 11 and 16 (Shivakumar et al., 2023; Annual Report, ICAR-IISR 2021).

Intergeneric Hybridization

There is no secondary gene pool (GP-2) in soybean; therefore, gene transfer has happened between accessions of GP1 and GP3. Interspecific, fertile hybrids between *G. max.* and *G. soja* (Sieb and Zucc.) have been observed successfully by several researchers worldwide, including Ahmad et al. (1977), Hadley and Hymowitz (1973), Broich (1978), Yashpal et al. (2015), Chandra et al. (2020), and between *G. max* and *G. gracilis* by Karasawa (1952). Intergeneric hybrids have also been obtained using embryo rescue and other innovative approaches. Intersubgeneric hybrids were obtained between *G. max* and *G. clandestina* Wendl; *G. max* and *G. tomentella* Hayata (Singh and Hymowitz, 1985; Singh et al., 1987); and between *G. max* and *G. canescens* (Broue et al., 1982).

Trait Specific and Molecular Breeding

Biotic Stress Resistance

Resistance to insect stress has been reported in wild-type soybean for soybean aphid and defoliant insects (Table 10.2) (Zhang et al., 2017; Annual Report, ICAR-IISR 2021). Three *G. soja* genotypes were found to have resistance to the soybean aphid (Hesler, 2013). Furthermore, through molecular mapping, researchers identified two new QTLs, Rag3c and Rag6, related to aphid resistance using linkage mapping (Zhang et al., 2017). Additionally, the foxglove aphid (*Aulacorthum solani*) resistance gene, *Raso2*, was reported in *G. soja* (Lee et al., 2015), followed by the *Raso2* gene.

G. soja was found to have resistance against several diseases, including Mungbean Yellow Mosaic India Virus (MYMIV). Two genes were reported to control MYMIV resistance, which was transferred to cultivated varieties through linked SSR markers

TABLE 10.2
Various Genomics and Bioinformatics Tools for Accessing, Managing, and Utilizing SWRs' Data and Metadata

S. No	Names	Websites
1	SoyBase	https://soyabase.org
2	GenBank	https://www.ncbi.nlm.nih.gov/genbank/
3	NCBI (National Center for Biotechnology Information)	https://www.ncbi.nlm.nih.gov/
4	DDBJ (DNA Data Bank of Japan)	https://www.ddbj.nig.ac.jp/
5	SOAPdenovo	http://soap.genomics.org.cn/soapdenovo.html
6	SPAdes	http://cab.spbu.ru/software/spades/
7	Velvet	https://www.ebi.ac.uk/~zerbino/velvet
8	MAKER	http://www.yandell-lab.org/software/maker.html
9	AUGUSTUS	http://bioinf.uni-greifswald.de/augustus/
10	BLASR	https://github.com/PacificBiosciences/blasr
11	Bowtie	http://bowtie-bio.sourceforge.net/index.shtml
12	OrthoMCL	https://orthomcl.org
13	Mauve	http://darlinglab.org/mauve/mauve.html
14	GATK (Genome Analysis Toolkit)	https://gatk.broadinstitute.org
15	SAMtools	http://www.htslib.org/
16	TASSEL	https://www.maizegenetics.net/tassel
17	PLINK	https://www.cog-genomics.org/plink/
18	HISAT2	http://daehwankimlab.github.io/hisat2/
19	StringTie	https://ccb.jhu.edu/software/stringtie/
20	DESeq2	https://bioconductor.org/packages/release/bioc/html/DESeq2.html
21	Gene Ontology (GO)	http://geneontology.org
22	KEGG (Kyoto Encyclopedia of Genes and Genomes)	https://www.kegg.jp
23	Genomes OnLine Database (GOLD)	http://www.genomesonline.org
24	GCViT	https://github.com/LegumeFederation/gcvit

to develop a resistant version of the variety JS335 (Rani et al., 2018; Annual Report, ICAR-IISR 2021). Soybean cyst nematode (SCN) is one of the most disturbing pests in cultivated soybeans. *G. soja* has exhibited variability in resistance to different SCN populations (Kim et al., 2011; Zhang et al., 2016). Two novel SCN-resistance QTLs (cqSCN-006 and cqSCN-007) were identified in *G. soja* accession PI468916 (Kim et al., 2011). Further fine mapping studies for this aspect have identified SCN-resistant candidate genes, including the SNAP gene Glyma.15g191200 (Yu and Diers, 2017). *G. soja* accessions resistant to SCN HG type 2.5.7 were also reported by Zhang et al. (2016). Recently, Kofsky et al. (2021) identified candidate genes and associated pathways involved in SCN resistance for SCN in wild-type soybean. They reported the involvement of Jasmonic acid (JA)/Salicylic acid (SA) signaling genes for resistance against SCN. Thus, wild-type soybeans harbor genes for resistance to biotic stresses, which would help in developing soybean genotypes tolerant to emerging pests in changing climate situations (Ratnaparkhe et al., 2022a).

Abiotic Stress Resistance

The abiotic stress factors such as drought, high temperature, water logging, salinity, etc. are emerging problems in the changing climate regime. In India, more than 90% of soybean grown in rainfed situations are victims of abiotic stress factors. Besides the cultivated species, *G soja* can be a strong contributor to tolerance against drought stress in terms of root architecture and other traits (Prince et al., 2015; Valliyodan et al., 2017). *G soja* accession PI 483463 was found to have a better root angle for combating drought stress (Valliyodan et al., 2017). Prince et al. (2015) identified SNPs/genes, associated with shorter root or tap roots in wild soybeans, which are related to drought adaptations. Ning et al. (2017) identified drought tolerance gene GsWRKY20 from *G. soja*. Similarly, *G. soja* accessions, PI 467162, PI 479751, PI 407229, PI 597459C, PI 424082, PI 378699A, PI 424107A, and PI 366124, have tolerance against water logging stress (Valliyodan et al., 2017).

Cultivated soybeans are sensitive to salt stress. However, *G. soja* accessions are found to have resistance to salt stress, and several resistance genes such as *Ncl2*, *GmCHX1*, etc., have been identified (Qi et al., 2014). An aquaporin gene *GmTIP2;1* has also been found to be linked with salt tolerance in *G. soja* (Zhang et al., 2017a). It is believed that more aquaporin alleles may be available in G. soja conferring salt stress tolerance.

Food Grade Traits

Glycine soja has remained underutilized for its diversity of food grade traits. Cultivated soybean (G. max) seeds consist of low protein (38%–42%) and high oil (18%–22%) compared to G. soja, which has high protein (46%–48%) and low oil (8%–10%) (Kumar et al., 2021). Favorable alleles for high protein introgressed from *G. soja* accessions and located on Chr. 20 have been reported to increase seed protein and decrease seed oil content (Diers et al., 1992; Sebolt et al., 2000; Bolon et al., 2010). Two *G. soja* accessions, PI 407246 (South Korea) and PI 407301 (China), are reported to have high protein content (Kumar et al., 2021), so genetic diversity for these traits may be exploited in breeding programs for further improvement in soybean meal-based cultivars for industries.

Seed Related Traits

As in other grain legumes, seeds are the main product output in soybeans for consumption and processing. The cultivated soybean has a bold seed size and more yield as compared to the wild types; however, different traits from wild-type soybean can contribute to broadening genetic base and developing better idiotype for changing climate conditions (Yashpal, 2019). Various researchers (Concibido et al., 2003; Liu et al., 2007; Yashpal et al., 2019) identified QTLs/genes for different yield attributes, including seed size using *G. soja* × *G. max* crosses.

Hard seededness in soybeans can maintain high quality and seed viability through lengthy storage; therefore, it is valuable for the food economy in general and germplasm preservation in particular (Zhou et al., 2010). The wild-type soybean, which primarily produces black-colored seeds, also differs in the extent of hard seededness and viability from cultivated soybean (Mullin, 2000; Chandra et al., 2017; Kumar et al., 2019a). Typically, wild-type soybean seeds remain viable for a longer duration than cultivated type due to varying degrees of hard seededness, which can be introgressed into cultivated type for the development of lines with medium-term storability (12–36 months) (Kilen and Hartwig, 1978; Chandra et al., 2017). However, optimum thickness of seed coat and permeability are required for food industries (Kebede et al., 2014) for processing and making its use for food. On the other side, wild species of soybean such as *G. tomentella* and *G. soja* may be used as potential sources for seed storability Chandra et al. (2021). Chandra et al. (2021) and Kumar et al. (2019a) screened wild accessions of soybeans and subsequently developed introgression lines from *G. soja* × *G. max* and identified lines that maintain 75%–90% seed germination even after 3 years. Zhou et al. (2010) observed this variation in the ranges of 100%–28% and 100%–9% germination in *G. soja* accessions DS and QN, respectively, after 10 years of storage. Thus, *G. soja* can be a potential resource for developing breeding materials having better germination (>70%) even after longer ambient storage.

STRATEGIES FOR RAPID GENERATION ADVANCEMENTS OF SWRs

G. soja and *G. max* have the same chromosome number, $2n=40$, and are easily crossable to produce fertile hybrids, making it possible to transfer useful genes from *G. soja* to *G. max* through conventional breeding. However, introgression of traits from *G. soja* into *G. max* may result in linkage drag. Unfavorable characteristics like lower yield, shattering, lodging, climbing habit, etc., are frequently carried by linkage drag. There are now approaches like marker-assisted backcrossing (MABC), which can increase the effectiveness of making use of SWRs and get over some of these obstacles (Iftekharuddaula et al., 2011; Hasan et al., 2015) and genomic selection (e.g., Duhnen et al., 2017). The rapid advancement of biotechnology, including enhanced genetic transformation and cutting-edge genome editing (e.g., Kim et al., 2017), can overcome these constraints. Precise genome editing now enables the introduction of desirable allele-conferring wild-derived genes into elite backgrounds without the need for lengthy introgression regimes, overcoming the linkage drag and reduced fertility hurdles that frequently hinder the use of SWRs (Bohra et al., 2021). These cutting-edge strategies, which take advantage of genomic and genome editing advancements, offer potential ways to address persistent problems and bring SWRs to the forefront of crop invention. Some of such approaches are described below:

Speed Breeding

Speed breeding involves the manipulation of the temperature and photoperiod in a controlled growing environment to grow several crop generations quickly each year (Watson et al., 2018). It is one of the most viable approaches for the utilization of the SWRs in breeding programs. Speed breeding allows genotypes' genetic backgrounds to be fixed more rapidly than it would normally require through years of inbreeding. For crops like barley, canola, chickpea, pea, rice, sorghum, and wheat, speed breeding has been tested and successfully produced several generations in a single year (Espósito et al., 2012; Rizal et al., 2014; Watson et al., 2018; Nagatoshi and Fujita, 2019; Rana et al., 2019). Thus, it allows selection against unfavorable characteristics in a much shorter period. Speed breeding might also be advantageous for non-CRISPR (Clustered Regularly Interspaced Palindromic Repeat) methods for domesticating SWRs, such as germplasm conversion (Klein et al., 2016; Stephens et al., 1967; Rosenow et al., 1997), as it would simply require knowledge of the genomic region that confers the domestication trait(s), rather than specific knowledge of the target sequences. Thus, introducing the SWRs to conditions for speed breeding could reduce the amount of time needed to pass through several generations to effectively convert their germplasm into commercially viable crops (Bhatta et al., 2021).

Genomics and Genome Editing

Genomics provides ways for the exploration of a genetically diverse spectrum of SWRs and identification of the agronomically useful genes or QTL. The *de novo* assembly followed by SWR sequencing may result in reference assemblies that enable later applications like targeted genome editing and functional characterization of genes. Several CWRs assemblies are now becoming accessible, including relatives of barley, rice, soybean, tomato, and wheat (Brozynska et al., 2016; Bohra et al., 2022). These genome assemblies have made it possible to identify some significant genes and QTL from CWRs, such as numerous disease-resistance genes in wheat (Yahiaoui et al., 2009; Periyannan et al., 2013; Saintenac et al., 2013) and QTL associated with oil content in soybean (Zhou et al., 2015). This has frequently been done when combined with high-throughput phenotyping. For reference, genomes in important crops and high-quality assemblies based on third-generation long-read sequencing are increasingly becoming standard. The development of high-quality long-read assemblies in a wide range of CWRs will be essential for unlocking a reservoir of beneficial CWR genetic variation that is ready to be exploited for crop improvement. Advances in long-read sequencing in terms of expanded accessibility and reduced price points will be crucial for this.

Novel traits or missing traits in cultivated crops may be developed through naturally available or induced mutations. However, in the current era, genetic engineering and gene editing (genome editing) are the techniques that enable precise and targeted modifications for desired traits (Chandra et al., 2022c). Now, gene-editing technologies are gaining momentum for crop genetic enhancement. CRISPR/Cas9 gene editing is particularly useful in deciphering the traits. As researchers characterize more genes related to key domestication traits in model or major crops and high-quality CWR genome assemblies are generated, the potential for editing these genes in CWRs skyrockets, leading to the possible creation of new crops through *de novo* domestication. Furthermore, simultaneously identifying and cataloging agronomically beneficial traits in CWRs will greatly enhance our ability to exploit wild genetic diversity for the development of climate-resilient crops. Moreover, CWRs are also difficult to regenerate, further complicating the transformation process (Zhu et al., 2020). Several alternative approaches for reagent delivery, which were initially developed in animal cells, are being explored in plants (Ghogare et al., 2021).

GENOMIC AND BIOINFORMATICS TOOLS FOR ACCESSING, MANAGING, AND UTILIZING SWRS DATA AND METADATA

SWRs are valuable genetic resources that hold immense potential for enhancing cultivated soybean varieties (Li et al., 2014). The utilization of SWRs in breeding programs requires efficient tools for accessing, managing, and utilizing their genomic data and associated metadata. Now, a wide range of genomic and bioinformatics tools (Figure 10.3) are available for studying SWRs, from data retrieval to analysis and interpretation.

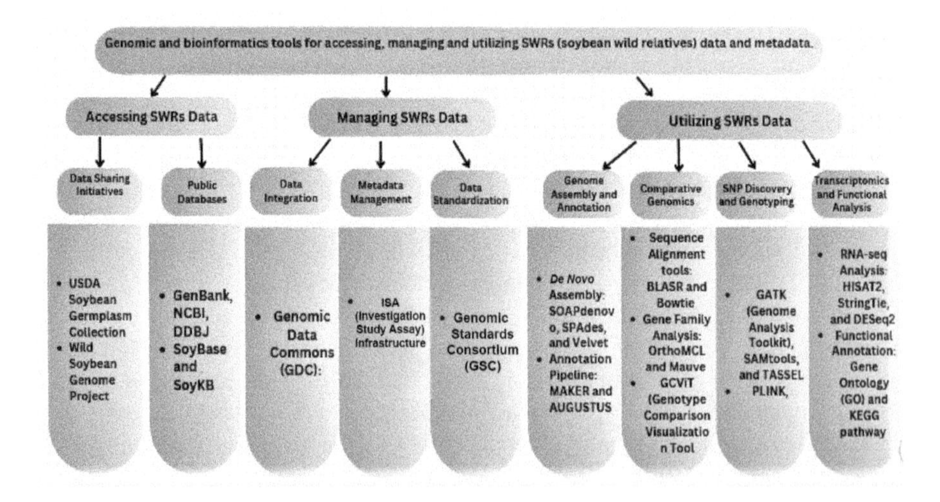

FIGURE 10.3 Flow chart depicting the various genomics and bioinformatics tools for accessing, managing, and utilizing SWRs' data and metadata.

Accessing SWRs Data

Accessing SWRs data is crucial for researchers, breeders, and conservationists who aim to utilize the genetic resources and information associated with these wild species. SWRs data provides valuable insights into the genetic diversity, traits, and potential applications in crop improvement and genetic resource conservation.

Data-Sharing Initiatives

Several collaborative projects such as USDA Soybean Germplasm Collection aim to collect, preserve, and share SWRs data and genetic resources. Researchers can access these datasets through online portals and data-sharing platforms (Diego, 2016). Similarly, the Wild Soybean Genome Project focuses on the genomic characterization of wild soybean accessions and provides open access to the generated data for further analysis and utilization (Xie et al., 2019).

Public Databases

Several publicly accessible databases such as GenBank, NCBI, and DDBJ host a vast collection of genomic data, including SWRs. Researchers can search and retrieve relevant data using keywords, accession numbers, or sequence similarity (Benson et al., 2002; Mashima et al., 2015). Similarly, specialized databases such as SoyBase and SoyKB provide curated genomic information specifically focused on soybeans and its wild relatives, enabling targeted searches and data retrieval (Joshi et al., 2014; Brown et al., 2020).

Managing SWRs Data

Managing SWRs data is essential for organizing, storing, and utilizing the vast amount of information associated with these valuable genetic resources. SWRs data management involves capturing, cataloging, and making the data accessible for research, breeding programs, and conservation efforts. Researchers and breeders can leverage SWR data for crop improvement, research, and conservation efforts.

Data Integration

Several digital platforms facilitate data integration for further analysis. Genomic Data Commons (GDC) is a platform that facilitates the integration and harmonization of diverse genomic datasets. It enables researchers to merge and analyze genomic data, annotations, and associated metadata to create unified datasets for downstream analysis (Jensen et al., 2017; Petereit et al., 2022).

Metadata Management

Managing a huge data set is a challenge in the digital era. The ISA (Investigation Study Assay) infrastructure is a framework that enables the capture, storage, and management of metadata associated with SWRs datasets. It ensures that essential information about the characteristics and properties of the genomic data is properly documented, enhancing data discoverability and reproducibility.

Data Standardization

Standardization of the genomic data is required for further analysis and harnessing the real meaning of it. The Genomic Standards Consortium (GSC) is a consortium that has developed guidelines and protocols for organizing and annotating genomic data, promoting data standardization and interoperability across different datasets. Adhering to these standards ensures compatibility and facilitates data integration (Dawn et al., 2011).

Utilizing SWRs Data

Utilizing SWRs data plays a crucial role in harnessing the genetic potential of these wild species for enhancing soybean cultivation and conserving genetic resources. By collecting, characterizing, and utilizing SWRs data, researchers and breeders can identify valuable traits, introgress desirable genetic variations, and enhance soybean's adaptation to diverse environmental conditions. Moreover, the effective utilization of SWRs data contributes to the conservation of genetic resources, ensuring the preservation of valuable wild relatives for future generations.

Genome Assembly and Annotation

***De Novo* assembly**: Tools such as SOAPdenovo, SPAdes, and Velvet facilitate *de novo* genome assembly, which is essential for constructing the reference genome of SWRs. These tools enable the reconstruction of complete or draft genomes by assembling short DNA sequences generated from high-throughput sequencing technologies (Zerbino et al., 2008; Luo et al., 2012).

ANNOTATION PIPELINE

Tools like MAKER and AUGUSTUS aid in the annotation of SWRs genomes by predicting genes, identifying regulatory elements, and assigning functional annotations to genomic regions. These pipelines assist in understanding the structure and function of SWRs genomes (Cantarel et al., 2008).

Comparative Genomics

Comparative genomics helps in understanding the genetic diversity, evolutionary relationships, and potential functional variations between SWRs and cultivated soybean (Langmead et al., 2012; Chowdhury et al., 2021). Tools like BLASR and Bowtie enable sequence alignment between SWRs and cultivated soybean genomes, identifying shared and unique genomic features.

Gene Family Analysis

OrthoMCL and Mauve assist in the identification and analysis of gene families in SWRs. These tools help in identifying conserved regions, gene duplications, and functional divergence, providing insights into the genomic changes that have occurred during the evolution of SWRs (Wafula et al., 2023).

Genotype Comparison Visualization Tool

This is an advanced bioinformatics tool designed for visualizing and analyzing genotype data. It is particularly useful in the field of plant breeding, including research involving SWRs. Genotype Comparison Visualization Tool (GCViT) provides a powerful platform to compare and analyze the genotypes of different individuals or populations, allowing researchers to explore genetic variations, identify patterns, and make informed breeding decisions (Wilkey et al., 2020).

SNP Discovery and Genotyping

Tools such as GATK (Genome Analysis Toolkit), SAMtools, and TASSEL enable the discovery of single nucleotide polymorphisms (SNPs) in SWRs genomes. These tools employ variant calling algorithms to identify genetic variations, which are crucial for genetic mapping, association studies, and marker-assisted breeding.

After SNP discovery, genotyping tools like PLINK (Purcell et al., 2007) and TASSEL facilitate genotypic data analysis (Bradbury et al., 2007). These tools help in determining the genetic profiles of SWRs accessions and enable the identification of genomic regions associated with specific traits or phenotypes (Ma et al., 2015).

TRANSCRIPTOMICS AND FUNCTIONAL ANALYSIS

RNA-seq Analysis

Tools such as HISAT2, StringTie, and DESeq2 are used for RNA sequencing (RNA-seq) data analysis. These tools aid in transcriptome assembly, quantification of gene expression levels, and identification of differentially expressed genes. Transcriptomics provides insights into gene regulation, functional pathways, and molecular responses of SWRs to different environmental conditions (Love et al., 2014; Costa-Silva et al., 2017; Pertea et al., 2015).

Functional Annotation

Tools like Gene Ontology (GO) and KEGG pathway analysis help in annotating the functions of genes identified through transcriptomics. These tools enable the interpretation of gene function, enrichment analysis, and pathway mapping, providing a comprehensive understanding of the biological processes underlying SWR traits (Ashburner et al., 2000).

Genomes OnLine Database (GOLD)

It is a resource designed to catalog and monitor genetic studies globally. It aims to provide comprehensive information about ongoing and completed sequencing projects, along with curated metadata. The database's fifth version (v.5) introduces a redesigned scheme and web user interface, which includes new features like a four-level (meta) genome project classification system and an intuitive web interface for accessing reports and search tools.

ACHIEVEMENTS FROM SWRS IN SOYBEAN BREEDING

POD SHATTERING

Artificial selection for pod-shattering resistance is regarded as a milestone of crop domestication in soybeans. Loss of pod shattering in soybean is triggered by a transcription factor NAC (NAM, ATAF1/2, and CUC2) and *SHAT1-5* in lignified fiber cap cells (Dong et al., 2014). The domesticated allele of *SHAT1-5* activates 15-fold biosynthesis of the secondary wall, promoting the thickening of the secondary walls of fiber cap cells. During domestication, this locus experienced artificial selection and was probably derived from a single domestication event, making *SHAT1-5* a prime domestication gene of soybean (Dong et al., 2014). Another gene, *prephenate dehydrogenase 1 (Pdh1)*, encoding a dirigent-like protein, is also involved in lignification, affecting the soybean pod-shattering phenotype. Modern cultivars carry a single nucleotide substitution at the beginning of the coding sequence of *Pdh1* that produces a stop codon (Funatsuki et al., 2014).

SEED HARDNESS

Seed hardness is another important trait for soybean domestication, governing the water permeability of dry seeds and the hardness of cooked seeds. The gene *GmHs1-1*, encoding a calcineurin-like metallophosphoesterase transmembrane protein, is expressed in the epidermal layer of the seed coat (Sun et al., 2015). It is reported that most soybean accessions carrying a SNP resulting in an amino acid substitution, show low polymorphism in the c. 160 kb genomic region surrounding *GmHs1-1*, indicating a possible signature of artificial selection during soybean domestication (Sun et al., 2015).

STEM GROWTH HABIT

Stem growth habit (determinate) is an important agronomic trait associated with the domestication process of soybean. It is controlled by two major loci, namely Dt1 and Dt2, where Dt1 has an epistatic effect on Dt2 (Bernard, 1972). The gene of Dt1 locus, *GmTfl1*, is functionally similar to the Arabidopsis *TERMINAL FLOWER1* (*TFL1*) gene (Tian et al., 2010). During early vegetative growth, *GmTfl1* transcripts accumulate in the shoot apical meristem of both determinate and indeterminate lines. However, these transcripts are abruptly lost in the reproductive phase in determinate lines (Liu et al., 2010). In a subset of *G. max*, four independent single nucleotide substitutions were identified in the *GmTfl1* gene, but these substitutions were absent in *G. soja* (Tian et al., 2010), indicating that selection for determinacy took place during soybean diversification. The Dt2 suppresses the expression of the *GmTfl1* gene in the shoot apical meristem (SAM), promoting early conversion of SAM into reproductive inflorescence (Ping et al., 2014). The semi-determinate and determinate stem growth habits are rarely observed in *G. soja*. It is anticipated that the dominant Dt2 allele is a recent gain-of-function mutation that occurred after soybean domestication.

Photoperiodic Flowering

In the process of domestication, attention was given to the selection of soybeans suitable for a particular latitudinal photoperiod (Cober and Morrison, 2010; Kim et al., 2012a). Being a short-day flowering plant, soybean requires photoperiod insensitivity for its latitudinal expansion. *GmCRY1a* (Li et al., 2013) promotes floral initiation in soybean and its rhythmic expression correlates with flowering time and the latitudinal distribution of soybean cultivars under flowering-inhibitory long-day photoperiods (Zhang et al., 2008). Zhai et al. (2014) observed recessive alleles at the major maturity loci E1, E2, E3, and E4 and suggest that these maturity loci have contributed to diversification or local adaptation rather than soybean domestication. Among these E loci, E1 shows a predominant effect on photoperiodic control of flowering and maturation. The dominant E1 allele delays flowering and maturation and is further enhanced by long-day photoperiods, found mostly in both wild and cultivated soybeans in China (Langewisch et al., 2014). The recessive e1 allele is distributed mainly in high latitudinal regions, such as the United States and Canada, northeastern China, Japan, and Korea, where soybeans typically show shorter maturity periods compared with those in southern regions (Zhou et al., 2015b). The maturity locus E2 encodes *GmGIa*, a regulator of photoperiodic flowering (Watanabe et al., 2011), and the dominant E2 allele also delays flowering and maturity. Accessions with e2 alleles carry an SNP causing a premature stop codon that elevates the expression of *GmFT2a*, leading to early flowering. The E2 allele is mostly found in wild soybeans, whereas the e2 allele is mostly found in cultivated soybeans (Langewisch et al., 2014). E3 and E4 are known as photoperiod sensitivity alleles and the geographic distribution of the e4 allele is restricted to high latitudinal regions in Japan (Liu et al., 2008; Wu et al., 2013; Kanazawa et al., 2009).

Flower, Seed Coat, and Pod Color

The genes responsible for flower color (pinkish-white and white flowers) have been identified from wild soybeans. The W1 and inhibitor (I) loci were identified for flower, seed coat, and pod color by targeted GWAS studies (Zhou et al., 2015b). Different loci, such as W1, W3, W4, w1-s1, w1-s2, w1-lp, and w1-p2, have been reported to control white color and pinkish-white (Van Huan et al., 2005). The W1 locus having *flavonoid 3050-hydroxylase* (*F3050H*) gene encodes delphinidin-based anthocyanins and/or proanthocyanins in flower and seed coats (Zabala and Vodkin, 2007). The recessive allele from the white flower carries a small insertion (65 bp) of tandem repeats in exon three, resulting in a premature stop codon. The dominant alleles of the I locus have a cluster of duplicated and inverted *chalcone synthase (CHS)* genes (Tuteja et al., 2009), encoding the first enzyme in the flavonoid synthesis pathway. This enzyme creates a diverse set of secondary metabolites including the seed coat pigmentation of certain genotypes. Short interfering RNAs (siRNAs) initiated from the transcripts of this CHS inverted repeat target mRNAs of CHS genes on other genomic regions and silence these genes in a seed coat-specific manner.

Other Genes/QTLs

Many other useful QTLs/genes for various biotic and abiotic stress tolerance, biochemical pathways, and several yield-related traits have been characterized by using wild soybeans or genetic populations resulting from crosses between wild and cultivated soybeans (Zhao et al., 2004) for details see (Table 10.3). Many studies on seed antioxidants, phenolics, and flavonoid contents have identified *GmMATE1, 2, 4* genes (Li et al., 2016) (Table 10.2). In soy products, astringent taste is due to the presence of group-A saponins. Wild soybean mutant line CWS5095 carries the *sg-5* gene. The sg-5 protein oxygenates the C-21 position of soyasapogenol-B or another intermediate, resulting in the production of saponin A (Yano et al., 2017; Rehman et al., 2017). Korean wild soybean natural mutant also carries another gene *Sg-1* of this pathway which controls the Ab series of saponins (Park et al., 2016).

Wild soybean has been used to discover genes related to SCN resistance. About 10 SNPs and genes significantly associated with SCN resistance have been identified. Among those genes, a mitogen-activated protein kinase gene and a *v-Myb* myeloblastosis viral oncogene homolog (MYB) transcription factor (TF) were found to be strong candidates (Shi et al., 2015). Drought stress tolerance genes and transcription factors have also been identified in wild soybeans (Luo et al., 2013). The *GmCHX1*, a salt-tolerant gene, has been identified in wild soybean through a whole-genome *de novo* sequencing approach (Qi et al., 2014). A phosphatase 2C-1 (PP2C-1) allele known to be involved in seed weight and seed size has also been identified from wild soybeans (Li et al., 2007). Many QTLs related to linolenic acid production (Pantalone et al., 1997), yield, height, seed weight, seed filling period, maturity, and lodging (Zhang et al., 2004), soybean cyst nematode (Zhang et al., 2017), seed salt tolerance (Ha et al., 2013), sclerotinia stem rot (Iquira et al., 2015), root traits (Manavalan et al., 2015), oil and local breeding, shoot fresh weight (Asekova et al., 2016), and many other traits have been mapped and identified from wild soybeans or populations developed by crossing wild and cultivated soybean (Yashpal et al., 2019).

LIMITATION TO USES OF SWRS WITHIN BREEDING PROGRAM

The widespread utilization of SWRs as a source of superior alleles, however, necessitates a large investment of time, money, and labor (Dempewolf et al., 2017). The relatedness, compatibility, and crossability of SWRs to their cultivated corresponding types are major obstacles preventing the direct introduction of SWR characteristics through conventional breeding. A trait that is useful in a domesticated state (and selected for) may not be useful in the wild, and vice versa, owing to drastically different selection regimes in wild state or region compared to a domesticated state or region. This makes trait identification and selection potentially difficult and significantly influenced by the environment. Since SWRs can exhibit growth forms or traits such as *G. soja's* large lateral branching phenotype and uncertain growth habits, it is still time-consuming and labor-intensive to quantify agriculturally significant features in these plants that are challenging to manage. It can be difficult to accurately assess their yield. Even when advantageous wild-derived characteristics are introduced into elite material, linkage

TABLE 10.3
Application of SWRs for Improvement of Stress Tolerance, Food Grade Traits, Seed Quality and Yield in Soybean

Trait	Gene/locus reported	Methods	Significance	References
		Biotic Stress Resistance		
Foxglove aphid Resistance	*Raso2*	QTL mapping	Antixenosis and antibiosis resistance to foxglove aphid	Lee et al. (2015)
Aphid (*Aphis glycines*) Resistance	*Rag3c* and *Rag6*	QTL mapping	Genes confer antibiosis resistance to aphids	Zhang et al. (2017f)
Mungbean Yellow Mosaic India Virus	Markers at Chr 8 and Chr 14	Gene mapping through Bulk segregants analysis	Tight linkages of MYMIV resistance with SSR markers i.e. BARCSOYSSR_08_0867 BARCSOYSSR_14_1416 and BARCSOYSSR_14_1417	Rani et al. (2018)
Mungbean Yellow Mosaic India Virus				Kumar et al. (2012)
Soybean cyst nematode resistance	Marker A245_1 (chr. 18) and Satt598 (chr. 15)	QTL mapping	QTLs associated with race 3 resistance, validated in a backcrosses	Wang et al. (2001)
Soybean cyst nematode resistance	QTLs cqSCN-006 and cqSCN-007. Candidate gene Glyma.15g191200 in cqSCN-006	QTL mapping and positional Cloning	Increased resistance when in combination with *G. max* resistance alleles (rhg1, Rhg4, rhg1-b). Glyma.15g191200 encodes SNAP proteins, similar to Rhg1 function.	Kim et al. (2011); Kim and Diers (2013); Yu and Diers (2017)
Soybean cyst nematode resistance	MAPK (Glyma.18g106800) and CDPK (Glyma.18g064100)	GWAS	New candidate genes associated with H. glycines type 2.5.7 resistance	Zhang et al. (2016)
Soybean cyst nematode resistance	RLKs, CaMs, JA/SA signaling genes MAPKs, WRKYs-	Transcriptomics	Comprehensive regulatory network conferring HG type 2.5.7 resistance in G. soja	Zhang and Song (2017); Zhang et al. (2017d)
Soybean cyst nematode resistance	Jasmonic acid (JA)/Salicylic acid (SA) signaling genes	Transcriptomics	Novel candidate genes and associated pathways involved in SCN resistance	Kofsky et al. (2021)

(*Continued*)

TABLE 10.3 (*Continued*)
Application of SWRs for Improvement of Stress Tolerance, Food Grade Traits, Seed Quality and Yield in Soybean

Trait	Gene/locus reported	Methods	Significance	References
		Abiotic Stress Tolerance		
Drought tolerance	GsWRKY20	Transcriptomics and transgenicoverexpression	Overexpression of GSWRKY20 enhances tolerance to drought stress	Ning et al. (2017)
Root architecture	Glyma15g42220, Glyma06g46210, Glyma06g45910, Glyma06g45920, and Glyma07g32480	QTL mapping	SNPs in discovered genes are associated withshorter root or taproot in wild soybeans which is related to droughtadaptations	Prince et al. (2015)
Alkalinity tolerance	ALMT, LEA, ABC transporter, GLR, NRT/POT and SLAH gene(s)	Transcriptomics	Genes up regulated during $NaHCO_3$ stress	Zhang et al. (2016)
Salt tolerance	Ncl2	QTL mapping	First report of salt tolerance gene in soybeanseparate from the S-100 derived *G. max*salttolerance allele	Lee et al. (2009)
Salt tolerance	GmCHX1	Whole genome sequencing and QTL mapping	Salt tolerance via ion transporters role to maintain homeostasis during salt stress	Qi et al. (2014)
Salt tolerance	GmTIP2;1	Genome wide identification of aquaporins and expression analysis	Overexpression of candidate gene increases salt stress response, likely also associated with drought response	Zhang et al. (2017a)
		Food Grade Traits		
Seed protein content	Marker pA-245 (LG C)	QTL mapping	This *G. soja* allele is associated with increased seed protein content and dominant over the *G. max* allele	Diers et al. (1992)

(*Continued*)

TABLE 10.3 (*Continued*)
Application of SWRs for Improvement of Stress Tolerance, Food Grade Traits, Seed Quality and Yield in Soybean

Seed saturated fatty acid Content	SNPs ss71559532, ss715597684, ss715617910	GWAS	QTLs associated with lower palmitic acid levels	Leamy et al. (2017)
Seed saturated fatty acid Content	Glyma.14G121400, Glyma.16G068500	GWAS	Candidate genes associated with steric acid Production	Leamy et al. (2017)
Trait	**Gene/locus reported**	**Methods**	**Significance**	**References**
Seed saturated fatty acid Content	Glyma.14G121400, Glyma.16G068500	GWAS	Candidate genes associated with steric acid Production	Leamy et al. (2017)
Seed unsaturated fatty acid content	Glyma.16G014000, Glyma.07G112100	GWAS	Candidate gene associated with oleic acid levels and linoleic acid levels respectively	Leamy et al. (2017)
Oil, Protein and sucrose	5 QTLs for protein, 9 QTLs oil, 4 QTLs sucrose	QTL mapping	A major QTL for protein and oil were mapped on Chr. 20 (qPro_20) and suggested negative correlation between oil and protein	Patil et al. (2018)
Oil	22 QTLs	QTL mapping	One QTL, qPA10_1 (5.94–9.98 Mb) on Chr. 10 found as novel locus for palmitic acid	Yao et al. (2020)
		Yield-Related Traits		
Yield	QTL on chromosome 14, Satt168 (significant marker)	QTL mapping	9% yield advantage in *G. max* individuals carrying *G. soja* QTL	Concibido et al. (2003)
Seed size	*qSW-D2;* Satt154	QTL mapping	Significant and consistent QTL for 100 seed weight	Liu et al. (2007)
Yield and seed size	qHSW9-1, qHSW9-2, qHSW19, qYLD17	QTL mapping	Consistent QTLs for yield per plant and 100 seed weight identified for 3 years	Yashpal et al. (2019)

(*Continued*)

TABLE 10.3 (*Continued*)
Application of SWRs for Improvement of Stress Tolerance, Food Grade Traits, Seed Quality and Yield in Soybean

		Seed Quality Related Traits		
Seed longevity	4 QTLs; Satt282, Sat_198, Sat_216	QTL mapping	50% more germination reported in *G. soja* derived RILs	Kumar et al. (2019a, 2019b)
Hard seededness	*qHS1*	Fine mapping and expression analysis	A SNP an Endo-1,4-β-Glucanase gene controls seed coat permeability in Soybean	Jang et al. (2015)
Hard seededness	*GmHs1-1*	Map-based cloning and candidate gene association analysis	Gene codes for calcineurin-like protein, primarily expressed in the Malpighian layer of the seed coat	Sun et al. (2015)
Hard seededness	2 major and 9 minor QTLs; Sat_202, Satt703, Satt274, Satt686, Satt619 and sct_046	QTL mapping	Two consistent minor QTLs were identified, Semipermeable RILs identified	Chandra et al. (2020)

This table is adapted from Kofsky et al. (2018) with updated modifications.

drag can frequently have a detrimental impact on yield or attributes associated with yield. Unfavorable features like lower yield, propensity to shatter, lodging, etc., are frequently dragged by linkage drag (Brouwer and St. Clair, 2004; Summers and Brown, 2013). Furthermore, sterility issues, which are frequently found in the F_1 or BC_1 generation, may develop after introducing genetic material from CWRs into an elite background (Wang et al., 2020; Bohra et al., 2022). Although wild soybeans have been used to boost crops, some potential locations still need upgrading.

FUTURE PROSPECTS: A WAY AHEAD

SWRs have a significant deal of potential to offer novel genes and alleles for soybean and other species of legumes for crop development. Dissecting the genetic architecture underpinning the desired features and identifying the underlying molecular, physiological, and biochemical pathways are prerequisites for employing *G. soja* to enhance the agricultural potential of *G. max*. Approaches like GWAS and linkage mapping facilitated the dissection of potentially useful traits, and integrating these approaches with research into gene expression (transcriptomes), protein expression (proteomics), and metabolite profiling (metabolomes) can be helpful to understand the mechanisms involved in phenotypic differences. The identification and transfer of novel, valuable genes to soybeans can be greatly facilitated by high-throughput sequencing technologies and biotechnology. For instance, precise genome editing now makes it possible to introduce wild-derived genes resulting in desirable alleles into elite backgrounds without the need for time-consuming introgression regimes, overcoming the barriers of linkage drag and decreased fertility that frequently make the use of SWRs challenging (Bohra et al., 2021). In the near future, further discoveries from wild soybeans are anticipated. Along with other crop wild cousins, efforts should be made to preserve wild soybean populations (Khoury et al., 2010).

CONCLUSION

SWRs are the treasure of genes/QTLs. It can provide new allelic combinations needed to adjust the disease pressures, farming methods, market needs, and climatic conditions and can improve the adaptability of agricultural systems worldwide. As of now, a small fraction of the SWRs has been utilized to develop soybean varieties with tolerance to various biotic and abiotic stresses. A concerted global effort is therefore needed for the collection, conservation, and utilization of the precious wild types resources of soybeans for the welfare of mankind. Conventional and modern molecular biological tools and techniques with thorough utilization of the SWRs are expected to support the advancement of soybeans to make them suitable for the changing climatic regimes.

REFERENCES

Abe J, Xu D, Suzuki Y, Kanazawa A, Shimamoto Y. 2003. Soybean germplasm pools in Asia revealed by nuclear SSRs. *Theoretical and Applied Genetics* 106: 445–453.

Ahmad QN, Britten EJ, Byth DE. 1977. Inversion bridges and meiotic behaviour in species hybrids of soybeans. *Journal of Heredity* 68: 360–364.

Anne VB, Others. 2021. A new decade and new data at SoyBase, the USDA-ARS soybean genetics and genomics database. *Nucleic Acids Research* 49(D1), D1496–D1501. https://doi.org/10.1093/nar/gkaa1107

Annual Report 2021 ICAR-Indian Institute of Soybean Research (ICAR-IISR) Indore (MP), India, pp. 1–122.

Asekova S, Kulkarni KP, Kim M, Kim JH, Song JT, Shannon JG, Lee JD. 2016. Novel quantitative trait loci for forage quality traits in a cross between PI 483463 and 'Hutcheson' in soybean. *Crop Science* 56, 2600–2611. https://doi.org/10.2135/cropsci2016.02.0125.

Ashburner M, Ball CA, Blake JA, Botstein D, Butler H, Cherry JM, et al. 2000. Gene ontology: tool for the unification of biology. *Nature Genetics* 25, 25–29.

Benson DA, Karsch-Mizrachi I, Lipman DJ, Ostell J, Rapp BA, Wheeler DL. 2002. GenBank. *Nucleic Acids Research* 30(1), 17–20. https://doi.org/10.1093/nar/30.1.17. PMID: 11752243; PMCID: PMC99127.

Bernard RL 1972. Two genes affecting stem termination in soybeans. *Crop Sci*ence 12, 235–239.

Bhatt P, Rene ER, Kumar AJ, Gangola S, Kumar G, Sharma A, Zhang W, Chen S. 2021. Fipronil degradation kinetics and resource recovery potential of Bacillus sp. Strain FA4 isolated from a contaminated agricultural field in Uttarakhand, India. *Chemosphere* 276, 130156. https://doi.org/10.1016/j.chemosphere.2021.130156

Bhatta M, Sandro P, Smith MR, Delaney O, Voss-Fels KP, Gutierrez L, Hickey LT. 2021. Need for speed: manipulating plant growth to accelerate breeding cycles. *Current Opinion in Plant Biology* 60, 101986.

Bohra A, Kilian B, Sivasankar S, Caccamo M, Mba C, McCouch SR, Varshney RK. 2022. Reap the crop wild relatives for breeding future crops. *Trends in Biotechnology* 40(4), 412–431.

Bolon YT, Joseph B, Cannon SB, Graham MA, Diers BW, Farmer AD, May GD, Muehlbauer GJ, Specht JE, Tu ZJ, Weeks N. 2010. Complementary genetic and genomic approaches help characterize the linkage group I seed protein QTL in soybean. *BMC Plant Biology* 10(1), 1–24.

Bradbury P, Zhang Z, Kroon D, Casstevens T, Ramdoss Y, Buckler E. 2007. TASSEL: software for association mapping of complex traits in diverse samples. *Bioinformatics (Oxford, England)* 23, 2633–2635. https://doi.org/10.1093/bioinformatics/btm308

Broich SL. 1978. The systematic relationships within the genus *Glycine* Willd. subgenus Soja (Moench) F.J. Hermann. M.S. thesis. Iowa State University, Ames.

Broue P, Douglass J, Grace JP, Marshall DR. 1982. Interspecific hybridisation of soybeans and perennial *Glycine* species indigenous to Australia via embryo culture. *Euphytica* 31, 715–724.

Brouwer DJ, St.Clair DA. 2004. Fine mapping of three quantitative trait loci for late blight resistance in tomato using near isogenic lines (NILs) and sub-NILs. *Theoretical and Applied Genetics* 108, 28–638.

Brown A, Conners S, Huang W, Wilkey A, Grant D, Weeks N, Cannon S, Graham M, Nelson, R. 2020. A new decade and new data at SoyBase, the USDA-ARS soybean genetics and genomics database. *Nucleic Acids Research* 49. https://doi.org/10.1093/nar/gkaa1107.

Brozynska M, Furtado A, Henry RJ. 2016. Genomics of crop wild relatives: expanding the gene pool for crop improvement. *Plant Biotechnology Journal*, *14*(4), 1070–1085. https://doi.org/10.1111/pbi.12454

Cantarel BL, Korf I, Robb, SMC, et al. 2008. MAKER: an easy-to-use annotation pipeline designed for emerging model organism genomes. *Genome Research* 18(1), 188–196.

CarterWu F-Q, Fan C-M, Zhang X-M, Fu Y-F. 2013. The phytochrome gene family in soybean and a dominant negative effect of a soybean PHYA transgene on endogenous Arabidopsis PHYA. *Plant Cell Reports* 32, 1879–1890.

Chandra S, Yadav RR, Poonia S, Yashpal RDR, Kumar A, Lal SK, Talukdar A. 2017. Seed coat permeability studies in wild and cultivated species of soybean. *International Journal of Current Microbiology and Applied Science* 6(7), 2358–2363.

Chandra S, Taak Y, Rathod DR, Yadav RR, Poonia S, Sreenivasa V, Talukdar A. 2020. Genetics and mapping of seed coat impermeability in soybean using inter-specific populations. *Physiology and Molecular Biology of Plants* 26(11), 2291–2299.

Chandra S, Talukdar A, Taak Y, Yadav RR, Saini M, Sipani NS. 2021. Seed longevity studies in wild type, cultivated and inter-specific recombinant inbred lines (RILs) of soybean [*Glycine max* (L.) Merr.]. *Genetic Resources and Crop Evolution* 2022, 1–11.

Chandra S, Maranna S, Manisha S, Kumawat G, Nataraj V, Satpute GK, Rajesh V, Verma RK, Ratnaparkhe MB, Sanjay G, Akshay T. 2022a. Achievements, challenges and prospects of hybrid soybean. In: Bohra, A., Parihar, A. K., Naik, S. J. S., Chandra, A. (eds) *Plant Male Sterility Systems for Accelerating Crop Improvement*. Springer, Singapore, pp 167–194.

Chandra S, Choudhary M, Bagaria PK, Nataraj V, Kumawat G, Choudhary JR, Sonah H, Gupta S, Wani SH, Ratnaparkhe MB. 2022c. Progress and prospectus in genetics and genomics of Phytophthora root and stem rot resistance in soybean (Glycine max L.). *Frontiers in Genetics* 2022, 3170.

Chowdhury R, Nallusamy S, Shanmugam V, Loganathan A, Muthurajan R, Subramanian KS, Jeyaprakash P, Sudhakar D. 2021. Genome-wide understanding of evolutionary and functional relationships of rice Yellow Stripe-Like (YSL) transporter family in comparison with other plant species. *Biologia*. https://doi.org/10.1007/s11756-021-00924-5.

Chung G, Singh RJ. 2008. Broadening the genetic base of soybean: a multidisciplinary approach. *Critical Reviews in Plant Sciences* 27, 295–341.

Chung W-H, Jeong N, Kim J, Lee WK, Lee Y-G, Lee S-H, Yoon W, Kim J-H, Choi I-Y, Choi H-K. 2014. Population structure and domestication revealed by high-depth resequencing of Korean cultivated and wild soybean genomes. *DNA Research* 21, 153–167.

Cober ER, Morrison MJ. 2010. Regulation of seed yield and agronomic characters by photoperiod sensitivity and growth habit genes in soybean. *Theoretical and Applied Genetics* 120, 1005–1012.

Concibido V, La Vallee B, Mclaird P, Pineda N, Meyer J, Hummel L, Yang J, Wu K, Delannay X. 2003. Introgression of a quantitative trait locus for yield from Glycine soja into commercial soybean cultivars. *Theoretical and Applied Genetics* 106, 575–582.

Costa-Silva J, Domingues D, & Lopes FM. 2017. RNA-Seq differential expression analysis: an extended review and a software tool. *PLoS One* 12(12), e0190152. https://doi.org/10.1371/journal.pone.0190152

Dawn F, Amaral-Zettler L, Cochrane GR, Cole's J, Peter D, Garrity G, Jack G, Glöckner F, Hirschman L, Karsch-Mizrachi I, Klenk HP, Knight R, Kottmann R, Kyrpides N, Meyer F, Gil I, Sansone SA, Lynn S, Sterk P, Wooley J. 2011. The genomic standards consortium. *PLoS biology* 9, e1001088. https://doi.org/10.1371/journal.pbio.1001088

Dempewolf H, Baute G, Anderson J, Kilian B, Smith C, Guarino L. 2017. Past and future use of wild relatives in crop breeding. *Crop Science* 57(3), 1070–1082. https://doi.org/10.2135/cropsci2016.10.0885

Diego J. 2016. Prospects of genomic prediction in the USDA soybean germplasm collection: historical data creates robust models for enhancing selection of accessions. *G3 Genes Genomes Genetics* 6(8), 2329–2341. https://doi.org/10.1534/g3.116.031443

Diers BW, Keim P, Fehr WR, Shoemaker RC. 1992. RFLP analysis ofsoybean seed protein and oil content. *Theoretical and Applied Genetics* 83, 608–612.

Dong Y, Yang X, Liu J, Wang B-H, Liu B-L, Wang Y-Z. 2014. Pod shattering resistance associated with domestication is mediated by a NAC gene in soybean. *Nature Communications* 5, 3352.

Duhnen A, Gras A, Teyssèdre S, Romestant M, Claustres B, Daydé J, Mangin B. 2017. Genomic selection for yield and seed protein content in soybean: a study of breeding program data and assessment of prediction accuracy. *Crop Science* 57(3), 1325–1337. https://doi.org/10.2135/cropsci2016.06.0496

Esposito K, Chiodini P, Colao A, Lenzi A, Giugliano D. 2012. Metabolic syndrome and risk of cancer: a systematic review and meta-analysis. *Diabetes Care*. 35(11), 2402–2411. https://doi.org/10.2337/dc12-0336. PMID: 23093685; PMCID: PMC3476894.

Espósito MA, Almiron P, Gatti I, Cravero VP, Lopez Anido FS, Cointry Peix EL. 2012. A rapid method to increase the number of F1 plants in pea (Pisum sativum) breeding programs. *Fundacao de Pesquisas Científicas de Riberao Preto; Genetics and Molecular Research.* 11(3), 2729–2732.

FAO. 1999. What is happening to agrobiodiversity? Available at: https://www.fao.org/3/y5609e/y5609e02.htm.

Funatsuki H, Suzuki M, Hirose A, Inaba H, Yamada T, Hajika M, Komatsu K, Katayama T, Sayama T, Ishimoto M, Fujino K. 2014. Molecular basis of a shattering resistance boosting global dissemination of soybean. *Proceedings of the National Academy of Sciences of the United States of America* 111(50), 17797–17802. https://doi.org/10.1073/pnas.1417282111.

Ghogare R, Ludwig Y, Bueno GM, Slamet-Loedin IH, Dhingra A. 2021. Genome editing reagent delivery in plants. *Transgenic Research* 30, 321–335.

Gregory PJ, Johnson SN, Newton AC, Ingram, JS. 2009. Integrating pests and pathogens into the climate change/food security debate. *Journal of Experimental Botany* 60(10), 2827–2838. https://doi.org/10.1093/jxb/erp080

Guo J, Wang Y, Song C, Zhou J, Qiu L, Huang H, Wang Y. 2010. A single origin and moderate bottleneck during domestication of soybean (Glycine max): implications from microsatellites and nucleotide sequences. *Annals of Botany* 106, 505–514.

Ha BK, Vuong TD, Velusamy V, Nguyen HT, Shannon JG, Lee JD. 2013. Genetic mapping of quantitative trait loci conditioning salt tolerance in wild soybean (Glycine soja) PI 483463. *Euphytica* 193, 79–88.

Hadley HH, Hymowitz T. 1973. Speciation and cytogenetics. *Soybeans: Improvement, Production, and Uses* 16, 97–136.

Han YP, Zhao X, Liu DY, Li YH, Lightfoot DA, Yang ZJ, Zhao L, Zhou G, Wang ZK, Huang L et al. 2016. Domestication footprints anchor genomic regions of agronomic importance in soybeans. *New Phytologist* 209, 871–884.

Hasan MM, Rafii MY, Ismail MR, Mahmood M, Rahim HA, Alam MA, Latif MA. 2015. Marker-assisted backcrossing: a useful method for rice improvement. *Biotechnology & Biotechnological Equipment* 29(2), 237–254.

Hesler LS 2013. Resistance to soybean aphid among wild soybean lines under controlled conditions. *Crop Protection* 53, 139–146.

Heuzé V, Tran G, Giger-Reverdin S, Lebas F. 2015. Perennial soybean (Neonotonia wightii). Feedipedia, a programme by INRA, CIRAD, AFZ and FAO. https://www.feedipedia.org/node/293. Last updated on September 30, 2015, 15:09.

Hymowitz T. 1995. Evaluation of wild perennial glycine species and crosses for resistance to phakopsora. In: Sinclair JB, Hartman GL (eds.) *Proceedings of the Soybean Rust Workshop*. Urbana, IL: National Soybean Research Laboratory. pp. 33–37.

Iftekharuddaula KM, Newaz MA, Salam MA, Ahmed HU, Mahbub MAA, Septiningsih EM, Mackill DJ. 2011. Rapid and high-precision marker assisted backcrossing to introgress the SUB1 QTL into BR11, the rainfed lowland rice mega variety of Bangladesh. *Euphytica* 178, 83–97.

Innes RW, Ameline-Torregrosa C, Ashfield T, Cannon E, et al. 2008. Differential accumulation of retroelements and diversification of NB-LRR disease resistance genes in duplicated regions following polyploidy in the ancestor of soybean. *Plant Physiology* 148, 1740–1759.

Iquira E, Humira S, François B. 2015. Association mapping of QTLs for sclerotinia stem rot resistance in a collection of soybean plant introductions using a genotyping by sequencing (GBS) approach. *BMC Plant Biology* 15, 5. https://doi.org/10.1186/s12870-014-0408-y

Jensen MA, Ferretti V, Grossman RL, Staudt LM. 2017. The NCI genomic data commons as an engine for precision medicine. *Blood* 130(4), 453–459. https://doi.org/10.1182/blood-2017-03-735654.

Joshi T, Fitzpatrick MR, Chen S, Liu Y, Zhang H, Endacott RZ, Gaudiello EC, Stacey G, Nguyen HT, Xu D. 2014. Soybean knowledge base (SoyKB): a web resource for integration of soybean translational genomics and molecular breeding. *Nucleic Acids Research*. 42(Database issue), D1245–52. https://doi.org/10.1093/nar/gkt905.

Kanazawa A, Liu B, Kong F, Arase S, Abe J. 2009. Adaptive evolution involving gene duplication and insertion of a novel Ty1/copia-like retrotransposon in soybean. *Journal of Molecular Evolution* 69, 164– 175.

Karasawa, K. 1952. Crossing experiments with *Glycine soja* and *G. gracilis*. *Genetica* 26, 357–358.

Kebede H, Smith JR, Ray JD. 2014. Identification of a single gene for seed coat impermeability in soybean PI 594619. *Theoretical and Applied Genetics* 127(9), 1991–2003.

Khoury CK, Brush S, Costich DE, Curry HA, de Haan S, Engels JMM, et al. 2022. Crop genetic erosion: understanding and responding to loss of crop diversity. *New Phytologist* 233(1), 84–118. https://doi.org/10.1111/nph.17733

Kiang Y, Gorman M. 1982. Soybean. *Developments in Plant Genetics and Breeding* 1, 295–328. https://doi.org/10.1016/B978-0-444-42227-9.50017-2

Kilen TC, Hartwig EE. 1978. An inheritance study of impermeable seed in soybean. *Field Crops Research* 1, 65–70.

Kim H, Kim ST, Ryu, J, Kang BC, Kim JS, and Kim SG. 2017. CRISPR/Cpf1-mediated DNA-free plant genome editing. *Nature Communications* 8, 14406. https://doi.org/10.1038/ncomms14406.

Kim MY, Lee S, Van K, Kim T-H, Jeong S-C, Choi I-Y, Kim D-S, Lee Y-S, Park D, Ma J. 2010b. Whole-genome sequencing and intensive analysis of the undomesticated soybean (Glycine soja Sieb. and Zucc.) genome. *Proceedings of the National Academy of Sciences* 107, 22032–22037.

Kim M, Hyten DL, Niblack TL, Diers BW. 2011. Stacking resistance alleles from wild and domestic soybean sources improves soybean cyst nematode resistance. *Crop Science* 51, 2301–2301. https://doi.org/10.2135/cropsci2010.08.0459

Kim MY, Shin JH, Kang YJ, Shim SR, Lee S-H. 2012a. Divergence of flowering genes in soybean. *Journal of Biosciences* 37, 857–870.

Klein RR, Miller FR, Bean S, Klein PE. 2016. Registration of 40 converted germplasm sources from the reinstated sorghum conversion program. *Journal of Plant Registrations* 10(1), 57–61.

Klein SL, Flanagan KL. 2016. Sex differences in immune responses. *Nature Reviews Immunology* 16(10), 626–638. https://doi.org/10.1038/nri.2016.90.

Kofsky J, Zhang H, Song BH. 2021. Novel resistance strategies to soybean cyst nematode (SCN) in wild soybean. *Scientific Reports* 11, 7967. https://doi.org/10.1038/s41598-021-86793-z.

Kong L, Wang J, Zhao S, Gu X, Luo J, Gao G. 2012. ABrowse-a customizable next-generation genome browser framework. *BMC Bioinform*, 13, 1–8. https://doi.org/10.1186/1471-2105-13-2.

Kumar A, Chandra S, Talukdar A, Yadav RR, Saini M, Poonia S, Lal SK. 2019a. Genetic studies on seed coat permeability and viability in RILs derived from an inter-specific cross of soybean [Glycine max (L.) Merrill]. *Indian Journal of Genetics and Plant Breeding* 79(1), 48–55.

Kumar, A., Talukdar, A., Yadav, R.R., Poonia, S., Ranjan, R. and Lal, S.K. 2019b. Identification of QTLs for seed viability in soybean [Glycine max (L.) Merill]. *Indian Journal of Genetics and Plant Breeding* 79(4), 713–718.

Kumar V, Vats S, Kumawat S, Bisht A, Bhatt V, Shivaraj SM, Padalkar G, Goyal V, Zargar S, Gupta S, Kumawat G. 2021. Omics advances and integrative approaches for the simultaneous improvement of seed oil and protein content in soybean (*Glycine max* L.). *Critical Reviews in Plant Sciences*, 40(5), 398–421.

Langewisch T, Zhang H, Vincent R, Joshi T, Xu D, Bilyeu K. 2014. Major soybean maturity gene haplotypes revealed by SNPViz analysis of 72 sequenced soybean genomes. *PLoS One* 9, e94150.

Langmead B, Salzberg SL. 2012. Fast gapped-read alignment with Bowtie 2. *Nature Methods* 9, 357–359.

Larson G, Piperno DR, Allaby RG, Purugganan MD, Andersson L, Arroyo-Kalin M, Barton L, Climer Vigueira C, Denham T, Dobney K, Doust AN, Gepts P, Gilbert MT, Gremillion KJ, Lucas L, Lukens L, Marshall FB, Olsen KM, Pires JC, Richerson PJ, Rubio de Casas R, Sanjur OI, Thomas MG, Fuller DQ. 2014. Current perspectives and the future of domestication studies. *Proceedings of the National Academy of Sciences of the United States of America* 111, 6139–6146.

Lee G-A, Crawford GW, Liu L, Sasaki Y, Chen X. 2011. Archaeological soybean (Glycine max) in East Asia: does size matter? *PLoS One* 6, e26720.

Lee, JY, Min HJ, Jinkyo, Bilyeu Kristin Lee Jeong Dong, Kang Sungtaeg. 2015. Detection of novel QTLs for foxglove aphid resistance in soybean. *Theoretical and Applied Genetics*. 128. https://doi.org/10.1007/s00122-015-2519-8.

Li D, Pfeiffer TW, Cornelius PL. 2007. Soybean QTL for yield and yield components associated with *Glycine soja* alleles. *Crop Science* 48, 571–581. https://doi.org/10.2135/cropsci2007.06.0361

Li YH, Li W, Zhang C, Yang L, Chang RZ, Gaut BS, Qiu LJ. 2010. Genetic diversity in domesticated soybean (Glycine max) and its wild progenitor (Glycine soja) for simple sequence repeat and single-nucleotide polymorphism loci. *New Phytologist* 188, 242–253.

Li Y-H, Zhao S-C, Ma J-X, Li D, Yan L, Li J, Qi X-T, Guo X-S, Zhang L, He W-M. 2013. Molecular footprints of domestication and improvement in soybean revealed by whole genome re-sequencing. *BMC Genomics* 14, 579.

Li Yh, Zhou G, Ma J. et al. 2014. *De novo* assembly of soybean wild relatives for pan-genome analysis of diversity and agronomic traits. *Nature Biotechnology* 32, 1045–1052. https://doi.org/10.1038/nbt.2979

Li MW, Muñoz NB, Wong CF, Wong FL, Wong KS, Wong JW, Qi X, Li KP, Ng MS, Lam HM. 2016. QTLs regulating the contents of antioxidants, phenolics, and flavonoids in soybean seeds share a common genomic region. *Frontiers in Plant Science* 7, 854. https://doi.org/10.3389/fpls.2016.00854.

Liu L. 2008. BEST: Bayesian estimation of species trees under the coalescent model. *Bioinformatics* 24(21), 2542–2543. https://doi.org/10.1093/bioinformatics/btn484

Liu B, Fujita T, Yan ZH, Sakamoto S, Xu D, Abe J. 2007. QTL mapping of domestication-related traits in soybean (Glycine max). *Annals of Botany* 100(5), 1027–1038.

Liu B, Watanabe S, Uchiyama T, Kong F, Kanazawa A, Xia Z, Nagamatsu A, Arai M, Yamada T, Kitamura K. 2010. The soybean stem growth habit gene Dt1 is an ortholog of arabidopsis terminal flower1. *Plant Physiology* 153, 198–210.

Love MI, Huber W, Anders S. 2014. Moderated estimation of fold change and dispersion for RNA-seq data with DESeq2. *Genome Biology* 15(12), 550. PMID: 25516281.

Luo R, Liu B, Xie Y, et al. 2012. SOAPdenovo2: an empirically improved memory-efficient short-read de novo assembler, *GigaScience* 1(1), 18.

Luo X, Bai X, Sun X, Zhu D, Liu B, Ji W, Cai H, Cao L, Wu J, Hu M, Liu X, Tang L, Zhu Y. 2013. Expression of wild soybean WRKY20 in arabidopsis enhances drought tolerance and regulates ABA signalling. *Journal of Experimental Botany* 64, 2155–2169. https://doi.org/10.1093/jxb/ert073

Ma Q, Huang L, Ma L, Jin Q, Li F, Wang K, Zheng K, Yao Z, Chen L. 2015. Large-scale SNP discovery and genotyping for constructing a high-density genetic map of tea plant using specific-locus amplified fragment sequencing (SLAF-seq). *PLoS One* 10(6), e0128798. https://doi.org/10.1371/journal.pone.0128798

Manavalan LP, Prince SJ, Musket TA, Chaky J, Deshmukh R, Vuong TD, Song L, Cregan PB, Nelson JC, Shannon JG, Specht JE, Nguyen HT. 2015. Identification of novel QTL governing root architectural traits in an interspecific soybean population. *PLoS One* 10, e0120490. https://doi.org/10.1371/journal.pone.0120490

Maranna S, Nataraj V, Kumawat G, Chandra S, Rajesh V, Ramteke R, Patel RM, Ratnaparkhe MB, Husain SM, Gupta S, Khandekar N. 2021. Breeding for higher yield, early maturity, wider adaptability and waterlogging tolerance in soybean (*Glycine max* L.): a case study. *Scientific Reports* 24(1), 1–6.

Maryana R, Ma'rifatun D, Wheni A, Satriyo K, Rizal WA. 2013. Alkaline pretreatment on sugarcane bagasse for bioethanol production. *Energy Procedia* 47, 250–254. https://doi.org/10.1016/j.egypro.2014.01.221

Mashima J, Kodama Y, Kosuge T, Fujisawa T, Katayama T, Nagasaki H, Okuda Y, Kaminuma E, Ogasawara O, Okubo K, Nakamura Y, Takagi T. 2016. DNA data bank of Japan (DDBJ) progress report. *Nucleic Acids Research* 44(D1), D51–57. https://doi.org/10.1093/nar/gkv1105. PMID: 26578571; PMCID: PMC4702806.

Mittler R, Blumwald, E. 2010. Genetic engineering for modern agriculture: challenges and perspectives. *Annual Review of Plant Biology* 61, 443–462. https://doi.org/10.1146/annurev-arplant-042809-112116.

Mullin WJ, Xu WL. 2000. A study of the intervarietal differences of cotyledon and seed coat carbohydrates in soybean. *Food Research International* 33, 883–891.

Nagatoshi Y, Fujita Y. 2019. Accelerating soybean breeding in a CO_2-supplemented growth chamber. *Plant and Cell Physiology* 60(1), 77–84.

Oecd Food and Nations AOOTU. 2016. *OECD-FAO Agricultural Outlook 2016–2025*. Paris: OECD Food and Nations AOOTU.

Pantalone VR, Rebetzke GJ, Burton JW, Wilson RF. 1997. Genetic regulation of linolenic acid concentration in wild soybean Glycine soja accessions. *Journal of the American Oil Chemists' Society* 74, 159–163.

Periyannan S, Moore J, Ayliffe M, Bansal U, Wang X, Huang L, Deal K, Luo M, Kong X, Bariana H, Mago R, McIntosh R, Dodds P, Dvorak J, Lagudah E. 2013. The gene Sr33, an ortholog of barley Mla genes, encodes resistance to wheat stem rust race Ug99. *Science* 341(6147), 786–788. https://doi.org/10.1126/science.1239028

Pertea M, Pertea GM, Antonescu CM, Chang TC, Mendell JT, Salzberg SL. 2015. String Tie enables improved reconstruction of a transcriptome from RNA-seq reads. *Nature Biotechnology*, 33:290–295.

Petereit J, Marsh JI, Bayer PE, Danilevicz MF, Thomas WJW, Batley J, Edwards D. 2022. Genetic and genomic resources for soybean breeding research. *Plants (Basel)* 11(9), 1181. https://doi.org/10.3390/plants11091181

Pimentel D, Wilson C, McCullum C, Huang R, Dwen P, Flack J, et al. 1997. Economic and environmental benefits of biodiversity. *BioScience* 47(11), 747–757. https://doi.org/10.2307/1313097

Ping J, Liu Y, Sun L, Zhao M, Li Y, She M, Sui Y, Lin F, Liu X, Tang Z. 2014. Dt2 is a gain-of-function MADS-domain factor gene that specifies semideterminacy in soybean. *Plant Cell* 26, 2831–2842.

Purcell S, Neale B, Todd-Brown K, Thomas L, Ferreira MA, Bender D, Maller J, Sklar P, de Bakker PI, Daly MJ, Sham PC. 2007. PLINK: a tool set for whole-genome association and population-based linkage analyses. *American Journal of Human Genetics* 81(3), 559–575. https://doi.org/10.1086/519795. PMID: 17701901; PMCID: PMC1950838.

Qi X, Li MW, Xie M, Liu X, Ni M, Shao G, Song C, Yim AKY, Tao Y, Wong FL, Isobe S, Wong CF, Wong KS, Xu C, Li C, Wang Y, Guan R, Sun F, Fan G, Ziao Z, Zhou F, Phang TH, Liu X, Tong SW, Chang TF, Yiu SM, Tabata S, Wang J, Xu X, Lam HM. 2014. Identification of a novel salt tolerance gene in wild soybean by whole-genome sequencing. *Nature Communications* 5, 4340. https://doi.org/10.1038/ncomms5340

Rana SK, Price TD, Qian H. 2019. Plant species richness across the Himalaya driven by evolutionary history and current climate. *Ecosphere* 10(11), e02945. https://doi.org/10.1002/ecs2.2945

Rana MM, Takamatsu T, Baslam M, Kaneko K, Itoh K, Harada N, Sugiyama T. et al. 2019. Salt tolerance improvement in rice through efficient SNP marker-assisted selection coupled with speed-breeding. *International Journal of Molecular Sciences* 20(10), 2585.

Ranganathan J, Vennard D, Waite R, Dumas P, Lipinski B Searchinger T. 2016. *Shifting Diets for a Sustainable food Future*. Washington, DC: World Resour. Institute.

Rathod DR, Chandra S, Kumar A, Yadav RR, Talukdar A. 2019. Deploying inter-specific recombinant inbred lines to map QTLs for yield-related traits in soybean. *Indian Journal of Genetics and Plant Breeding* 79(04), 693–703.

Ray DK, Mueller ND, West PC, Foley JA. 2013. Yield trends are insufficient to double global crop production by 2050. *PLoS One* 8, e66428. https://doi.org/10.1371/journal.pone.0066428

Raza A, Razzaq A, Mehmood SS, Zou X, Zhang X, Lv Y, et al. 2019. Impact of climate change on crops adaptation and strategies to tackle its outcome: a review. *Plants* 8(2), 34. https://doi.org/10.3390/plants8020034

Rehman HM, Nawaz MA, Shah ZH, Yang SH, Chung G. 2017. Functional characterization of naturally occurring wild soybean mutant (sg-5) lacking astringent saponins using whole genome sequencing approach. *Plant Science* 267, 148–156. https://doi.org/10.1016/j.plantsci.2017.11.014

Rehman HM, Nawaz MA, Shah ZH, Yang SH, Chung G. 2018. Functional characterization of naturally occurring wild soybean mutant (sg-5) lacking astringent saponins using whole genome sequencing approach. *Plant Science*, 267, 148–156. https://doi.org/10.1016/j.plantsci.2017.11.014

Rizal G, Karki S, Alcasid M, Montecillo F, Acebron K, Larazo N, Garcia R, Slamet-Loedin IH, Quick WP. 2014. Shortening the breeding cycle of sorghum, a model crop for research. *Crop Science* 54(2), 520–529.

Rosenow DT, Dahlberg JA, Stephens JC, Miller FR, Barnes DK, Peterson GC, Johnson JW, Schertz KF. 1997. Registration of 63 converted sorghum germplasm lines from the sorghum conversion program. *Crop Science* 37(4), 1399–1400.

Rosenow DT, Clark LE, Peterson GC, Odvody GN, Rooney WL. 2021. Registration of Tx3440 through Tx3482 sorghum germplasm. *Journal of Plant Registrations* 15(2), 379–387. https://doi.org/10.1002/plr2.20082

Saintenac C, Zhang W, Salcedo A, Rouse MN, Trick HN, Akhunov E, Dubcovsky J. 2013. Identification of wheat gene Sr35 that confers resistance to Ug99 Stem rust race group. *Science* 341(6147), 783–786. https://doi.org/10.1126/science.1239022

Schouten HJ, Tikunov Y, Verkerke W, Finkers R, Bovy A, Bai Y, Visser RG. 2019. Breeding has increased the diversity of cultivated tomato in the Netherlands. *Frontiers in Plant Science* 10, 1606.

Sebolt AM, Shoemaker RC, Diers BW. 2000. Analysis of a quantitative trait locus allele from wild soybean that increases seed protein concentration in soybean. *Crop Science* 40(5), 1438–1444.

Sedivy EJ, Wu F, Hanzawa Y. 2017. Soybean domestication: the origin, genetic architecture and molecular bases. *New Phytologist* 214, 539–553. https://doi.org/10.1111/nph.14418

Sherman-Broyles S, Bombarely A, Powell AF, Doyle JL, Egan AN, Coate JE, Doyle JJ. 2014. The wild side of a major crop: soybean's perennial cousins from down under. *American Journal of Botany* 101, 1651–1665. https://doi.org/10.3732/ajb.1400121

Shi Z, Liu S, Noe J, Arelli P, Meksem K, Li Z. 2015. SNP identification and marker assay development for high-throughput selection of soybean cyst nematode resistance. *BMC Genomics* 16(1), 314. https://doi.org/10.1186/s12864-015-1531-3

Shivakumar M, Kumawat G, Nataraj V, Gill BS et al., 2023. Genetic enhancement for grain yield and Mungbean Yellow Mosaic India Virus MYMIV) resistance through introgressions from Glycine soja. In: Paper presentation in International Conference on Vegetable Oils 2023 (ICVO 2023) on 'Research, Trade, Value Chain and Policy' Hyderabad during January 17–21, 2023.

Singh RJ. 2019. Cytogenetics and genetic introgression from wild relatives in soybean. *Nucleus* 62, 3–14. https://doi.org/10.1007/s13237-019-00263-6

Singh RJ, Kollipara KP, Hymowitz T. 1987. Inter-subgeneric hybridisation of soybeans with a wild perennial species, *Glycine* clandestina Wendl. *Theoretical and Applied Genetics* 74, 391–396.

Singh K, Kumar S, Kumar SR, Singh M, Gupta K. 2019. Plant genetic resources management and pre-breeding in genomics era. *Indian Journal of Genetics and Plant Breeding* 79(Sup-01), 117–130.

Stephens JC, Miller FR, Rosenow, DT. 1967. Conversion of alien sorghums to early combine genotypes 1. *Crop Science* 7(4), 396–396. https://doi.org/10.2135/cropsci1967.0011183X000700040036x

Sun L, Miao Z, Cai C, Zhang D, Zhao M, Wu Y, Zhang X, Swarm SA, Zhou L, Zhang ZJ, Nelson RL, Ma J. 2015. *GmHs1-1*, encoding a calcineurin-like protein, controls hard-seededness in soybean. *Nature Genetics* 47(8), 939–943. https://doi.org/10.1038/ng.3339.

Talukdar A, Harish GD, Shivakumar M, Kumar B, Verma K, Lal SK, Sapra RL, Singh KP. 2013. Genetics of yellow mosaic virus (YMV) resistance in cultivated soybean [*Glycine max* (L) Merr.]. *Legume Research* 36(3), 263–267.

Tian Z, Wang X, Lee R, Li Y, Specht JE, Nelson RL, McClean PE, Qiu L, Ma J. 2010. Artificial selection for determinate growth habit in soybean. *Proceedings of the National Academy of Sciences* 107(19), 8563–8568. https://doi.org/10.1073/PNAS.1000088107.

Tiwari, SP. 2017. Emerging trends in soybean industry. *Soybean Research* 15(1), 1–17.

Trupti, J, others, 2014. Soybean knowledge base (SoyKB): a web resource for integration of soybean translational genomics and molecular breeding. *Nucleic Acids Research* 42(1), D1245–D1252. https://doi.org/10.1093/nar/gkt905

USDA. 2016. Soybeans: Acreage planted, harvested, yield, production, value, and loan rate, U.S., 1960–2015. In Oil Crops Yearbook 2016, ed ERSUSDO (Agriculture United States Department of Agriculture; Economic Research Service, USDA). Available online at: https://www.ers.usda.gov/data-products/ oil-crops-yearbook/

Van Huan N, Sugimoto H, Harada K. 2005. Genetic variation of local varieties of soybean in the western part of the Shikoku Mountains in Japan. *Breeding Science*, 55, 441–446.

Wafula EK, Zhang H, Von Kuster GHJ, Honaas LA, dePamphilis CW. 2023. PlantTribes2: Tools for comparative gene family analysis in plant genomics. *Frontiers in Plant Science* 13, 1011199. https://doi.org/10.3389/fpls.2022.1011199

Wang X, Li Y, Zhang H, Sun G, Zhang W, Qiu L. 2015. Evolution and association analysis of *GmCYP78A10* gene with seed size/weight and pod number in soybean. *Molecular Biology Reports* 42, 489–496.

Watson A, Ghosh S, Williams MJ, Cuddy WS, Simmonds J, Rey M, Hinchliffe A, Steed A, Reynolds D, Adamski NM, Breakspear A, Korolev A, Rayner T, Dixon LE, Riaz A, Martin W, Ryan M, Edwards D, Batley J, Hickey LT. 2017. Speed breeding is a powerful tool to accelerate crop research and breeding. *Nature Plants* 4(1), 23–29. https://doi.org/10.1038/s41477-017-0083-8

Wilkey AP, Brown AV, Cannon, SB. et al. 2020. GCViT: a method for interactive, genome-wide visualization of resequencing and SNP array data. *BMC Genomics* 21, 822. https://doi.org/10.1186/s12864-020-07217-2

Wu F-Q, Fan C-M, Zhang X-M, Fu Y-F. 2013. The phytochrome gene family in soybean and a dominant negative effect of a soybean PHYA transgene on endogenous Arabidopsis PHYA. *Plant Cell Reports* 32, 1879–1890.

Xie M, Chung CY, Li MW, Wong FL, Wang X, Liu A, Wang Z, Leung AK, Wong TH, Tong SW, Xiao Z, Fan K, Ng MS, Qi X, Yang L, Deng T, He L, Chen L, Fu A, Ding Q, He J, Chung G, Isobe S, Tanabata T, Valliyodan B, Nguyen HT, Cannon SB, Foyer CH, Chan TF, Lam HM. 2019. A reference-grade wild soybean genome. *Nature Communications* 10(1), 1216. https://doi.org/10.1038/s41467-019-09142-9.

Xu D, Abe J, Gai J, Shimamoto Y. 2002. Diversity of chloroplast DNA SSRs in wild and cultivated soybeans: evidence for multiple origins of cultivated soybean. *Theoretical and Applied Genetics* 105, 645–653.

Yahiaoui N, Kaur N, Keller B. 2009. Independent evolution of functional Pm3 resistance genes in wild tetraploid wheat and domesticated bread wheat. *The Plant Journal* 57(5), 846–856. https://doi.org/10.1111/j.1365-313X.2008.03731.x

Yano R, Takagi K, Takada Y, Mukaiyama K, Tsukamoto C, Sayama T, Kaga A, Anai T, Sawai S, Ohyama K, Saito K, Ishimoto M. 2017. Metabolic switching of astringent and beneficial triterpenoid saponins in soybean is achieved by a loss of function mutation in cytochrome P450 72A69. *The Plant Journal* 89, 527–539.

Yashpal Y, Rathod DR, Jyoti D, Anil K, Keya M, Deepika C, Subhash C, Lal SK, Akshay T. 2015. Genomic variation studies in *Glycine max* and *Glycine soja* using SSR markers. *Current Science* 109(11), 1929–1931.

Yashpal Y, Singh NP, Saroj SK. 2015. Combining ability, heterosis and inbreeding depression in inter specific hybrids involving greengram [Vigna radiata L. Wilczek] and blackgram [Vigna mungo L. Hepper]. *Electronic Journal of Plant Breeding*, 6(1), 87–92.

Yashpal Y, Rathod DR, Chandra S, Kumar A, Yadav RR, Talukdar A. 2019. Deploying inter-specific recombinant inbred lines to map QTLs for yield-related traits in soybean. *Indian Journal of Genetics and Plant Breeding* 79(04), 693–703.

Yukari N, Yasunari F. 2019. Accelerating soybean breeding in a CO2-supplemented growth chamber. *Plant and Cell Physiology* 60(1), 77–84. https://doi.org/10.1093/pcp/pcy189

Zabala G, Vodkin LO. 2006. A rearrangement resulting in small tandem repeats in the f3′5′h gene of white flower genotypes is associated with the soybean w1 locus. *Crop Science* 47, S-113–S-124. https://doi.org/10.2135/cropsci2006.12.0838tpg

Zerbino DR, Birney E. 2008. Velvet: algorithms for de novo short read assembly using de Bruijn graphs. *Genome Research* 18(5), 821–829.

Zhai H, Zhao Y, Li W, Chen Q, Bai H, Hu H, Piazza ZA, Tian W, Lu H, Wu Y, Mu Y, Wei G, Liu Z, Li J, Li S, Wang L. 2014. Observation of an all-boron fullerene. *Nature Chemistry* 6(8), 727–731. https://doi.org/10.1038/nchem.1999

Zhang WK, Wang YJ, Luo GZ, Zhang JS, He CY, Wu XL, Gai JY, Chen SY. 2004. QTL mapping of ten agronomic traits on the soybean (*Glycine max* L. Merr.) genetic map and their association with EST markers. *Theoretical and Applied Genetics* 108, 1131–1139.

Zhang Q, Li H, Li R, Hu R, Fan C, Chen F, Wang Z, Liu X, Fu Y, Lin C. 2008. Association of the circadian rhythmic expression of GmCRY1a with a latitudinal cline in photoperiodic flowering of soybean. *Proceedings of the National Academy of Sciences* 105, 21028–21033.

Zhao TJ, Gai JY. 2004. The origin and evolution of cultivated soybean [Glycine max (L.) Merr.]. *Scientia Agricultura Sinica* 37(7), 954–962.

Zhou S, Sekizaki H, Yang Z, Sawa S, Pan J. 2010. Phenolics in the seed coat of wild soybean (Glycine soja) and their significance for seed hardness and seed germination. *Journal of Agricultural and Food Chemistry* 58(20), 10972–10978.

Zhou Z, Jiang Y, Wang Z, Gou Z, Lyu J, Li W, Yu Y, Shu L, Zhao Y, Ma Y. 2015b. Resequencing 302 wild and cultivated accessions identifies genes related to domestication and improvement in soybean. *Nature Biotechnology* 33, 408–414.

Zhu H, Li C, Gao C. 2020. Applications of CRISPR-Cas in agriculture and plant biotechnology. *Nature Reviews Molecular Cell Biology* 21(11), 661–677.

Zhu X, Ge Y, Wu T, Zhao K, Chen Y, Wu B, Zhu F, Zhu B, Cui L. 2020. Co-infection with respiratory pathogens among COVID-2019 cases. *Virus Research* 285, 198005. https://doi.org/10.1016/j.virusres.2020.198005

Index

For Product Safety Concerns and Information please contact our EU
representative GPSR@taylorandfrancis.com
Taylor & Francis Verlag GmbH, Kaufingerstraße 24, 80331 München, Germany

www.ingramcontent.com/pod-product-compliance
Ingram Content Group UK Ltd.
Pitfield, Milton Keynes, MK11 3LW, UK
UKHW021829240425
457818UK00006B/126

* 9 7 8 1 0 3 2 5 6 2 2 3 0 *